S0-EHD-907

THE LOEB CLASSICAL LIBRARY
FOUNDED BY JAMES LOEB, LL.D.

EDITED BY

†T. E. PAGE, C.H., LITT.D.

†E. CAPPS, PH.D., LL.D. †W. H. D. ROUSE, LITT.D.

L. A. POST, L.H.D.   E. H. WARMINGTON, M.A., F.R.HIST.SOC.

LETTERS TO ATTICUS
III

CICERO
LETTERS TO ATTICUS

WITH AN ENGLISH TRANSLATION BY

E. O. WINSTEDT, M.A.
OF MAGDALEN COLLEGE, OXFORD

IN THREE VOLUMES

III

CAMBRIDGE, MASSACHUSETTS
HARVARD UNIVERSITY PRESS
LONDON
WILLIAM HEINEMANN LTD
MCMLXI

First printed 1918
Reprinted 1925, 1945, 1953, 1961

Printed in Great Britain

CONTENTS

Introduction	*Page* vii
Letters to Atticus Book XII	1
Letters to Atticus Book XIII	109
Letters to Atticus Book XIV	217
Letters to Atticus Book XV	293
Letters to Atticus Book XVI	369
Chronological Order of the Letters	445
Index of Names	449

INTRODUCTION

THE letters contained in this volume begin with one written just after Caesar's final victory over the remains of the Pompeian party at Thapsus in April, 46 B.C., and cover three of the last four years of Cicero's life. When they open, Cicero was enjoying a restful interval after the troublous times of the Civil War. He had made his peace with Caesar and reconciled himself to a life of retirement and literary activity. In the Senate he never spoke except to deliver a speech pleading for the return from exile of his friend Marcellus; and his only other public appearance was to advocate the cause of another friend, Ligarius. In both he was successful; and, indeed, so he seems also to have been in private appeals to Caesar on behalf of friends. But their relations were never intimate,[1] and Cicero appears always to have felt ill at ease in Caesar's society,[2] disliking and fearing him as a possible tyrant or at least an anomaly in a Republican state. He evidently felt, too, some natural qualms at being too much of a turn-coat, as he dissuaded his son from joining Caesar's expedition to Spain at the end of the year on that ground, and persuaded him to go to Athens to study instead.[3] No doubt he considered that it was more consonant with the dignity which he was always claiming for himself to take no part in public affairs at all than to play a secondary part where he had once been first. Consequently he spent the year 46 peacefully engaged in writing and in his

[1] XIV. 1 and 2.   [2] XIII. 52.   [3] XII. 7.

INTRODUCTION

private affairs; and even of those we hear little, as he was at Rome the greater part of the time. Somewhat under protest he wrote, apparently at the suggestion of the Caesarian party,[1] with most of whom he was on good terms, a work on Cato, which satisfied neither friend nor foe, as Brutus thought it necessary to write another himself, and Caesar composed an *Anti-Cato*. Of his other writings, two rhetorical works, the *Brutus* and the *Orator*, and one philosophical, the *Paradoxa*, fall in this year. In the early part of it he divorced Terentia, and at the end of it married his rich and youthful ward Publilia; but he soon separated from her. The unhappy marriage between his daughter Tullia and her profligate husband, Dolabella, was dissolved at much the same time, but she only survived for a few months. Her death, which occurred in February, 45 B.C., seems to have prostrated Cicero with grief, and a long series of daily letters, from March to August of that year, are largely filled with reiterations of his grief and projects for the erection of a shrine in her honour. They are interesting for the light they cast on Atticus' treatment of Cicero when he was unstrung and excited. Atticus evidently disapproved entirely of the project; but from Cicero's answers one infers that he kept on humouring him and at the same time delaying action on his part by continual suggestions of a fresh site for the shrine, knowing that Cicero's ardour would cool and the scheme drop through, as it did.

Much is said, too, in these letters about the literary work to which Cicero turned with more eagerness than ever to assuage his grief; and the output was enormous. A book on consolation in

[1] XII. 4.

INTRODUCTION

times of sorrow, a general introduction to the philosophical works which followed, the *De Finibus*, the *Academica*—rewritten three times [1]—and a small rhetorical treatise, the *Partitiones Oratoriae*, were published during the year, while the *Tusculanae Disputationes*, the *De Natura Deorum* and the *De Senectute* were projected and begun. Certainly Cicero was right in saying that he had no lack of words! [2]

Of political affairs little is said; indeed, in Caesar's absence there was not much to say. But there are occasional sneers at the honours paid to him [3] and at his projected extension of Rome.[4] For the latter part of the year, after Caesar's return from Spain, there are no letters in this collection except two amusing letters in December, one describing a conversation with his nephew, who was trying to make peace with his relatives after a violent quarrel,[5] and the other Cicero's entertainment of Caesar at Puteoli.[6]

Not long afterwards came the murder of Caesar, at which Cicero to his regret was not present, though he was in Rome and hastened to the Capitol to lend his support to the murderers. He found, however, the cold Brutus hard to stir into action, and after Antony's speech at the funeral he thought it wiser to retire from Rome. The letters written at the time are full of rejoicing at the death of a man, towards whom he never seems to have felt any attraction, in spite of the kindness he had received at his hands. But he soon realised the hopelessness of the Republican cause, which lacked both a leader and a following. He himself regained something

---

[1] XIII. 13 and 16. [2] XII. 52.
[3] XII. 45; XIII. 27 and 44. [4] XIII. 35.
[5] XIII. 42. [6] XIII. 52.

INTRODUCTION

of his old position, and we find him not only consulted by Brutus and the rest of his party, but politely addressed by Antony in a note, asking his permission to recall Cicero's old enemy Clodius.[1] Cicero, taking the request as a demand, returned an equally polite note of assent;[2] but what he thought of the request and of Antony is shown by a letter sent to Atticus simultaneously.[3] For a while there are occasional bursts of hope in a revival of the old constitution, for instance when Dolabella threw down the column erected in the forum in honour of Caesar;[4] but despair at the inactivity of Brutus and his friends and at Antony's growing influence and the respect shown for Caesar's enactment after his death prevail; and Cicero contemplated crossing to Greece to visit his son and escape from the war he foresaw. Octavian's arrival and opposition to Antony did not comfort him much, in spite of attentions paid to himself by the future emperor, as he mistrusted Octavian's youth, his abilities and his intentions. But, when just on the point of sailing, news reached him that there was a chance of Antony giving way and peace with something of the old conditions being restored; and he hurried back to Rome to take his part in its restoration.[5] There he found little chance of peace, but, once returned, he recovered sufficient courage to take the lead in the Senate and deliver his first *Philippic* against Antony. After that there are only a few letters written towards the end of the year. In them he still expresses great mistrust for Octavian, who was continually appealing to him for his support;[6] and, in spite of his renewed entry into public affairs, one

[1] XIV. 13a.   [2] XIV. 13b.   [3] XIV. 13.
[4] XIV. 15.   [5] XVI. 7.   [6] XVI. 9 and 11.

INTRODUCTION

is rather surprised to find that he was still working at his philosophical treatises, writing the *De Officiis* to dedicate to his son,[1] and even eager to turn to history at the suggestion of Atticus.[2] Such is the last glimpse we get of him in the *Letters to Atticus*. Shortly afterwards he returned to Rome, and for some six months led the senatorial party in its opposition to Antony; but, when Octavian too turned against the party and the struggle became hopeless, he retired to Tusculum, where he lived until he was proscribed by the Triumvirs early in December. Then he contemplated flight to Greece, but was killed at Astura before he had succeeded in leaving Italy.

I must again acknowledge my indebtedness in preparing the translation to Tyrrell's edition of the Letters and to Shuckburgh's translation, from both of which I have "conveyed" many a phrase. The text is as usual based on the Teubner edition, and textual notes have been mainly confined to passages where a reading not found in that edition was adopted. In those notes the following abbreviations are used:—

M = the *Codex Mediceus* 49, 18, written in the year 1389 A.D., and now preserved in the Laurentian Library at Florence. M^1 denotes the reading of the first hand, and M^2 that of a reviser.

Δ = the reading of M when supported by that of the *Codex Urbinas* 322, a MS. of the fifteenth century, preserved in the Vatican Library.

O = *Codex* 1, 5, 34 in the University Library at Turin, written in the fifteenth century. O^1 denotes the reading of the first hand, and O^2 that of a reviser.

[1] XVI. 11.   [2] XVI. 13b.

INTRODUCTION

C = the marginal readings in Cratander's edition of 1528, drawn from a MS. which is lost.

Z = the readings of the lost *Codex Tornaesianus*. Z^b, Z^l, Z^t, the readings of the same MS. when attested only by Bosius, Lambinus, or Turnebus respectively.

L(marg.) = readings in the margin of Lambinus' second edition.

Vict. = the *editio Petri Victori* (Venice, 1534–37).

CICERO'S LETTERS
TO ATTICUS
BOOK XII

M. TULLI CICERONIS
EPISTULARUM AD ATTICUM
LIBER DUODECIMUS

I

CICERO ATTICO SAL.

Scr. in Arpinati VIII K. Dec. a. 708

Undecimo die, postquam a te discesseram, hoc litterularum exaravi egrediens e villa ante lucem, atque eo die cogitabam in Anagnino, postero autem in Tusculano, ibi unum diem; v Kalend. igitur ad constitutum. Atque utinam continuo ad complexum meae Tulliae, ad osculum Atticae possim currere! Quod quidem ipsum scribe, quaeso, ad me, ut, dum consisto in Tusculano, sciam, quid garriat, sin rusticatur, quid scribat ad te; eique interea aut scribes salutem aut nuntiabis itemque Piliae. Et tamen, etsi continuo congressuri sumus, scribes ad me, si quid habebis.

Cum complicarem hanc epistulam, noctuabundus ad me venit cum epistula tua tabellarius; qua lecta de Atticae febricula scilicet valde dolui. Reliqua, quae exspectabam, ex tuis litteris cognovi omnia; sed, quod scribis "igniculum matutinum γεροντικόν," γεροντικώτερον est memoriola vacillare. Ego enim IIII Kal. Axio dederam, tibi III, Quinto, quo die venissem, id est v Kal. Hoc igitur habebis, novi nihil.

CICERO'S LETTERS TO ATTICUS
BOOK XII

I

CICERO TO ATTICUS, GREETING.

On the eleventh day after parting from you I have scribbled these few lines while leaving my country house before daybreak. I am thinking of stopping to-day at my place at Anagnia, to-morrow at Tusculum and staying there one day. On the 26th then to our tryst; and I only wish I could run straight to the embraces of my Tullia and the lips of Attica. What those little lips are prattling, please write and let me know, while I am at Tusculum, or, if she is in the country, what she is writing to you: and in the meantime pay my respects by letter or in person to her, and to Pilia too. And all the same, though we are to meet at once, write to me, if you have anything to say.

Arpinum, Nov. 23, B.C. 46

As I was folding up this letter, a messenger came in the night to me with a letter of yours, and on reading it I was naturally very sorry to hear of Attica's slight attack of fever. Everything else I was wanting to hear, I learn from your letter. You say it is a sign of old age to want a bit of fire in the morning: it's a worse sign of old age to be a bit weak in your memory. I had arranged for the 27th with Axius, the 28th with you, and the 26th, the day I arrive, with Quintus. So please count on

MARCUS TULLIUS CICERO

Quid ergo opus erat epistula? Quid, cum coram sumus et garrimus, quicquid in buccam? Est profecto quiddam λέσχη, quae habet, etiamsi nihil subest, collocutione ipsa suavitatem.

II

CICERO ATTICO SAL.

Scr. Romae ante med. m. Apr. a. 708

Hic rumores tamen Murcum perisse naufragio, Asinium delatum vivum in manus militum, L navis delatas Uticam reflatu hoc, Pompeium non comparere nec in Balearibus omnino fuisse, ut Paciaecus adfirmat. Sed auctor nullius rei quisquam. Habes, quae, dum tu abes, locuti sint. Ludi interea Praeneste. Ibi Hirtius et isti omnes. Et quidem ludi dies VIII. Quae cenae, quae deliciae! Res interea fortasse transacta est. O miros homines! At Balbus aedificat; τί γὰρ αὐτῷ μέλει; Verum si quaeris, homini non recta, sed voluptaria quaerenti nonne βεβίωται? Tu interea dormis. Iam explicandum est πρόβλημα, si quid acturus es. Si quaeris, quid putem, ego fructum[1] puto. Sed quid multa? Iam te videbo,

---

[1] fructum *MSS.*: peractum *Moser*: confectum *Schütze*: eluctum *Ellis*.

---

[1] Or, as Tyrrell suggests, "There's tit for tat. I have no news."

[2] Statius Murcus, an officer in Caesar's army. He is mentioned again later in *Fam.* XII. 11, 1.

[3] *i.e.* soldiers of Pompey, Asinius Pollio being another adherent of Caesar.

LETTERS TO ATTICUS XII. 1-2

that: there is no new arrangement.[1] What's the use of writing then? What's the use of our meeting and chattering about everything that comes into our heads? A bit of gossip is something after all, and, even if there is nothing in our talk, the mere fact of talking together has some charms.

II

CICERO TO ATTICUS, GREETING.

All the same there are reports here that Murcus[2] has been lost at sea, that Asinius reached shore alive to fall into the soldiers'[3] hands, that 50 ships have been carried to Utica by this contrary wind, that Pompey[4] is nowhere to be found and never has been in the Baleares, as Paciaecus declares. But there is no definite authority for any of this. That is what people have been saying while you are away. Meanwhile there are the games at Praeneste. That's where Hirtius and all that crew are; and there are eight days of games! Picture their dinners and their extravagant goings on. Perhaps in the meantime the great question has been settled. What people they are! So Balbus is building: little he recks. But, if you ask me, is not life over and done with, when a man begins to look for pleasure rather than duty? In the meantime you slumber on. Now is the time the problem must be solved, if you mean to do anything. If you ask me what I think, I think "Gather ye roses."[5] But what's the good of going on? I shall see you at once, and I hope you

Rome, April, B.C. 46

[4] Cn. Pompeius, the eldest son of Pompey the Great.

[5] *Fructum* may be the first word of some proverb; but probably the word is corrupt, as the sentiment seems rather at variance with that expressed just above.

et quidem, ut spero, de via recta ad me. Simul enim et diem Tyrannioni constituemus, et si quid aliud.

III

CICERO ATTICO SAL.

Scr. in Tusculano III Id. Iun. a. 708

Unum te puto minus blandum esse quam me, et, si uterque nostrum est aliquando adversus aliquem, inter nos certe numquam sumus. Audi igitur me hoc ἀγοητεύτως dicentem. Ne vivam, mi Attice, si mihi non modo Tusculanum, ubi ceteroqui sum libenter, sed μακάρων νῆσοι tanti sunt, ut sine te sim tot dies. Quare obduretur hoc triduum, ut te quoque ponam in eodem πάθει; quod ita est profecto. Sed velim scire, hodiene statim de auctione, et quo die venias. Ego me interea cum libellis; ac moleste fero Vennoni me historiam non habere. Sed tamen, ne nihil de re, nomen illud, quod a Caesare, tres habet condiciones, aut emptionem ab hasta (perdere malo, etsi praeter ipsam turpitudinem hoc ipsum puto esse perdere) aut delegationem a mancipe annua die (quis erit, cui credam, aut quando iste Metonis annus veniet?) aut

---

[1] To read a book he had written, possibly on accents. Cf. *Att.* XII. 6.

[2] Or, as Tyrrell and Shuckburgh, "whether you are coming to-day or, if not, on what day you are coming." But Cicero does not seem to have anticipated Atticus' arrival before three days.

[3] Probably a debt owed to Cicero by some proscribed Pompeian.

LETTERS TO ATTICUS XII. 2-3

will come straight from the road to me. For we will arrange a day for Tyrannio at the same time,[1] and anything else there is to do.

III

CICERO TO ATTICUS, GREETING.

Tusculum, June 11, B.C. 46

You are the only person I know less given to flattery than myself, and, if we both fall into it sometimes in the case of other people, certainly we never use it to one another. So listen to what I am saying with all sincerity. On my life, Atticus, I don't count even the Isles of the Blest, let alone my place at Tusculum—though in other respects I'm comfortable enough there—worth so long a separation from you. So let us harden our hearts for these three days—assuming that you are affected as I am, which I am sure is the case. But I should like to know whether you are starting to-day[2] immediately after the auction, and on what day you are coming. In the meantime I am buried in my books, and annoyed that I have not got Vennonius' history. But, not to neglect business altogether, for that debt that Caesar assigned to me[3] there are three means I might use. I could buy the property at a public auction; but I would rather lose it—it comes to the same thing in the end, besides the disgrace. I might transfer my rights for a bond payable a year hence by the buyer: but whom can I trust, and when would that "year of Meton"[4] come? Or I

[4] Meton, an Athenian mathematician, of the beginning of the 5th century B.C., discovered the solar cycle of 19 years. "Meton's year" was proverbially used for an indefinitely long period.

Vettieni condicione semissem. Σκέψαι igitur. Ac vereor, ne iste iam auctionem nullam faciat, sed ludis factis Ἀτύπῳ[1] subsidio currat, ne talis vir ἀλογηθῇ. Sed μελήσει. Tu Atticam, quaeso, cura et ei salutem et Piliae Tulliae quoque verbis plurimam.

IV

CICERO ATTICO SAL.

Scr. in Tusculano Id. Iun. a. 708

O gratas tuas mihi iucundasque litteras! Quid quaeris? restitutus est mihi dies festus. Angebar enim, quod Tiro ἐνερευθέστερον te sibi esse visum dixerat. Addam igitur, ut censes, unum diem.

Sed de Catone πρόβλημα Ἀρχιμήδειον est. Non adsequor, ut scribam, quod tui convivae non modo libenter, sed etiam aequo animo legere possint; quin etiam, si a sententiis eius dictis, si ab omni voluntate consiliisque, quae de re publica habuit, recedam; ψιλῶςque velim gravitatem constantiamque eius laudare, hoc ipsum tamen istis odiosum ἄκουσμα sit. Sed vere laudari ille vir non potest, nisi haec ornata sint, quod ille ea, quae nunc sunt, et futura viderit, et, ne fierent, contenderit, et, facta ne videret, vitam reliquerit. Horum quid est, quod Aledio probare possimus? Sed cura, obsecro, ut valeas, eamque, quam ad omnes res adhibes, in primis ad convalescendum adhibe prudentiam.

[1] Ἀτύπῳ *Popma* : clypo *M* : Olympo *m*.

---

[1] A banker (cf. *Att.* x. 5) who proposed to take over the debt, in return for present payment of half the sum owed.

might accept Vettienus'[1] proposal and take half paid down. So look into the matter. The fact is I am afraid Caesar may not hold any auction now, but, as soon as his games are over, may run off to the aid of his stammering friend,[2] not to slight so important a person. But I will attend to the matter. Pray take care of Attica and give her and Pilia and Tullia my kindest greetings.

IV

CICERO TO ATTICUS, GREETING.

How glad I was of your delightful letter! Why, it made my day a red-letter day after all. For I was anxious because Tiro had said you looked to him rather flushed. So I will stay another day, as you suggest.

Tusculum, June 13, B.C. 46

But about Cato, that would puzzle a Philadelphia lawyer. I cannot manage to write anything that your boon companions could read, I won't say with pleasure, but even without annoyance. If I steer clear of his utterances in the House and of his entire political outlook and policy, and content myself with simply eulogizing his unwavering constancy, even that would be no pleasant hearing for them. But he is a man who cannot properly be eulogized, unless these points are fully treated, that he foresaw the present state of affairs, and tried to prevent it, and that he took his own life by preference to seeing it come about. Can I win Aledius' approval of any of that? But pray be careful of yourself and devote the common sense you devote to other things, before all to recovering your health.

[2] Balbus, if the reading is right.

MARCUS TULLIUS CICERO

V

CICERO ATTICO SAL.

Scr. in Tusculano in. m. Quint. a. 708

Quintus pater quartum vel potius millesimum nihil sapit, qui laetetur Luperco filio et Statio, ut cernat duplici dedecore cumulatam domum. Addo etiam Philotimum tertium. O stultitiam, nisi mea maior esset, singularem! quod autem os in hanc rem ἔρανον a te! Fac non ad "διψῶσαν κρήνην," sed ad Πειρήνην eum venisse, "ἄμπνευμα σεμνὸν Ἀλφειοῦ" in te "κρήνῃ," ut scribis, haurire, in tantis suis praesertim angustiis. Ποῖ ταῦτα ἄρα ἀποσκήψει; Sed ipse viderit.

Cato me quidem delectat, sed etiam Bassum Lucilium sua.

Va

CICERO ATTICO SAL.

Scr. in Tusculano prid. K. Iun. a. 709

De Caelio tu quaeres, ut scribis; ego nihil novi. Noscenda autem est natura, non facultas modo. De Hortensio et Verginio tu, si quid dubitabis. Etsi, quod magis placeat, ego quantum aspicio, non facile inveneris. Cum Mustela, quem ad modum scribis, cum venerit Crispus. Ad Avium scripsi, ut ea, quae

---

[1] A quotation from a verse of Ennius, *Quintus pater quartum fit consul*, preserved in Aulus Gellius x. 1.

[2] Caesar had restored the ancient priestly corporation of

LETTERS TO ATTICUS XII. 5–5a

V

CICERO TO ATTICUS, GREETING.

"Quintus the elder for the fourth time"[1] or *Tusculum,* rather for the thousandth time is a fool to rejoice *July,* B.C. *46* in his son's new office[2] and in Statius, that he may see a double disgrace heaped on his house. I may add Philotimus as a third disgrace. His folly would be unparalleled, if my own had not been greater. But what cheek of him to ask you for a contribution towards it! Even suppose he had not come to a "fount athirst," but to a Pirene or "the hallowed spot where Alpheus took breath,"[3] to think of his drawing on you as his fountain, to use your word, especially when he is in such straits! Where will such conduct end? But that is his own look out.

Myself I am delighted with my Cato: but then Lucilius Bassus is delighted with his works too.

Va

CICERO TO ATTICUS, GREETING.

About Caelius you must make enquiries, as you *Tusculum,* say: I know nothing. But one must get to know *May 31,* his character as well as his resources. If you have B.C. *45* any doubts about Hortensius and Verginius, look into the matter: though, so far as I can see, you are not likely to find anything that will suit better. Deal with Mustela as you say, when Crispus has arrived. I have written to Avius to tell Piso all he

Luperci and the celebration of the Lupercalia on the Palatine hill on February 15.

[3] From Pindar, *Nem.* 1, 1, where it is used of the Arethusa at Syracuse, which was popularly believed to be connected with the river Alpheus in the Peloponnese.

bene nosset de **auro**, Pisoni demonstraret. Tibi enim sane adsentior, et istud nimium diu duci et omnia nunc undique contrahenda. Te quidem nihil agere, nihil cogitare aliud, nisi quod ad me pertineat, facile perspicio, meisque negotiis impediri cupiditatem tuam ad me veniendi. Sed mecum esse te puto, non solum quod meam rem agis, verum etiam quod videre videor, quo modo agas. Neque enim ulla hora tui mihi est operis ignota.

Vb

CICERO ATTICO SAL.

Scr. in Tusculano III aut II Id. Iun. a. 709

Tubulum praetorem video L. Metello, Q. Maximo consulibus. Nunc velim, P. Scaevola, pontifex maximus, quibus consulibus tribunus pl. Equidem puto proximis, Caepione et Pompeio; praetor enim L. Furio, Sex. Atilio. Dabis igitur tribunatum et, si poteris, Tubulus quo crimine. Et vide, quaeso, L. Libo, ille qui de Ser. Galba, Censorinone et Manilio an T. Quinctio, M'. Acilio consulibus tribunus pl. fuerit. Conturbabat enim me [epitome Bruti Fanniana][1] in Bruti epitoma Fannianorum [scripsi][1] quod erat in extremo, idque ego secutus hunc Fannium, qui scripsit historiam, generum esse scripseram Laeli. Sed tu me γεωμετρικῶς refelleras, te autem nunc Brutus et

[1] *The words in brackets are deleted by most editors as glosses.*

[1] 142 B.C.   [2] 136 B.C.
[3] For taking a bribe, when presiding at a murder trial (Cicero, *de Finibus*, 2, § 54).

knows about the gold: for I quite agree with you, I have delayed too long already and must get in all I can from every source. I quite realize that you are doing nothing and thinking of nothing except my concerns, and that your longing to come to me is prevented by my business. But in my imagination you are with me, not only because you are managing my affairs, but because I seem to see how you are managing them, for I know what you are doing in every single one of your working hours.

Vb

CICERO TO ATTICUS, GREETING.

Tusculum, June 11 or 12, B.C. 45

I see Tubulus was praetor in the consulship of L. Metellus and Q. Maximus.[1] Now I should like to know when P. Scaevola the Pontifex Maximus was tribune. I think it was in the next year, under Caepio and Pompey, as he was praetor under L. Furius and Sex. Atilius.[2] So please give me the date of his tribunate, and, if you can, the charge on which Tubulus was tried.[3] Pray look and see too whether L. Libo, who brought in the bill about Ser. Galba, was tribune in the consulship of Censorinus and Manilius or in that of T. Quinctius and M'. Acilius.[4] For I was confused by a passage at the end of Brutus' epitome of Fannius' history. Following that I made Fannius, the author of the history, son-in-law of Laelius. But you refuted me by rule and line; now however Brutus and Fannius have refuted

[4] 150 or 149 B.C. Libo impeached Galba in 147 B.C. for selling the Lusitani, who had surrendered on promise of freedom, as slaves.

MARCUS TULLIUS CICERO

Fannius. Ego tamen de bono auctore, Hortensio, sic acceperam, ut apud Brutum est. Hunc igitur locum expedies.

Vc

CICERO ATTICO SAL.

Scr. in Tus-culano prid. Id. Iun. a. 708

Ego misi Tironem Dolabellae obviam. Is ad me Idibus revertetur. Te exspectabo postridie. De Tullia mea tibi antiquissimum esse video, idque ita ut sit, te vehementer rogo. Ergo ei in integro omnia; sic enim scribis. Mihi, etsi Kalendae vitandae fuerunt Nicasionumque ἀρχέτυπα fugienda conficiendaeque tabulae, nihil tamen tanti, ut a te abessem, fuit. Cum Romae essem et te iam iamque visurum me putarem, cotidie tamen horae, quibus exspectabam, longae videbantur. Scis me minime esse blandum; itaque minus aliquanto dico, quam sentio.

VI

CICERO ATTICO SAL.

Scr. in Tus-culano m. interc. post. a. 708

De Caelio vide, quaeso, ne quae lacuna sit in auro. Ego ista non novi. Sed certe in collubo est detrimenti satis. Huc aurum si accedit—sed quid loquor? Tu videbis. Habes Hegesiae genus, quod Varro lau-

---

[1] *Brutus*, § 101.
[2] Interest was payable on the 1st of the month.
[3] Before the alteration of the calendar made by Caesar in the next year, two months, of 29 and 28 days respectively, were inserted between November and December, 46 B.C., to set the calendar right.

LETTERS TO ATTICUS XII. 5b–6

you. I had followed a good authority, Hortensius, for my statement in *Brutus*.[1] So please set the matter straight.

Vc

CICERO TO ATTICUS, GREETING.

I have sent Tiro to meet Dolabella. He will return on the 13th. I shall expect you on the next day. I see you are putting Tullia before everything, and I earnestly beg you to do so. So her dowry is untouched: for that is what you say. For myself, though I had to avoid pay-day,[2] to keep off the money-lenders' precious books, and make up my accounts, there was nothing to compensate for my absence from you. When I was at Rome and expected to see you every minute, still the hours I spent in expecting you every day seemed long. You know I am nothing of a flatterer, and so I rather understate my feelings.

Tusculum, June 12, B.C. 46

VI

CICERO TO ATTICUS, GREETING.

As to Caelius, please see that there is nothing lacking in the gold. I know nothing about that. But anyhow there is loss enough in the exchange. If there is anything wrong with the gold on the top of that—but what's the use of my talking? You will see to it. There is a specimen of Hegesias' style,[4] of which Varro approves. Now I come to

Tusculum, intercalary month,[3] B.C. 46

---

[4] Hegesias of Magnesia introduced the Asiatic school of rhetoric. Abrupt breaks such as that in the last sentence were one of its features.

MARCUS TULLIUS CICERO

dat. Venio ad Tyrannionem. Ain tu? verum hoc fuit, sine me? At ego quotiens, cum essem otiosus, sine te tamen nolui? Quo modo hoc ergo lues? Uno scilicet, si mihi librum miseris; quod ut facias, etiam atque etiam rogo. Etsi me non magis liber ipse delectabit, quam tua admiratio delectavit. Amo enim πάντα φιλειδήμονα teque istam tam tenuem θεωρίαν tam valde admiratum esse gaudeo. Etsi tua quidem sunt eius modi omnia. Scire enim vis; quo uno animus alitur. Sed, quaeso, quid ex ista acuta et gravi refertur ad τέλος?

Sed longa oratio est, et tu occupatus es in meo fortasse aliquo negotio. Et pro isto asso sole, quo tu abusus es in nostro pratulo, a te nitidum solem unctumque repetemus. Sed ad prima redeo. Librum, si me amas, mitte. Tuus est enim profecto, quoniam quidem est missus ad te.

VIa

CICERO ATTICO SAL.

Scr. in Tusculano m. interc. post. a. 708

"Chremés, tantumne ab ré tua est otí tibi," ut etiam Oratorem legas? Macte virtute¹ Mihi quidem gratum, et erit gratius, si non modo in tuis libris, sed etiam in aliorum per librarios tuos "Aristophanem" reposueris pro "Eupoli." Caesar autem

---

¹ Atticus had read the book of Tyrannio, which was referred to in XII. 2.
² Cicero refers to the introduction of Atticus in his *Brutus* (24) *in pratulo propter Platonis statuam*; but his meaning is

16

LETTERS TO ATTICUS XII. 6–6a

Tyrannio. Do you really mean it?[1] How unfair, without me! Think how often, even when I had plenty of time, I refused without you. How are you going to atone for your crime then? There is only one way: you must send me the book. I earnestly entreat you to do so; though the book itself will not delight me more than your admiration of it has. For I love everyone who loves learning and I am glad you admired so strongly an essay on so minute a point. But that is you all over. You want knowledge, which is the only mental food. But please tell me what there was in that acute and grave treatise which contributed to your *summum bonum*.

However I'm making a long story of it, and you may be busy about some of my business. And in return for that dry basking in the sun, in which you revelled in my meadow, I shall claim from you a richer and a warmer glow.[2] But to return to my first point. If you love me, send me the book: for it is yours of course, as it was sent to you.

VIa

CICERO TO ATTICUS, GREETING.

Tusculum, intercalary month, B.C. 46

"What, so much leisure from your own affairs"[3] that you have found time to read the *Orator* too. Bravo! I am pleased to hear it, and shall be still more pleased if you will get your copyists to alter Eupolis to Aristophanes[4] not only in your own copy but in others too. Caesar seemed to me to be amused

not very clear. Probably he only means that he is expecting to enjoy Atticus' hospitality soon.
[3] Terence, *Heaut.* 75.
[4] In the quotation from Aristophanes, *Ach.* 530, in *Orat.* 29.

MARCUS TULLIUS CICERO

mihi irridere visus est "quaeso" illud tuum, quod erat et εὐπινὲς et urbanum. Ita porro te sine cura esse iussit, ut mihi quidem dubitationem omnem tolleret. Atticam doleo tam diu; sed, quoniam iam sine horrore est, spero esse, ut volumus.

VII

CICERO ATTICO SAL.

Scr. in Tusculano m. interc. post. a. 708

Quae desideras, omnia scripsi in codicillis eosque Eroti dedi; breviter, sed etiam plura, quam quaeris, in iis de Cicerone; cuius quidem cogitationis initium tu mihi attulisti. Locutus sum cum eo liberalissime; quod ex ipso velim, si modo tibi erit commodum, sciscitere. Sed quid differo? Exposui te ad me detulisse, et quid vellet et quid requireret. Velle Hispaniam, requirere liberalitatem. De liberalitate dixi, quantum Publilius, quantum flamen Lentulus filio. De Hispania duo attuli, primum idem quod tibi, me vereri vituperationem. Non satis esse, si haec arma reliquissemus? etiam contraria? Deinde fore ut angeretur, cum a fratre familiaritate et omni gratia vinceretur. Vellem magis liberalitate uti mea quam sua libertate. Sed tamen permisi; tibi enim intellexeram non nimis displicere. Ego etiam atque etiam cogitabo, teque, ut idem facias, rogo. Magna res est; simplex est manere, illud anceps. Verum videbimus.

---

[1] There was a danger of Atticus' land at Buthrotum being confiscated, as Caesar was thinking of planting a colony there.

at your use of *quaeso*, as rather quaint and cockneyfied. He bade you have no anxiety in such a way that I had no doubts left.[1] I am sorry Attica's attack lasts so long: but, as she has lost her shivering fits now, I hope it will be all right.

VII

CICERO TO ATTICUS, GREETING.

I have scribbled a note with all you want on a *Tusculum*, tablet, and given it to Eros—quite shortly, but *intercalary* more than you ask for. In it I have spoken about *month*, B.C. my son, of whose intentions you gave me the first 46 hint. I took a most liberal tone with him, and, if you think it convenient, I should like you to ask him about that. But why put it off? I pointed out that you had told me what he wished to do and what he wanted: "he wished to go to Spain, and wanted a liberal allowance." As for the allowance, I said I would give him as much as Publilius or Lentulus the flamen gave their sons. Against Spain I brought forward two arguments, the first, the one I used to you, that I was afraid of adverse criticism: "Was it not enough that we abandoned one side? Must we take the other?" The second that he would be annoyed, if his cousin enjoyed Caesar's intimacy and general goodwill more than he did. I should prefer him to make use of my liberal offer rather than of his liberty. However I gave him permission; for I saw you did not really dislike the idea. I shall think the matter over carefully, and I hope you will too. It is an important point: to stay is simple, to go risky. But we shall see.

MARCUS TULLIUS CICERO

De Balbo et in codicillis scripseram et ita cogito, simul ac redierit. Sin ille tardius, ego tamen triduum, et, quod praeterii, Dolabella etiam mecum.

VIII

CICERO ATTICO SAL.

Scr. in Tusculano m. interc. post. a. 708

De Cicerone multis res placet. Comes est idoneus. Sed de prima pensione ante videamus. Adest enim dies, et ille currit. Scribe, quaeso, quid referat Celer egisse Caesarem cum candidatis, utrum ipse in fenicularium an in Martium campum cogitet. Et scire sane velim, numquid necesse sit comitiis esse Romae. Nam et Piliae satis faciendum est et utique Atticae.

IX

CICERO ATTICO SAL.

Scr. Asturae VI K. Sext. a. 709

Ne ego essem hic libenter atque id cotidie magis, ni esset ea causa, quam tibi superioribus litteris scripsi. Nihil hac solitudine iucundius, nisi paulum interpellasset Amyntae filius. Ὦ ἀπεραντολογίας ἀηδοῦς! Cetera noli putare amabiliora fieri posse

---

[1] Cicero wished to send him to Athens with L. Tullius Montanus.

[2] *i.e.* will he appoint the magistrates in Spain or let the

LETTERS TO ATTICUS XII. 7–9

About Balbus I have written in the tablet, and I think of doing as you say, as soon as he comes back. If he is rather slow about it, still I shall be three days there; and, I forgot to say, Dolabella will be with me too.

VIII

CICERO TO ATTICUS, GREETING.

My plan for my son meets with general approval. *Tusculum.* I have found a suitable companion.[1] But let us first *intercalary* see to the payment of an instalment of Tullia's *month,* B.C. dowry. The time is near and Dolabella is in a *46* hurry. Please write and tell me what Celer says Caesar has settled about the candidates, whether he thinks of going to the field of Fennel or the field of Mars.[2] I should much like to know too whether I must come to Rome for the elections. For I must do my duty by Pilia and anyhow by Attica.

IX

CICERO TO ATTICUS, GREETING.

I should be perfectly comfortable here and become *Astura, July* more and more so every day, if it weren't for the *27,* B.C. *45* reason I mentioned in my former letter. Nothing could be pleasanter than this solitude, except for the occasional interruptions of Amyntas' son.[3] How his chatter does bore one! All the rest is more charming than you can imagine, the villa, the shore,

elections at Rome take place? The *campus Fenicularius* was near Tarraco.

[3] *i.e.* L. Marcius Philippus, jestingly referred to as Philip, king of Macedonia.

villa, litore, prospectu maris, tumulis, his rebus omnibus. Sed neque haec digna longioribus litteris, nec erat, quod scriberem, et somnus urgebat.

X

CICERO ATTICO SAL.

Scr. Asturae V K. Sext. a. 709

Male mehercule de Athamante. Tuus autem dolor humanus is quidem, sed magno opere moderandus. Consolationum autem multae viae, sed illa rectissima: impetret ratio, quod dies impetratura est. Alexin vero curemus, imaginem Tironis, quem aegrum Romam remisi, et, si quid habet collis ἐπιδήμιον, ad me cum Tisameno[1] transferamus. Tota domus vacat superior, ut scis. Hoc puto valde ad rem pertinere.

XI

CICERO ATTICO SAL.

Scr. in Tusculano m. interc. post. a. 708

Male de Seio. Sed omnia humana tolerabilia ducenda. Ipsi enim quid sumus, aut quam diu haec curaturi sumus? Ea videamus, quae ad nos magis pertinent nec tamen multo, quid agamus de senatu. Et, ut ne quid praetermittam, Caesonius ad me litteras misit Postumiam Sulpici domum ad se venisse. De Pompei Magni filia tibi rescripsi nihil me hoc tempore cogitare; alteram vero illam, quam tu scribis,

[1] Tisameno Z^b, testamento *other MSS*.

the sea view, the hillocks and everything. But they don't deserve a longer letter, and I have nothing else to say, and I'm very sleepy.

X

CICERO TO ATTICUS, GREETING.

I am very sorry to hear about Athamas. But your grief, though it is a kindly weakness, should be kept well in check. There are many roads to consolation, but this is the straightest: let reason bring about what time is sure to bring about. Let us take care of Alexis, the living image of Tiro, whom I have sent back to Rome ill, and, if there is any epidemic on the hill,[1] send him to my place with Tisamenus. The whole of the upper story is vacant as you know. This I think is an excellent suggestion.

Astura, July 28, B.C. 45

XI

CICERO TO ATTICUS, GREETING.

I am sorry to hear about Seius. But one has to learn to put up with all human troubles. For what are we ourselves and how long will they be bothering us? Let us look to a thing that is more in our power, though not very much,—what we are to do about the Senate. And, before I forget it, Caesonius sent me word that Sulpicius' wife Postumia had paid him a visit. As to Pompey's daughter I answered you saying I was not thinking of her at present. I suppose you know the other lady you

Tusculum, intercalary month, B.C. 46

[1] Atticus' house was on the Quirinal hill.

puto, nosti. Nihil vidi foedius. Sed adsum. Coram igitur.

Obsignata epistula accepi tuas. Atticae hilaritatem libenter audio. Commotiunculis συμπάσχω.

XII

CICERO ATTICO SAL.

Scr. Asturae XVII K. Apr. a. 709

De dote tanto magis perpurga. Balbi regia condicio est delegandi. Quoquo modo confice. Turpe est rem impeditam iacere. Insula Arpinas habere potest germanam ἀποθέωσιν; sed vereor, ne minorem τιμὴν habere videatur ἐκτοπισμός. Est igitur animus in hortis; quos tamen inspiciam, cum venero.

De Epicuro, ut voles; etsi μεθαρμόσομαι in posterum genus hoc personarum. Incredibile est, quam ea quidam requirant. Ad antiquos igitur; ἀνεμέσητον γάρ. Nihil habeo, ad te quod scribam, sed tamen institui cotidie mittere, ut eliciam tuas litteras, non quo aliquid ex iis exspectem, sed nescio quo modo tamen exspecto. Quare, sive habes quid sive nil habes, scribe tamen aliquid teque cura.

---

[1] The first sentence refers to the repayment of Tullia's dowry; the second to Cicero's debt to his divorced wife. This Terentia had made over to Balbus in order to enforce

LETTERS TO ATTICUS XII. 11–12

write about. The ugliest thing I ever saw. But I am coming to town at once: so we will discuss it together.

When I had sealed this letter I received yours. I am very glad to hear Attica is so cheerful; and I'm grieved about the slight indisposition.

XII

CICERO TO ATTICUS, GREETING.

With regard to the dowry make all the more effort to clear the business up. To make over the debt to Balbus is a high-handed proceeding.[1] Get it settled anyhow. It is disgraceful to let the thing hang fire. The island at Arpinum would be an excellent place for a shrine, but I'm afraid it's too far out of the way to convey much honour. So my mind is set on the garden: however I'll have a look at it, when I arrive.

Astura, March 16, B.C. 45

About Epicurus you shall have your way:[2] but in the future I shall change my plan as regards the persons in my dialogues. You'd never believe how eager some people are for a place. So I shall confine myself to the ancients: that avoids invidious distinctions. I have nothing to say; but I've made up my mind to write every day to draw letters from you, not that there is anything I expect from them, still somehow or other I do expect something. So whether you have any news or not, anyhow write something; and take care of yourself.

payment quicker. The shrine mentioned below was intended to be in honour of Tullia.

[2] Apparently Atticus had asked to have the Epicurean view in the *De Finibus* put in the mouth of some friend of his.

MARCUS TULLIUS CICERO

XIII
CICERO ATTICO SAL.

Scr. Asturae Commovet me Attica; etsi adsentior Cratero.
Non. Mart. Bruti litterae scriptae et prudenter et amice multas
a. 709 mihi tamen lacrimas attulerunt. Me haec solitudo
minus stimulat quam ista celebritas. Te unum desidero; sed litteris non difficilius utor, quam si domi
essem. Ardor tamen ille idem urget et manet, non
mehercule indulgente me, sed tamen repugnante.

Quod scribis de Appuleio, nihil puto opus esse
tua contentione, nec Balbo et Oppio; quibus quidem
ille receperat mihique etiam iusserat nuntiari se
molestum omnino non futurum. Sed cura, ut excuser
morbi causa in dies singulos. Laenas hoc receperat.
Prende C. Septimium, L. Statilium. Denique nemo
negabit se iuraturum, quem rogaris. Quod si erit
durius, veniam et ipse perpetuum morbum iurabo.
Cum enim mihi carendum sit conviviis, malo id lege
videri facere quam dolore. Cocceium velim appelles.
Quod enim dixerat, non facit. Ego autem volo
aliquod emere latibulum et perfugium doloris mei.

XIV
CICERO ATTICO SAL.

Scr. Asturae De me excusando apud Appuleium dederam ad te
VIII Id. pridie litteras. Nihil esse negotii arbitror. Quem-
Mart. a. 709 cumque appellaris, nemo negabit. Sed Septimium

XIII
CICERO TO ATTICUS, GREETING.

I am upset about Attica, though I agree with Craterus. Brutus' letter, though full of wise saws and friendliness, drew from me many tears. This solitude stirs my grief less than your crowded city. You are the only person I miss; but I find no more difficulty about my literary work than if I were at home. Still the old anguish oppresses me and will not leave me, though I give you my word I do not give way to it, but fight against it.

Astura, March 7, B.C. 45

As to what you say about Appuleius, I don't think you need exert yourself, or trouble Balbus and Oppius. He has promised them and told them to let me know that he will not bother me at all. But take care that my plea of ill-health is put in every day. Laenas promised to certify. Add C. Septimius, and L. Statilius. Indeed anyone you ask will pass his word for it. But if there is any difficulty, I will come and swear myself to chronic ill-health. Since I am going to miss the banquets,[1] I would rather seem to do so according to the rules than on account of grief. Please dun Cocceius. He hasn't fulfilled his promise: and I am wanting to buy a hiding-place and a refuge for my sorrow.

XIV
CICERO TO ATTICUS, GREETING.

I wrote to you yesterday about offering my excuses to Appuleius. I don't think there will be any bother. Any one you apply to is sure not to

Astura, March 8, B.C. 45

---

[1] Apparently an augur had to bring evidence of ill-health attested by three other augurs to escape attendance on regular meetings and inaugural banquets.

vide et Laenatem et Statilium; tribus enim opus est Sed mihi Laenas totum receperat.

Quod scribis a Iunio te appellatum, omnino Cornificius locuples est; sed tamen scire velim, quando dicar spopondisse et pro patre anne pro filio. Neque eo minus, ut scribis, procuratores Cornifici et Appuleium praediatorem videbis.

Quod me ab hoc maerore recreari vis, facis ut omnia; sed me mihi non defuisse tu testis es. Nihil enim de maerore minuendo scriptum ab ullo est, quod ego non domi tuae legerim. Sed omnem consolationem vincit dolor. Quin etiam feci quod profecto ante me nemo, ut ipse me per litteras consolarer. Quem librum ad te mittam, si descripserint librarii. Adfirmo tibi nullam consolationem esse talem. Totos dies scribo, non quo proficiam quid, sed tantisper impedior; non equidem satis (vis enim urget), sed relaxor tamen omnique vi nitor non ad animum, sed ad vultum ipsum, si queam, reficiendum, idque faciens interdum mihi peccare videor, interdum peccaturus esse, nisi faciam. Solitudo aliquid adiuvat, sed multo plus proficeret, si tu tamen interesses. Quae mihi una causa est hinc discedendi; nam pro malis recte habebat. Quamquam id ipsum doleo. Non enim iam in me idem esse poteris. Perierunt illa, quae amabas.

De Bruti ad me litteris scripsi ad te antea. Pru-

refuse. But see Septimius, Laenas and Statilius. There must be three. However Laenas undertook the whole matter for me.

You say you have been dunned by Junius. Well anyhow Cornificius is rich enough to pay: but I should like to know when they say I went bail for him, and whether it was for the father or the son. Still for all that, do as you say, and see Cornificius' agents and Appuleius the estate agent.

You are as kind as usual in wishing that I could get some relief from my grief; but you can bear witness that it is no fault of mine. For every word that has been written by anyone on the subject of assuaging grief I read at your house. But my sorrow is beyond any consolation. Why, I have done what no one has ever done before, tried to console myself by writing a book. I will send it to you as soon as it is copied out. I assure you no other consolation equals it. I write the whole day long, not that it does any good, but it acts as a temporary check: not very much of that, for the violence of my grief is too strong; but still I get some relief and try with all my might to attain some composure of countenance, if not of mind. In so doing sometimes I think I am doing wrong, and sometimes that I should be doing wrong, if I were not to do it. Solitude helps a little, but it would have much more effect, if you at any rate could be with me: and that is my only reason for leaving, for the place is as right as any could be under the circumstances. However even the idea of seeing you upsets me: for now you can never feel the same towards me. I have lost all you used to love.

I have mentioned Brutus' letter to me before:

denter scriptae, sed nihil, quod me adiuvarent. Quod
ad te scripsit, id vellem, ut ipse adesset. Certe aliquid, quoniam me tam valde amat, adiuvaret. Quodsi
quid scies, scribas ad me velim, maxime autem, Pansa
quando. De Attica doleo, credo tamen Cratero.
Piliam angi veta. Satis est maerere pro omnibus.

XV

CICERO ATTICO SAL.

Scr. Asturae
VII Id.
Mart. a. 709

Apud Appuleium, quoniam in perpetuum non
placet, in dies ut excuser, videbis. In hac solitudine
careo omnium colloquio, cumque mane me in silvam
abstrusi densam et asperam, non exeo inde ante
vesperum. Secundum te nihil est mihi amicius solitudine. In ea mihi omnis sermo est cum litteris.
Eum tamen interpellat fletus; cui repugno, quoad
possum, sed adhuc pares non sumus. Bruto, ut
suades, rescribam. Eas litteras cras habebis. Cum
erit, cui des, dabis.

XVI

CICERO ATTICO SAL.

Scr. Asturae
VI Id. Mart.
a. 709

Te tuis negotiis relictis nolo ad me venire, ego
potius accedam, si diutius impediere. Etsi ne dis-
cessissem quidem e conspectu tuo, nisi me plane nihil
ulla res adiuvaret. Quodsi esset aliquod levamen, id

it was full of wise saws, but nothing that could help me. To you he wrote asking if I should like his company. Yes, it would do me some good, as he has so great an affection for me. If you have any news, please write and let me know, especially when Pansa is going.[1] I am sorry about Attica, but I believe Craterus. Tell Pilia not to worry: my sorrow is enough for all.

XV

CICERO TO ATTICUS, GREETING.

Astura, March 9, B.C. 45

See that my excuses are paid to Appuleius every day, since you do not approve of one general excuse. In this solitude I don't speak to a soul. In the morning I hide myself in a dense and wild wood, and I don't come out till the evening. After you I have not a greater friend than solitude. In it my only converse is with books, though tears interrupt it. I fight against them as much as I can: but as yet I am not equal to the struggle. I will answer Brutus as you suggest. You shall have the letter to-morrow. Give it to a messenger, when you have one.

XVI

CICERO TO ATTICUS, GREETING.

Astura, March 10, B.C. 45

I do not wish you to neglect your business to come to me. I would rather go to you, if you are delayed any longer. However I should never even have come out of sight of you, if it were not that I absolutely could not get relief from anything. If there were any alleviation for my sorrow, it would

[1] To his province in Cisalpine Gaul.

esset in te uno, et, cum primum ab aliquo poterit esse, a te erit. Nunc tamen ipsum sine te esse non possum. Sed nec tuae domi probabatur, nec meae poteram, nec, si propius essem uspiam, tecum tamen essem. Idem enim te impediret, quo minus mecum esses, quod nunc etiam impedit. Mihi nihil adhuc aptius fuit hac solitudine; quam vereor ne Philippus tollat. Heri enim vesperi venerat. Me scriptio et litterae non leniunt, sed obturbant.

XVII

CICERO ATTICO SAL.

Scr. Asturae IV Id. Mart. a. 709

Marcianus ad me scripsit me excusatum esse apud Appuleium a Laterense, Nasone, Laenate, Torquato, Strabone. Iis velim meo nomine reddendas litteras cures gratum mihi eos fecisse. Quod pro Cornificio me abhinc amplius annis xxv spopondisse dicit Flavius, etsi reus locuples est et Appuleius praediator liberalis, tamen velim des operam, ut investiges ex consponsorum tabulis, sitne ita (mihi enim ante aedilitatem meam nihil erat cum Cornificio. Potest tamen fieri; sed scire certum velim), et appelles procuratores, si tibi videtur. Quamquam quid ad me? Verum tamen. Pansae profectionem scribes, cum scies. Atticam salvere iube et eam cura, obsecro, diligenter. Piliae salutem.

be in you alone, and, as soon as any will be possible from anyone, it will come from you. Yet at this very moment I cannot bear your absence. But it did not seem right to stay in your house and I could not stay at my own house; and, if I stayed somewhere nearer, still I should not be with you, for you would be prevented from being with me by the same reason that you are now. For myself, this solitude has suited me better than anything so far, though I am afraid Philippus will destroy it. He came yesterday evening. Writing and reading do not soften my feelings, they only distract them.

XVII

CICERO TO ATTICUS, GREETING.

Marcianus has written to tell me that my excuses were made with Appuleius by Laterensis, Naso, Laenas, Torquatus and Strabo. Please send them a letter on my behalf, thanking them for what they have done. As for what Flavius says, that more than 25 years ago I went bail for Cornificius, though the defendant is well off, and Appuleius is a respectable estate agent, I should be glad, if you would verify the truth of that statement from the account books of the other sureties; for before my aedileship I had no dealings with Cornificius. It may be so: but I should like to know for certain. And please demand payment from his agents, if you think it right. However it's of no importance: but still—. Let me know when Pansa departs, when you know yourself. Pay my respects to Attica, and pray look after her well. Greet Pilia for me.

Astura, March 12, B.C. 45

MARCUS TULLIUS CICERO

XVIII

CICERO ATTICO SAL.

Scr. Asturae
V Id. Mart.
a. 709

Dum recordationes fugio, quae quasi morsu quodam dolorem efficiunt, refugio ad te admonendum. Quod velim mihi ignoscas, cuicuimodi est. Etenim habeo non nullos ex iis, quos nunc lectito auctores, qui dicant fieri id oportere, quod saepe tecum egi et quod a te approbari volo, de fano illo dico, de quo tantum, quantum me amas, velim cogites. Equidem neque de genere dubito (placet enim mihi Cluati) neque de re (statutum est enim), de loco non numquam. Velim igitur cogites. Ego, quantum his temporibus tam eruditis fieri potuerit, profecto illam consecrabo omni genere monimentorum ab omnium ingeniis sumptorum et Graecorum et Latinorum. Quae res forsitan sit refricatura vulnus meum. Sed iam quasi voto quodam et promisso me teneri puto, longumque illud tempus, cum non ero, magis me movet quam hoc exiguum, quod mihi tamen nimium longum videtur. Habeo enim nihil temptatis rebus omnibus, in quo acquiescam. Nam, dum illud tractabam, de quo ad te ante scripsi, quasi fovebam dolores meos; nunc omnia respuo nec quicquam habeo tolerabilius quam solitudinem; quam, quod eram veritus, non obturbavit Philippus. Nam, ut heri me salutavit, statim Romam profectus est.

Epistulam, quam ad Brutum, ut tibi placuerat, scripsi, misi ad te. Curabis cum tua perferendam.

XVIII

CICERO TO ATTICUS, GREETING.

In trying to escape from the painful sting of recollection I take refuge in recalling something to your memory. Whatever you think of it, please pardon me. The fact is I find that some of the authors over whom I am poring now, consider appropriate the very thing that I have often discussed with you, and I hope you approve of it. I mean the shrine. Please give it all the attention your affection for me dictates. For my part I have no doubt about the design (I like Cluatius' design), nor about the erection (on that I am quite determined); but I have some doubts about the place. So please consider it. I shall use all the opportunities of this enlightened age to consecrate her memory by every kind of memorial borrowed from the genius of all the masters, Greek and Latin. Perhaps it will only gall my wound: but I consider myself pledged by a kind of vow or promise; and I am more concerned about the long ages, when I shall not be here, than about my short day, which, short though it is, seems all too long to me. I have tried everything and find nothing that gives me rest. For, while I was engaged on the essay I mentioned before, I was to some extent fostering my grief. Now I reject everything and find nothing more tolerable than solitude. Philippus has not disturbed it as I feared: for after paying me a visit yesterday he returned at once to Rome.

I have sent you the letter I have written at your suggestion to Brutus. Please have it delivered with your own. However I have sent you a copy of it,

Astura, March 11, B.C. 45

MARCUS TULLIUS CICERO

Eius tamen misi ad te exemplum, ut, si minus placeret, ne mitteres.

Domestica quod ais ordine administrari, scribes, quae sint ea. Quaedam enim exspecto. Cocceius vide ne frustretur. Nam, Libo quod pollicetur, ut Eros scribit, non incertum puto. De sorte mea Sulpicio confido et Egnatio scilicet. De Appuleio quid est quod labores, cum sit excusatio facilis?

Tibi ad me venire, ut ostendis, vide ne non sit facile. Est enim longum iter, discedentemque te, quod celeriter tibi erit fortasse faciendum, non sine magno dolore dimittam. Sed omnia, ut voles. Ego enim, quicquid feceris, id cum recte tum etiam mea causa factum putabo.

XVIIIa

CICERO ATTICO SAL.

Scr. Asturae III Id. Mart. a. 709

Heri, cum ex aliorum litteris cognovissem de Antoni adventu, admiratus sum nihil esse in tuis. Sed erant pridie fortasse scriptae quam datae. Neque ista quidem curo; sed tamen opinor propter praedes suos accucurrisse.

Quod scribis Terentiam de obsignatoribus mei testamenti loqui, primum tibi persuade me istaec non curare neque esse quicquam aut parvae curae aut novae loci. Sed tamen quid simile? Illa eos non

---

[1] Antony had bought Pompey's confiscated property, but had not paid for it, and his sureties were in danger of an

LETTERS TO ATTICUS XII. 18–18a

so that, if you don't approve of it, you may not send it.

You say my private affairs are being properly managed. Write and tell me what they are; for there are some things I am expecting to hear about. See that Cocceius does not disappoint me: for I count Libo's promise, of which Eros writes, as trustworthy. My capital of course I leave in Sulpicius' and Egnatius' hands. Why trouble yourself about Appuleius, when my excuse is so easily made?

About coming to me as you suggest, take care not to inconvenience yourself. It is a long way, and it will cost me many a pang to let you go again, when you want to go, which may happen very quickly. But just as you please. Whatever you do, I shall count it right and know you have done it for my sake.

XVIIIa

CICERO TO ATTICUS, GREETING.

Astura, March 13, B.C. 45

When I learned yesterday from other people's letters of Antony's arrival I wondered there was nothing in yours. But perhaps it was written a day earlier than it was dated. It does not matter a bit to me; but I suppose he has rushed back to save his sureties.[1]

You say Terentia is talking about the witnesses to my will. In the first place bear in mind that I am not troubling my head about those things, and this is no time for any new or unimportant business. But anyhow are the two cases parallel? She did

execution on their property. Hence he returned in haste from Narbo, whither he had gone on his way to joining Caesar in Spain. Cf. the second *Philippic*, 76, 77.

adhibuit, quos existimavit quaesituros, nisi scissent,
quid esset.   Num id etiam mihi periculi fuit?   Sed
tamen faciat illa quod ego.   Dabo meum testamentum legendum, cui voluerit; intelleget non potuisse
honorificentius a me fieri de nepote, quam fecerim.
Nam, quod non advocavi ad obsignandum, primum
mihi non venit in mentem, deinde ea re non venit,
quia nihil attinuit.   Tute scis, si modo meministi,
me tibi tum dixisse, ut de tuis aliquos adduceres.
Quid enim opus erat multis?   Equidem domesticos
iusseram.   Tum tibi placuit, ut mitterem ad Silium.
Inde est natum, ut ad Publilium; sed necesse neutrum fuit.   Hoc tu tractabis, ut tibi videbitur.

XIX

CICERO ATTICO SAL.

Scr. Asturae   Est hic quidem locus amoenus et in mari ipso, qui
prid. Id.   et Antio et Circeiis aspici possit; sed ineunda nobis
Mart. a. 709 ratio est, quem ad modum in omni mutatione dominorum, quae innumerabiles fieri possunt in infinita
posteritate, si modo haec stabunt, illud quasi conse-
cratum remanere possit.   Equidem iam nihil egeo
vectigalibus et parvo contentus esse possum.   Cogito
interdum trans Tiberim hortos aliquos parare et quidem ob hanc causam maxime : nihil enim video, quod
tam celebre esse possit.   Sed quos, coram videbimus,
ita tamen, ut hac aestate fanum absolutum sit.   Tu
tamen cum Apella Chio confice de columnis.

LETTERS TO ATTICUS XII. 18a–19

not invite anyone she thought would ask questions, if they did not know the contents of the will. Was I likely to be afraid of anything of the kind? However let her do what I do. I will hand over my will to anyone she likes, to read. She will find I could not have treated my grandson more handsomely than I have. As to my not calling certain people as witnesses, in the first place it never entered my mind, and in the second the reason why it never entered it, was because it was of no importance. You know, if you remember, that I told you to bring some of your friends. What need was there of many? I had asked members of my household. Then you thought I ought to send for Silius. Hence it came about that I sent for Publilius. But neither of them was necessary. Manage the point as you think fit.

XIX

CICERO TO ATTICUS, GREETING.

This is certainly a delightful place, right on the sea and within sight of Antium and Circeii. But we must remember how it may change hands an infinite number of times in the countless years to come, if our empire last, and must arrange that that shrine may remain as consecrated ground. For my part I don't want a large income now and can be contented with little. I think at times of buying some gardens across the Tiber, especially for this reason: I don't see any other place that can be so much frequented. But what gardens, we will consider together; provided only that the shrine must be completed this summer. However settle with Apella of Chios about the columns.

Astura, March 14, B.C. 45

MARCUS TULLIUS CICERO

De Cocceio et Libone quae scribis, approbo, maxime quod de iudicatu meo. De sponsu si quid perspexeris, et tamen quid procuratores Cornifici dicant, velim scire, ita ut in ea re te, cum tam occupatus sis, non multum operae velim ponere. De Antonio Balbus quoque ad me cum Oppio conscripsit, idque tibi placuisse, ne perturbarer. Illis egi gratias. Te tamen, ut iam ante ad te scripsi, scire volo me neque isto nuntio esse perturbatum nec iam ullo perturbatum iri. Pansa si hodie, ut putabas, profectus est, posthac iam incipito scribere ad me, de Bruti adventu quid exspectes, id est quos ad dies. Id, si scies, ubi iam sit, facile coniectura adsequere.

Quod ad Tironem de Terentia scribis, obsecro te, mi Attice, suscipe totum negotium. Vides et officium agi meum quoddam, cui tu es conscius, et, ut non nulli putant, Ciceronis rem. Me quidem id multo magis movet, quod mihi est et sanctius et antiquius, praesertim cum hoc alterum neque sincerum neque firmum putem fore.

XX

CICERO ATTICO SAL.

Scr. Asturae
Id. Mart. a.
709

Nondum videris perspicere, quam me nec Antonius commoverit, nec quicquam iam eius modi possit commovere. De Terentia autem scripsi ad te eis litteris, quas dederam pridie. Quod me hortaris idque a ceteris desiderari scribis, ut dissimulem me tam

LETTERS TO ATTICUS XII. 19–20

What you say about Cocceius and Libo I approve, especially as regards my serving on juries. If you have ascertained anything about my guarantee, I should like to know, and anyhow, what Cornificius' agents say, though I don't want you to take much trouble about the matter, when you are so busy. About Antony, Balbus and Oppius too have written to me saying you wished them to write, to save me from anxiety. I have thanked them. I should wish you to know however, as I have said before, that I was not disturbed at that news and shall never be disturbed at any again. If Pansa has set out to-day, as you thought, henceforth begin to tell me in your letters what you expect about Brutus' return, I mean about what day. That you can easily guess, if you know where he is at the time of writing.

As regards your letter to Tiro about Terentia, I beg you, Atticus, to undertake the whole matter. You see there is a question of my duty concerned, and you know all about that: besides, some think there is my son's interest. With me it is the first point that weighs most, as being the more sacred and the more important: especially as I don't think she is either sincere or reliable about the second.

XX

CICERO TO ATTICUS, GREETING.

Astura, March 15, B.C. 45

You don't seem yet to see how little Antony disturbed me nor how little anything of that kind ever can disturb me now. About Terentia I wrote to you in the letter I sent yesterday. You exhort me and you say others want me to hide the depth of

MARCUS TULLIUS CICERO

graviter dolere, possumne magis, quam quod totos dies consumo in litteris? Quod etsi non dissimulationis, sed potius leniendi et sanandi animi causa facio, tamen, si mihi minus proficio, simulationi certe facio satis.

Minus multa ad te scripsi, quod exspectabam tuas litteras ad eas, quas pridie dederam. Exspectabam autem maxime de fano, non nihil etiam de Terentia. Velim me facias certiorem proximis litteris, Cn. Caepio, Serviliae Claudi pater, vivone patre suo naufragio perierit an mortuo, item Rutilia vivone C. Cotta, filio suo, mortua sit an mortuo. Pertinent ad eum librum, quem de luctu minuendo scripsimus.

XXI

CICERO ATTICO SAL.

Scr. Asturae
XVI K.
Apr. a. 709

Legi Bruti epistulam eamque tibi remisi sane non prudenter rescriptam ad ea, quae requisieras. Sed ipse viderit. Quamquam illud turpiter ignorat. Catonem primum sententiam putat de animadversione dixisse, quam omnes ante dixerant praeter Caesarem, et, cum ipsius Caesaris tam severa fuerit, qui tum praetorio loco dixerit, consularium putat leniores fuisse, Catuli, Servili, Lucullorum, Curionis, Torquati, Lepidi, Gelli, Volcaci, Figuli, Cottae, L. Caesaris, C. Pisonis, M'. Glabrionis, etiam Silani, Murenae, designatorum consulum. Cur ergo in sententiam Catonis? Quia verbis luculentioribus et pluribus rem eandem comprehenderat. Me autem hic laudat, quod rettulerim,

---

[1] Cotta was a celebrated orator, and held the consulship in 75 B.C. His mother Rutilia survived him, according to Seneca (*Consol. ad Helviam*, 16, 7).

my grief. Can I do so better than by spending all my days in writing? Though I do it, not to hide, but rather to soften and to heal my feelings, still, if I do myself but little good, I certainly keep up appearances.

My letter is shorter than it might be, because I am expecting your answer to mine of yesterday. I am most anxious about the shrine and a little about Terentia too. Please let me know in your next letter whether Cn. Caepio, father of Claudius' wife Servilia, perished by shipwreck during his father's life or after his death, and whether Rutilia died before or after her son C. Cotta.[1] They concern the book I have written on the lightening of grief.

XXI

CICERO TO ATTICUS, GREETING.

Astura, March 17, B.C. 45

I have read Brutus' letter and am sending it back to you. It is not at all a sensible answer to the points in which you found him wanting. But that is his look out: though in one thing it shows disgraceful ignorance on his part. He thinks Cato was the first to deliver a speech for the punishment of the conspirators, though everybody except Caesar had spoken before him: and that, though Caesar's speech, delivered from the praetorian bench, was so severe, those of the ex-consuls, Catulus, Servilius, the Luculli, Curio, Torquatus, Lepidus, Gellius, Volcacius, Figulus, Cotta, L. Caesar, C. Piso, M'. Glabrio, and even the consuls elect Silanus and Murena, were milder. Why then was the division taken on Cato's proposal? Because he had summed up the same matter in clearer and fuller words. My merit according to Brutus lay in bringing the affair

non quod patefecerim, quod cohortatus sim, quod denique ante, quam consulerem, ipse iudicaverim. Quae omnia quia Cato laudibus extulerat in caelum perscribendaque censuerat, idcirco in eius sententiam est facta discessio. Hic autem se etiam tribuere multum mi putat, quod scripserit "optimum consulem." Quis enim ieiunius dixit inimicus? Ad cetera vero tibi quem ad modum rescripsit! Tantum rogat, de senatus consulto ut corrigas. Hoc quidem fecisset, etiamsi a librario admonitus esset. Sed haec iterum ipse viderit.

De hortis quoniam probas, effice aliquid. Rationes meas nosti. Si vero etiam a Faberio recedit, nihil negotii est. Sed etiam sine eo posse videor contendere. Venales certe sunt Drusi, fortasse etiam Lamiani et Cassiani. Sed coram.

De Terentia non possum commodius scribere, quam tu scribis. Officium sit nobis antiquissimum. Si quid nos fefellerit, illius malo me quam mei paenitere. Oviae C. Lolli curanda sunt HS c̄. Negat Eros posse sine me, credo, quod accipienda aliqua sit et danda aestimatio. Vellem, tibi dixisset. Si enim res est, ut mihi scripsit, parata, nec in eo ipso mentitur, per te confici potuit. Id cognoscas et conficias velim.

---

[1] *Recedit* is generally altered by editors. But for this rare sense of the word Reid compares *Pro Quinctio*, 38.

[2] 100,000 sesterces.

before the House, not in finding it out, nor in urging them to take steps, nor yet in making up my own mind before I took the House's opinion. And it was because Cato extolled those actions of mine to the skies and moved that they should be put on record, that the vote was taken on his motion. Brutus again seems to think he is giving me high praise by calling me an "excellent consul." Why, has anyone, even a personal enemy, ever used a more grudging term? To the rest of your criticisms too what a poor answer he has given! He only asks you to alter the point about the decree of the Senate. He would have done as much as that at the suggestion of a clerk. But that again is his own look out.

Since you approve of the garden idea, manage it somehow. You know my resources. If I get something back[1] from Faberius, there will be no difficulty. But I think I can manage even without that. Drusus' gardens are certainly for sale, and I think those of Lamianus and Cassianus too. But, when we meet.

About Terentia I cannot say anything more suitable than you do in your letter. Duty must be my first consideration. If I have made a mistake, I would rather have to repent for her sake than for my own. C. Lollius' wife Ovia has to be paid 900 guineas.[2] Eros says it can't be done without me, I suppose because some property has to pass between us at a valuation.[3] I wish he had told you. For, if, as he said, the matter is arranged, and that is not precisely where he is deceiving me, it could be managed through you. Please find out and finish it.

[3] *Aestimatio* = land made over by a debtor to a creditor at a valuation.

MARCUS TULLIUS CICERO

Quod me in forum vocas, eo vocas, unde etiam bonis meis rebus fugiebam. Quid enim mihi foro sine iudiciis, sine curia, in oculos incurrentibus iis, quos aequo animo videre non possum? Quod autem a me homines postulare scribis ut Romae sim neque mihi ut absim concedere,[1] aut aliquatenus[2] eos mihi concedere, iam pridem scito esse, cum unum te pluris quam omnes illos putem. Ne me quidem contemno meoque iudicio multo stare malo quam omnium reliquorum. Neque tamen progredior longius, quam mihi doctissimi homines concedunt; quorum scripta omnia, quaecumque sunt in eam sententiam, non legi solum, quod ipsum erat fortis aegroti, accipere medicinam, sed in mea etiam scripta transtuli, quod certe adflicti et fracti animi non fuit. Ab his me remediis noli in istam turbam vocare, ne recidam.

XXII

CICERO ATTICO SAL.

Scr. Asturae XV K. Apr. a. 709

De Terentia quod mihi omne onus imponis, non cognosco tuam in me indulgentiam. Ista enim sunt ipsa vulnera, quae non possum tractare sine maximo gemitu. Moderare igitur, quaeso, ut potes. Neque enim a te plus, quam potes, postulo. Potes autem, quid veri sit, perspicere tu unus. De Rutilia quoniam videris dubitare, scribes ad me, cum scies, sed quam primum, et num Clodia D. Bruto consulari, filio suo, mortuo vixerit. Id de Marcello aut certe

---

[1] ut Romae . . . concedere *added by old editors.*
[2] aliquatenus *Andresen*: quatenus *MSS.*

In calling me back to the forum, you call me to a place I shunned even in my happy days. What have I to do with a forum, where there are no law-courts, no Senate, and where people are continually obtruding themselves on my sight, whom I cannot endure to see? You say people are demanding my presence at Rome, and will not allow me to be absent, or at any rate only for a certain time. Rest assured that I have long held you at a higher value than them all. Myself too I do not underrate, and I far prefer to trust my own judgment than that of all the rest. However I am not going further than the wisest heads allow. I have not only read all their writings on the point, which in itself shows I am a brave invalid and take my medicine, but I have transferred them to my own work; and that certainly does not argue a mind crushed and enfeebled. Do not call me back from these remedies into that busy life, for fear I relapse.

XXII

CICERO TO ATTICUS, GREETING.

Astura, March 18, B.C. 45

About Terentia, I do not recognise your usual consideration for me in throwing the whole weight of the matter on me. For those are the very wounds I cannot touch without deep groans. So please spare me, if you can. For I am not asking you more than you can do. You and you only can see what is fair. About Rutilia, as you seem to have doubts, write and let me know as soon as you know, but let that be as soon as possible: and also whether Clodia survived her son D. Brutus the ex-consul. The latter you can find out from Marcellus, or at any

de Postumia sciri potest, illud autem de M. Cotta aut de Syro aut de Satyro.

De hortis etiam atque etiam rogo. Omnibus meis eorumque, quos scio mihi non defuturos, facultatibus (sed potero meis) enitendum mihi est. Sunt etiam, quae vendere facile possim. Sed ut non vendam eique usuram pendam, a quo emero, non plus annum, possum adsequi, quod volo, si tu me adiuvas. Paratissimi sunt Drusi; cupit enim vendere. Proximos puto Lamiae; sed abest. Tu tamen, si quid potes, odorare. Ne Silius quidem quicquam utitur suis et is¹ usuris facillime sustentabitur. Habe tuum negotium, nec, quid res mea familiaris postulet, quam ego non curo, sed quid velim, existima.

XXIII

CICERO ATTICO SAL.

Scr. Asturae XIV K. Apr. a. 709

Putaram te aliquid novi, quod eius modi fuerat initium litterarum, "quamvis non curarem, quid in Hispania fieret, tamen te scripturum"; sed videlicet meis litteris respondisti ut de foro et de curia. Sed domus est, ut ais, forum. Quid ipsa domo mihi opus est carenti foro? Occidimus, occidimus, Attice, iam pridem nos quidem, sed nunc fatemur, posteaquam unum, quo tenebamur, amisimus. Itaque solitudinem sequor, et tamen, si qua me res isto adduxerit, enitar, si quo modo potero (potero autem), ut praeter te nemo dolorem meum sentiat, si ullo modo poterit, ne tu

¹ suis et is *Wesenberg*: et iis *MSS*.

rate from Postumia, the former from M. Cotta or Syrus or Satyrus.

About the gardens I earnestly entreat your aid. I must employ all my own resources and those of friends, who I know will not desert me: but I can manage with my own. There are things I could sell easily too. But without selling anything, if I pay interest to the person from whom I buy for no more than a year, I can get what I want, if you assist me. The most available are those of Drusus, as he wants to sell. The next I think are Lamia's; but he is away. However scent out anything you can. Silius again never uses his at all, and he will very easily be satisfied with the interest. Regard it as your own business, and don't consider what suits my purse, for that I don't care, but what suits me.

XXIII

CICERO TO ATTICUS, GREETING.

From the beginning of your letter "though I did not care what happened in Spain, still you would write," I thought you had some news from me: but I see you have answered my letter only as regards the forum and the Senate. But, you say, my house at Rome is a forum. What is the good of the house alone to me, if I have not the forum? I am dead and done for, Atticus, and have been this long while: but now I confess it, when I have lost the one link that bound me to life. So what I want is solitude. Still if in my despite anything drags me to Rome, I shall strive, if possible (and I will make it possible), to keep my grief from all eyes but yours, and, if it is anyhow possible, even from yours.

Astura, March 19, B.C. 45

quidem. Atque etiam illa causa est non veniendi. Meministi, quid ex te Aledius quaesierit. Qui etiam nunc molesti sunt, quid existimas, si venero?

De Terentia ita cura, ut scribis, meque hac ad maximas aegritudines accessione non maxima libera. Et, ut scias me ita dolere, ut non iaceam, quibus consulibus Carneades et ea legatio Romam venerit, scriptum est in tuo annali: haec nunc quaero quae causa fuerit. De Oropo, opinor, sed certum nescio. Et, si ita est, quae controversiae. Praeterea, qui eo tempore nobilis Epicureus fuerit Athenisque praefuerit hortis, qui etiam Athenis πολιτικοὶ fuerint illustres. Quae etiam ex Apollodori puto posse inveniri.

De Attica molestum, sed, quoniam leviter, recte esse confido. De Gamala dubium non mihi erat. Unde enim tam felix Ligus pater? Nam quid de me dicam, cui ut omnia contingant, quae volo, levari non possum?

De Drusi hortis, quanti licuisse tu scribis, id ego quoque audieram, et, ut opinor, heri ad te scripseram; sed quantiquanti, bene emitur, quod necesse est. Mihi, quoquo modo tu existimas (scio enim, ego ipse quid de me existimem), levatio quaedam est, si minus doloris, at officii debiti.

Ad Siccam scripsi, quod utitur L. Cotta. Si nihil conficietur de Transtiberinis, habet in Ostiensi Cotta

Besides there is this reason for not coming. You remember the questions Aledius asked you. They are annoying to me even now. What do you suppose they will be, if I come?

Arrange about Terentia as you say, and rid me of this addition—though not the weightiest—to my weighty griefs and sorrows. To show you that my sorrow is not prostration, you have entered in your Chronicle the date of the visit of Carneades and that famous embassy to Rome:[1] I want to know now the cause of its coming. I think it was about Oropus: but I am not certain. And, if that is so, what was the point in question? Further, who was the most distinguished Epicurean of the time and the head of the Garden at Athens; also who were the famous politicians there? I think you can find all those things in Apollodorus' book.

It is annoying about Attica; but, as it is a mild attack, I expect it will be all right. About Gamala I had no doubt. For why was his father Ligus so fortunate? Need I mention my own case, when I am incapable of getting relief, though everything I wish were to happen.

The price you mention for Drusus' gardens I too had heard, and had written about it to you, yesterday I think. Whatever the price is, what is necessary is cheap. In my eyes, whatever you may think—for I know what I think of myself—it relieves my mind of a bounden duty, if not of sorrow.

I have written to Sicca, because he is intimate with L. Cotta. If nothing can be managed about gardens across the Tiber, Cotta has some at Ostia in

[1] Three celebrated philosophers, Carneades, Diogenes, and Critolaus, came to Rome in 155 B.C. to plead against the fine of 500 talents imposed on Athens for raiding Oropus.

celeberrimo loco, sed pusillum loci, ad hanc rem
tamen plus etiam quam satis. Id velim cogites. Nec
tamen ista pretia hortorum pertimueris. Nec mihi
iam argento nec veste opus est nec quibusdam amoenis locis; hoc opus est. Video etiam, a quibus
adiuvari possim. Sed loquere cum Silio; nihil enim
est melius. Mandavi etiam Siccae. Rescripsit constitutum se cum eo habere. Scribet igitur ad me,
quid egerit, et tu videbis.

XXIV

CICERO ATTICO SAL.

Scr. Asturae
XIII K.
Apr. a. u. c.
709

Bene fecit A. Silius, qui transegerit. Neque enim
ei deesse volebam et, quid possem, timebam. De
Ovia confice, ut scribis. De Cicerone tempus esse
iam videtur; sed quaero, quod illi opus erit, Athenis
permutarine possit an ipsi ferendum sit, de totaque
re, quem ad modum et quando placeat, velim consideres. Publilius iturusne sit in Africam et quando,
ex Aledio scire poteris. Quaeras et ad me scribas
velim. Et, ut ad meas ineptias redeam, velim me
certiorem facias, P. Crassus, Venuleiae filius, vivone
P. Crasso consulari, patre suo, mortuus sit, ut ego
meminisse videor, an post. Item quaero de Regillo,
Lepidi filio, rectene meminerim patre vivo mortuum.
Cispiana explicabis itemque Preciana. De Attica
optime. Et ei salutem dices et Piliae.

a very public place. They are cramped for room, but more than sufficient for this purpose. Please think of that. But don't be afraid of the price you mention for the gardens. I don't want plate or raiment or any pleasant places now: I want this. I see, too, who can help me. But speak to Silius; you can't do better. I have given Sicca a commission too. He answered that he has made an appointment with him. So he will write and tell me what he has done, and you will see to it.

XXIV

CICERO TO ATTICUS, GREETING.

I am glad Silius has settled the business: for I did not want to fail him and was afraid I might not be able to manage it. Settle about Ovia as you say. As to my son it seems high time now; but I want to know whether he can get a draft for his allowance changed at Athens or whether he must take it with him; and as regards the whole matter please consider how and when you think he ought to go. Whether Publilius is going to Africa and when, you can find out from Aledius. Please enquire and let me know. And, to return to my own nonsense, please inform me whether P. Crassus, the son of Venuleia, died in the lifetime of his father, P. Crassus the ex-consul, as I seem to remember, or after his death. I also want to know whether my recollection is right that Regillus, son of Lepidus, died in his father's lifetime. You must settle the business about Cispius and Precius. As to Attica, bravo! Pay my respects to her and to Pilia.

Astura, March 20, B.C. 45

MARCUS TULLIUS CICERO

XXV

CICERO ATTICO SAL.

Scr. Asturae
XII K. Apr.
a. 709

Scripsit ad me diligenter Sicca de Silio seque ad te rem detulisse; quod tu idem scribis. Mihi et res et condicio placet, sed ita, ut numerato malim quam aestimatione. Voluptarias enim possessiones nolet Silius; vectigalibus autem ut his possum esse contentus, quae habeo, sic vix minoribus. Unde ergo numerato? HS $\overline{DC}$ exprimes ab Hermogene, cum praesertim necesse erit, et domi video esse HS $\overline{DC}$. Reliquae pecuniae vel usuram Silio pendemus, dum a Faberio, vel cum aliquo, qui Faberio debet, repraesentabimus. Erit etiam aliquid alicunde. Sed totam rem tu gubernabis. Drusianis vero hortis multo antepono, neque sunt umquam comparati. Mihi crede, una me causa movet, in qua scio me τετυφῶ-σθαι. Sed, ut facis, obsequere huic errori meo. Nam, quod scribis " ἐγγήραμα," actum iam de isto est; alia magis quaero.

XXVI

CICERO ATTICO SAL.

Scr. Asturae
XI K. Apr.
a. 709

Sicca, ut scribit, etiamsi nihil confecerit cum A. Silio, tamen se scribit x Kal. esse venturum. Tuis

---

[1] 600,000 sesterces.

LETTERS TO ATTICUS XII. 25-26

XXV

CICERO TO ATTICUS, GREETING.

Sicca has written to me in detail about Silius, and says he has reported the matter to you; and you say the same in your letter. I am pleased with the property and the conditions, except that I would rather pay money down than assign property at a valuation. Silius will not want show places and I can make myself contented on the income I have, though hardly on less. So where can I get ready money? You can extort 5,000 guineas[1] from Hermogenes, especially as it will be necessary; and I find I have another 5,000 by me. For the rest of the money I will either pay interest to Silius, until I get it from Faberius, or get the money to pay with at once from some debtor of Faberius. There will be some coming in too from other quarters. But you can take charge of the whole matter. I much prefer them to Drusus' gardens; indeed the two have never been compared. Believe me I am actuated by one single motive. I know I have gone silly about it; but continue to bear with my folly. For it is no use your talking about a place to grow old in[2]; that is all over. There are other things I want more.

Astura, March 21, B.C. 45

XXVI

CICERO TO ATTICUS, GREETING.

According to his letter Sicca is coming to me on the 23rd, even if he has not settled anything with A. Silius. You I excuse on the score of business,

Astura, March 22, B.C. 45

[2] For ἐγγήραμα cf. XII. 29; others take it to mean a "solace for old age."

occupationibus ignosco, eaeque mihi sunt notae. De
voluntate tua, ut simul simus, vel studio potius et
cupiditate non dubito. De Nicia quod scribis, si ita
me haberem, ut eius humanitate frui possem, in
primis vellem illum mecum habere. Sed mihi solitudo et recessus provincia est. Quod quia facile
ferebat Sicca, eo magis illum desidero. Praeterea
nosti Niciae nostri imbecillitatem, mollitiam, consuetudinem victus. Cur ergo illi molestus esse velim,
cum mihi ille iucundus esse non possit? Voluntas
tamen eius mihi grata est. Unam rem ad me scripsisti, de qua decrevi nihil tibi rescribere. Spero
enim me a te impetrasse, ut privares me ista molestia.
Piliae et Atticae salutem.

XXVII

CICERO ATTICO SAL.

Scr. Asturae
X K. Apr.
a. 709

De Siliano negotio, etsi mihi non est ignota condicio, tamen hodie me ex Sicca arbitror omnia cogni-
turum. Cottae quod negas te nosse, ultra Silianam
villam est, quam puto tibi notam esse, villula sordida
et valde pusilla, nil agri, ad nullam rem loci satis
nisi ad eam, quam quaero. Sequor celebritatem.
Sed, si perficitur de hortis Sili, hoc est si perficis (est
enim totum positum in te), nihil est scilicet, quod de
Cotta cogitemus.

De Cicerone, ut scribis, ita faciam; ipsi permittam

---

[1] A grammarian of Cos. Cf. VII. 3.

knowing what your business is. I have no doubt of your wish, or rather your eager desire, to be with me. You mention Nicias.[1] If I were in a condition to enjoy his cultivated conversation, he is one of the first persons I should wish to have with me. But solitude and retirement are my proper sphere: and it is because Sicca can content himself with that, that I am the more eager for his visit. Besides you know how delicate our Nicias is, and his luxurious way of living. So why should I want to put him to inconvenience, when he cannot give me any pleasure? However I am grateful to him for wishing it. There is one point you wrote about, which I have made up my mind not to answer. For I hope I have prevailed upon you to relieve me from the burden.[2] My greetings to Pilia and Attica.

XXVII

CICERO TO ATTICUS, GREETING.

Astura, March 23, B.C. 45

As to the business with Silius, I know the terms well enough, but I expect to hear full details from Sicca to-day. Cotta's place, which you say you don't know, is beyond Silius' house, which I think you know. It is a shabby little house and very tiny, with no ground, and not big enough for anything except the purpose for which I require it. I am looking for a public position. But, if the matter is being settled about Silius' gardens,—that is, if you settle it, for it rests entirely with you—there is no reason for thinking of Cotta.

About my son I will do as you say. I will leave

[2] Cicero refers to the arrangement with Terentia for the repayment of her dowry.

MARCUS TULLIUS CICERO

de tempore. Nummorum quantum opus erit, ut permutetur, tu videbis. Ex Aledio, quod scribas, si quid inveneris, scribes. Et ego ex tuis animadverto litteris, et profecto tu ex meis, nihil habere nos quod scribamus, eadem cotidie, quae iam iamque ipsa contrita sunt. Tamen facere non possum, quin cotidie ad te mittam, ut tuas accipiam. De Bruto tamen, si quid habebis. Scire enim iam puto, ubi Pansam exspectet. Si, ut consuetudo est, in prima provincia, circiter Kal. adfuturus videtur. Vellem tardius; valde enim urbem fugio multas ob causas. Itaque id ipsum dubito, an excusationem aliquam ad illum parem; quod quidem video facile esse. Sed habemus satis temporis ad cogitandum. Piliae, Atticae salutem.

XXVIII

CICERO ATTICO SAL.

Scr. Asturae IX K. Apr. a. 709

De Silio nilo plura cognovi ex praesente Sicca quam ex litteris eius. Scripserat enim diligenter. Si igitur tu illum conveneris, scribes ad me, si quid videbitur. De quo putas ad me missum esse, sit missum necne, nescio; dictum quidem mihi certe nihil est. Tu igitur, ut coepisti, et, si quid ita conficies, quod equidem non arbitror fieri posse, ut illi probetur, Ciceronem, si tibi placebit, adhibebis. Eius aliquid interest videri illius causa voluisse, mea qui-

LETTERS TO ATTICUS XII. 27–28

the time to him. See that he is provided with a bill of exchange for as much as is necessary. If you have been able to get anything out of Aledius, as you say, write and tell me. I gather from your letter, and certainly you will from mine, that we have nothing to say to each other—the same old things day after day, though they are long ago worn threadbare. Still I cannot help sending to you every day to get a letter from you. However tell me about Brutus, if you have any information. For I suppose he knows now where to expect Pansa. If, as is generally the case, on the border of his province, he ought to be here about the first of the month. I wish it were later; for there are plenty of reasons why I shun the city. So I am even wondering whether I should make some excuse to him. I could do so easily enough. But there is plenty of time to think about it. My greetings to Pilia and Attica.

XXVIII

CICERO TO ATTICUS, GREETING.

Astura, March 24, B.C. 45

About Silius I have learned nothing more from Sicca now he is here than from his letter, for he had written quite fully. So if you meet him, write and tell me your views. As to the matter on which you think a message has been sent to me, I don't know whether one has been sent or not; certainly not a word has been said to me. So go on as you have begun, and, if you come to any arrangement that satisfies her, which I don't think at all likely, take my son with you to her, if you like. It is to his interest to appear to have wanted to do something to

dem nihil nisi id, quod tu scis, quod ego magni aestimo.

Quod me ad meam consuetudinem revocas, fuit meum quidem iam pridem rem publicam lugere, quod faciebam, sed mitius; erat enim, ubi acquiescerem. Nunc plane nec ego victum nec vitam illam colere possum, nec in ea re, quid aliis videatur, mihi puto curandum; mea mihi conscientia pluris est quam omnium sermo. Quod me ipse per litteras consolatus sum, non paenitet me, quantum profecerim. Maerorem minui, dolorem nec potui nec, si possem, vellem.

De Triario bene interpretaris voluntatem meam. Tu vero nihil, nisi ut illi volent. Amo illum mortuum, tutor sum liberis, totam domum diligo. De Castriciano negotio, si Castricius pro mancipiis pecuniam accipere volet eamque ita[1] solvi, ut nunc solvitur, certe nihil est commodius. Sin autem ita actum est, ut ipsa mancipia abduceret, non mihi videtur esse aequum (rogas enim me, ut tibi scribam, quid mihi videatur); nolo enim negotii Quintum fratrem quicquam habere; quod videor mihi intellexisse tibi videri idem. Publilius, si aequinoctium exspectat, ut scribis Aledium dicere, navigaturus videtur. Mihi autem dixerat per Siciliam. Utrum et quando, velim scire. Et velim aliquando, cum erit tuum commodum, Lentulum puerum visas eique de mancipiis, quae tibi videbitur, attribuas. Piliae, Atticae salutem.

[1] ita *Tyrrell* : ei *MSS*.

LETTERS TO ATTICUS XII. 28

please her; I have no interest in the matter, except that you know of, which I consider important.

You call me back to my old way of life. Well, I have long been bewailing the loss of the Republic, and that was what I was doing, though less strongly; for I had one harbour of refuge. Now I positively cannot follow my old way of life and employment; nor do I think I ought to care what others think about that. My own conscience is more to me than all their talk. For the consolation I have sought in writing, I am not discontented with my measure of success. It has made me show my grief less; but the grief itself I could not lessen, nor would I, if I could.

About Triarius you interpret my wishes well. However do nothing without his family's consent. I love him, though he is dead: I am guardian to his children, and feel affection for all his household. As regards the business with Castricius, if he is willing to take money estimated at its present rate instead of the slaves, nothing could be more convenient. But, if things have gone so far that he is taking the slaves away, I don't think it is fair to him to ask him (you ask me to give you my real opinion); for I don't want my brother Quintus to have any bother, and I rather fancy you take the same view. If Publilius is waiting for the equinox, as you say Aledius tells you, I suppose he is going by sea; but he told me he was going by way of Sicily. I should like to know which it is and when. I should like you too some time at your convenience to pay a visit to little Lentulus[1] and assign him such of the household as you think fit. Love to Pilia and Attica.

---

[1] The son of Tullia and Dolabella, so called because Dolabella was adopted into the plebeian *gens* of the Lentuli in 49 B.C. in order to stand for the tribunate.

MARCUS TULLIUS CICERO

XXIX

CICERO ATTICO SAL.

Scr. Asturae
VIII K.
Apr. a. 709

Silius, ut scribis, hodie. Cras igitur, vel potius cum poteris, scribes, si quid erit, cum videris. Nec ego Brutum vito nec tamen ab eo levationem ullam exspecto; sed erant causae, cur hoc tempore istic esse nollem. Quae si manebunt, quaerenda erit excusatio ad Brutum, et, ut nunc est, mansurae videntur.

De hortis, quaeso, explica. Caput illud est, quod scis. Sequitur, ut etiam mihi ipsi quiddam opus sit; nec enim esse in turba possum nec a vobis abesse. Huic meo consilio nihil reperio isto loco aptius, et de hac re quid tui consilii sit. Mihi persuasum est, et eo magis, quod idem intellexi tibi videri, me ab Oppio et Balbo valde diligi. Cum his communices, quanto opere et quare velim hortos; sed id ita posse, si expediatur illud Faberianum; sintne igitur auctores futuri. Si qua etiam iactura facienda sit in repraesentando, quoad possunt, adducito; totum enim illud desperatum. Denique intelleges, ecquid inclinent ad hoc meum consilium adiuvandum. Si quid erit, magnum est adiumentum; si minus, quacumque ratione contendamus. Vel tu illud "ἐγγήραμα," quem ad modum scripsisti, vel ἐντάφιον putato. De illo Ostiensi nihil est cogitandum. Si hoc non assequimur (a Lamia non puto posse), Damasippi experiendum est.

XXIX

CICERO TO ATTICUS, GREETING.

You say you will see Silius to-day; so to-morrow, or as soon as you can, write, if anything comes of your meeting. I am not trying to avoid Brutus, though I don't expect to get any consolation from him. But there are reasons why I do not want to go there at this particular time. If those reasons continue to exist, I shall have to find some excuse to offer him, and by the look of things at present, I think they will continue.

Astura, March 25, B.C. 45

As for the gardens, please finish the business. The main point is what you know. A further consideration is that I myself want something of the kind; for I cannot exist in a crowd, nor can I be far from you. For my purpose I cannot see anything better adapted than that particular place, and I should like to know what your opinion is. I am quite sure, especially as I see you think so too, that Oppius and Balbus are very fond of me. Let them know how eager I am for the gardens and why; but that it is only possible, if the business with Faberius is settled; and ask whether they will go bail for the payment. Even if I must bear some loss in return for getting ready money, draw them on as far as they will go: for there is no chance of getting the full debt. In fact, find out if they show any inclination to assist my plan. If they do, it is a great assistance; if not, we must manage somehow or other. Look upon it as "a place to grow old in," to use your own phrase, or if you like as a burial place for me. It is no use thinking of the place at Ostium. If we don't get this, I feel sure, we shall not get Lamia's; so we must try for Damasippus' place.

MARCUS TULLIUS CICERO

XXX

CICERO ATTICO SAL.

Scr. Asturae
VI K. Apr.
a. 709

Quaero, quod ad te scribam, sed nihil est. Eadem cotidie. Quod Lentulum invisis, valde gratum. Pueros attribue ei, quot et quos videbitur. De Sili voluntate vendendi et de eo, quanti, tu vereri videris, primum ne nolit, deinde ne tanti. Sicca aliter; sed tibi adsentior. Quare, ut ei placuit, scripsi ad Egnatium. Quod Silius te cum Clodio loqui vult, potes id mea voluntate facere, commodiusque est quam, quod ille a me petit, me ipsum scribere ad Clodium. De mancipiis Castricianis commodissimum esse credo transigere Egnatium, quod scribis te ita futurum putare. Cum Ovia, quaeso, vide ut conficiatur. Quoniam, ut scribis, nox erat, in hodierna epistula plura exspecto.

XXXI

CICERO ATTICO SAL.

Scr. Asturae
IV K. Apr.
a. 709

Silium mutasse sententiam Sicca mirabatur. Equidem magis miror, quod, cum in filium causam conferret, quae mihi non iniusta videtur (habet enim, qualem vult), ais te putare, si addiderimus aliud, a quo refugiat, cum ab ipso id fuerit destinatum, venditurum. Quaeris a me, quod summum pretium

---

[1] Shuckburgh takes this as "I think Egnatius is making a very good bargain." But it seems difficult to get that out of the Latin. Cf. also XII. 32, 1.

[2] Others take *destinare* here in the Plautine sense of

LETTERS TO ATTICUS XII. 30-31

XXX

CICERO TO ATTICUS, GREETING.

I am trying to find something to say to you; but *Astura,* there is nothing. The same things every day. I am *March 27,* much obliged to you for paying a visit to Lentulus. B.C. *45* Assign him as many slaves as you like and select them yourself. As to Silius' inclination to sell and his price, you seem to fear first that he won't want to sell and secondly not at that price. Sicca thought differently; but I agree with you. So, as he suggested, I wrote to Egnatius. Silius wants you to speak to Clodius. You have my full consent to do so, indeed it is more convenient than for me to write to Clodius myself, as he wanted. As to Castricius' slaves I think it is most convenient that Egnatius should carry the matter through,[1] as you say you think he will. With Ovia please see that some arrangement is made. As you say it was night when you wrote, I expect more in to-day's letter.

XXXI

CICERO TO ATTICUS, GREETING.

Sicca is surprised that Silius has changed his *Astura,* mind. For my part I am more surprised that, when *March 29,* he makes his son the excuse—and it seems to me a B.C. *45* good enough excuse, as his son is all he could wish— you say you think he will sell, if we add one other thing, which he shrinks from mentioning, though he has set his heart on it.[2] You ask me to fix my

"buy"; and Shuckburgh translates the end of the sentence "if we should include something else, which he is anxious to get rid of, as he had of his own accord determined not to do so."

constituam et quantum anteire istos hortos Drusi.
Accessi numquam; Coponianam villam et veterem
et non magnam novi, silvam nobilem, fructum autem
neutrius, quod tamen puto nos scire oportere. Sed
mihi utrivis istorum tempore magis meo quam ratione
aestimandi sunt. Possim autem adsequi necne, tu
velim cogites. Si enim Faberianum venderem, explicare vel repraesentatione non dubitarem de Silianis, si modo adduceretur, ut venderet. Si venales
non haberet, transirem ad Drusum vel tanti, quanti
Egnatius illum velle tibi dixit. Magno etiam adiumento nobis Hermogenes potest esse in repraesentando. At tu concede mihi, quaeso, ut eo animo
sim, quo is debeat esse, qui emere cupiat, et tamen
ita servio cupiditati et dolori meo, ut a te regi
velim.

XXXII

CICERO ATTICO SAL.

Scr. Asturae
V K. Apr.
a. 709

Egnatius mihi scripsit. Is si quid tecum locutus
erit (commodissime enim per eum agi potest), ad me
scribes, et id agendum puto. Nam cum Silio non
video confici posse. Piliae et Atticae salutem.

Haec ad te mea manu. Vide, quaeso, quid agendum sit. Publilia ad me scripsit matrem suam, cum
Publilio videretur,[1] ad me cum illo venturam, et se
una, si ego paterer. Orat multis et supplicibus verbis, ut liceat, et ut sibi rescribam. Res quam molesta

[1] videretur *Klotz* : loqueretur *MSS.*

LETTERS TO ATTICUS XII. 31–32

outside price and say how much I prefer them to Drusus' gardens. I have never been in them; I know Coponius' country house is old and not very large and the wood a fine one; but I don't know what either brings in, and that I think we ought to know. But for me either of them should be reckoned rather by my need than by the market value. However please consider whether I can get them or not. If I were to sell my claim on Faberius, I should have no doubt about settling for Silius' gardens even with ready money, if only he could be induced to sell. If his are not for sale, I should have recourse to Drusus, even at the price Egnatius said he asked. Hermogenes too can be a great assistance to me in getting ready money. You must not mind my being eager, one ought to be when one is wanting to make a purchase. However I won't give way to my wishes and my grief so far as not to be ruled by you.

XXXII

CICERO TO ATTICUS, GREETING.

Egnatius has written to me. If he has spoken to you, write and tell me, for the matter can be arranged most conveniently through him, and I think that is what ought to be done. For I don't see any chance of settling with Silius. My greetings to Pilia and Attica.

Astura, March 28, B.C. 45

The rest I have written myself. Pray see what can be done. Publilia has written to me that her mother is coming to me with Publilius at his suggestion and that she will come too, if I will let her. She begs me urgently and humbly to allow her and to answer her. You see what a nuisance it is. I

sit, vides. Rescripsi mi etiam gravius esse quam tum, cum illi dixissem me solum esse velle. Quare nolle me hoc tempore eam ad me venire. Putabam, si nihil rescripsissem, illam cum matre venturam; nunc non puto. Apparebat enim illas litteras non esse ipsius. Illud autem, quod fore video, ipsum volo vitare, ne illae ad me veniant, et una est vitatio, ut ego avolem. Nollem, sed necesse est. Te hoc nunc rogo, ut explores, ad quam diem hic ita possim esse, ut ne opprimar. Ages, ut scribis, temperate.

Ciceroni velim hoc proponas, ita tamen, si tibi non iniquum videbitur, ut sumptus huius peregrinationis, quibus, si Romae esset domumque conduceret, quod facere cogitabat, facile contentus futurus erat, accommodet ad mercedes Argileti et Aventini, et, cum ei proposueris, ipse velim reliqua moderere, quem ad modum ex iis mercedibus suppeditemus ei, quod opus sit. Praestabo nec Bibulum nec Acidinum nec Messallam, quos Athenis futuros audio, maiores sumptus facturos, quam quod ex eis mercedibus recipietur. Itaque velim videas, primum conductores qui sint et quanti, deinde ut sit, qui ad diem solvat, et quid viatici, quid instrumenti satis sit. Iumento certe Athenis nihil opus est. Quibus autem in via utatur, domi sunt plura, quam opus erat, quod etiam tu animadvertis.

XXXIII

CICERO ATTICO SAL.

Scr. Asturae
VII K. Apr.
a. 709

Ego, ut heri ad te scripsi, si et Silius is fuerit, quem tu putas, nec Drusus facilem se praebuerit, Damasippum velim adgrediare. Is, opinor, ita partes

LETTERS TO ATTICUS XII. 32-33

answered that I was even worse than when I told her I wanted to be alone; so she must not think of coming to me at the present time. I thought, if I had not answered, she would come with her mother, now I don't think she will. For evidently that letter is not her own. But the thing that I see will happen—that they will come to me—is the very thing I want to avoid, and the one way of avoiding it is for me to flee. I don't want to, but I must. Now I want you to find out how long I can stay without being caught. Act as you say, with moderation.

Please suggest to my son, that is if you think it fair, that he should keep the expenses of this journey within the rents of my property in the Argiletum and the Aventine, with which he would have been quite contented, if he stayed in Rome and hired a house, as he was thinking of doing: and, when you have made the suggestion, I should like you to arrange the rest, so that we may supply him with what is necessary from those rents. I will guarantee that neither Bibulus nor Acidinus nor Messalla, who I hear are at Athens, will spend more than he will get out of those rents. So please see who the tenants are and what they pay, secondly that they are punctual payers, and what journey money and outfit will suffice. There is certainly no need of a carriage at Athens, while for what he wants on the journey, we have more than enough, as you also observe.

XXXIII

CICERO TO ATTICUS, GREETING.

As I said in my letter yesterday, if Silius is the sort of man you think him and Drusus is hard to deal with, I should like you to approach Damasippus. He

Astura, March 26, B.C. 45

fecit in ripa nescio quotenorum iugerum, ut certa
pretia constitueret; quae mihi nota non sunt. Scribes
ad me igitur, quicquid egeris.

Vehementer me sollicitat Atticae nostrae valetudo,
ut verear etiam, ne quae culpa sit. Sed et paedagogi
probitas et medici adsiduitas et tota domus in omni
genere diligens me rursus id suspicari vetat. Cura
igitur; plura enim non possum.

XXXIV

CICERO ATTICO SAL.

Scr. Asturae   Ego hic vel sine Sicca (Tironi enim melius est)
III K. Apr. facillime possem esse ut in malis, sed, cum scribas
a. 709 videndum mihi esse, ne opprimar, ex quo intellegam
te certum diem illius profectionis non habere, putavi
esse commodius me istuc venire; quod idem video
tibi placere. Cras igitur in Siccae suburbano. Inde,
quem ad modum suades, puto me in Ficulensi fore.
Quibus de rebus ad me scripsisti, quoniam ipse venio,
coram videbimus. Tuam quidem et in agendis nostris
rebus et in consiliis ineundis mihique dandis in ipsis
litteris, quas mittis, benevolentiam, diligentiam, prudentiam mirifice diligo. Tu tamen, si quid cum Silio,
vel illo ipso die, quo ad Siccam venturus ero, certiorem me velim facias, et maxime cuius loci detrac-
tionem fieri velit. Quod enim scribis "extremi,"
vide, ne is ipse locus sit, cuius causa de tota re, ut
scis, est a nobis cogitatum. Hirti epistulam tibi
misi et recentem et benevole scriptam.

I think, has divided up his property on the banks of the Tiber into lots of so and so many acres with fixed prices, which I don't know. So write and tell me, whatever you do.

I am much disturbed about dear Attica's ill-health, it almost makes me fear it is somebody's fault. But the good character of her tutor, the attention of her doctor, and the carefulness of the whole household in every way forbid me to entertain that suspicion. So take care of her. I can write no more.

XXXIV

CICERO TO ATTICUS, GREETING.

I could be very comfortable here considering my troubles even without Sicca—for Tiro is better; but, as you tell me to look out that I'm not caught, by which I am to understand you can't fix a day for the departure I mentioned, I thought the best thing would be to go to Rome. That I see is your opinion too. So to-morrow I shall go to Sicca's suburban place. Then I think I will stay at your place at Ficulea, as you suggest. The matters you have mentioned we will investigate together, as I am coming. Your kindness, diligence and good sense both in managing my affairs and in forming plans and suggesting them in your letters, goes to my heart wonderfully. However, if you do anything with Silius, even on the very day of my arrival at Sicca's place, please let me know, especially which part he wants to withdraw. You say "the far end." Take care that is not the very bit which, as you know, set me thinking about the thing at all. I am sending you a letter of Hirtius', which has just come. It is kindly expressed.

Astura, March 30, B.C. 45

MARCUS TULLIUS CICERO

XXXV

CICERO ATTICO SAL.

Scr. fort. in suburbano Siccae K. vesp. aut mane VI Non. Mai. a. 709

Antequam a te proxime discessi, numquam mihi venit in mentem, quo plus insumptum in monimentum esset quam nescio quid, quod lege conceditur, tantundem populo dandum esse. Quod non magno opere moveret, nisi nescio quomodo, ἀλόγως fortasse nollem illud ullo nomine nisi fani appellari. Quod si volumus, vereor, ne adsequi non possimus nisi mutato loco. Hoc quale sit, quaeso, considera. Nam, etsi minus urgeor meque ipse prope modum collegi, tamen indigeo tui consilii. Itaque te vehementer etiam atque etiam rogo, magis quam a me vis aut pateris te rogari, ut hanc cogitationem toto pectore amplectare.

XXXVI

CICERO ATTICO SAL.

Scr. Asturae V Non. Mai. a. 709

Fanum fieri volo, neque hoc mihi erui potest. Sepulcri similitudinem effugere non tam propter poenam legis studeo, quam ut maxime adsequar ἀποθέωσιν. Quod poteram, si in ipsa villa facerem; sed, ut saepe locuti sumus, commutationes dominorum reformido. In agro ubicumque fecero, mihi videor adsequi posse, ut posteritas habeat religionem. Hae meae tibi ineptiae (fateor enim) ferendae sunt; nam habeo ne me quidem ipsum, quicum tam audacter communicem quam tecum. Sin tibi res, si locus, si institu-

LETTERS TO ATTICUS XII. 35-36

XXXV

CICERO TO ATTICUS, GREETING.

It never occurred to me before I left you the *At Sicca's* other day, that if anything is spent on a monument *house, May* in excess of whatever it is that the law allows, one *1 or 2,* B.C. has to give an equal sum to the public funds. That *45* would not disturb me much, if it were not that somehow or other, perhaps without any good reason, I should be sorry for it to be called anything but a shrine. If I want that, I'm afraid I can't have it, unless I change the site. Please consider what there is in this point. For though I am less anxious and have almost recovered myself, still I want your advice. So I entreat you with more urgency than you wish or allow me to use, to give your whole mind to considering this question.

XXXVI

CICERO TO ATTICUS, GREETING.

I want it to be a shrine, and that idea cannot *Astura,* be rooted out of my mind. I am anxious to avoid *May 3,* B.C. its being taken for a tomb, not so much on account *45* of the legal penalty as to get as near to deification as possible. That would be possible, if it were in the actual house where she died; but, as I have often said, I am afraid of its changing hands. Wherever I build it in the open, I think I can contrive that posterity shall respect its sanctity. You must put up with these foolish fancies of mine, for such I confess they are; for there is no one, not even myself, with whom I talk so freely as with you. But, if you approve of the project, the place and

MARCUS TULLIUS CICERO

tum placet, lege, quaeso, legem mihique eam mitte. Si quid in mentem veniet, quo modo eam effugere possimus, utemur.

Ad Brutum si quid scribes, nisi alienum putabis, obiurgato eum, quod in Cumano esse noluerit propter eam causam, quam tibi dixit. Cogitanti enim mihi nihil tam videtur potuisse facere rustice. Et, si tibi placebit sic agere de fano, ut coepimus, velim cohortere et exacuas Cluatium. Nam, etiamsi alio loco placebit, illius nobis opera consilioque utendum puto. Tu ad villam fortasse cras.

XXXVII
CICERO ATTICO SAL.

Scr. Asturae IV Non. Mai. a. 709

A te heri duas epistulas accepi, alteram pridie datam Hilaro, alteram eodem die tabellario, accepique ab Aegypta liberto eodem die Piliam et Atticam plane belle se habere. Quod mihi Bruti litteras, gratum. Ad me quoque misit; quae litterae mihi redditae sunt tertio decimo die. Eam ipsam ad te epistulam misi et ad eam exemplum mearum litterarum.

De fano, si nihil mihi hortorum invenis, qui quidem tibi inveniendi sunt, si me tanti facis, quanti certe facis, valde probo rationem tuam de Tusculano. Quamvis prudens ad cogitandum sis, sicut es, tamen, nisi magnae curae tibi esset, ut ego consequerer id, quod magno opere vellem, numquam ea res tibi tam belle in mentem venire potuisset. Sed nescio quo pacto celebritatem requiro; itaque hortos mihi confi-

LETTERS TO ATTICUS XII. 36-37

the plan, please read the law and send it to me. If any means of avoiding it occurs to you, we will adopt it.

If you should be writing to Brutus and don't think it out of place, reproach him for refusing to stay in my house at Cumae for the reason he gave you. For when I come to think of it, I don't think he could have done anything ruder. If you think we ought to go on with our idea about the shrine, I should like you to speak to Cluatius and spur him on. For, even if we decide on another place, I think we must make use of his labour and advice. Perhaps you may be going to your country house to-morrow.

XXXVII

CICERO TO ATTICUS, GREETING.

Yesterday I received two letters from you, one given the day before to Hilarus, the other on the same day to a letter-carrier; and on the same day I heard from my freedman Aegypta that Pilia and Attica are quite well. Thanks for sending Brutus' letter. He sent one to me too, which only reached me on the thirteenth day. I am forwarding the letter itself and a copy of my answer.

Astura, May 4, B.C. 45

About the shrine, if you don't get me any gardens —and you ought, if you love me as much as I know you do—I approve highly of your scheme about the place at Tusculum. In spite of your acute powers of thought so bright an idea would never have come into your head, unless you had been very anxious for me to secure what I was very much bent on having. But somehow or other I want a public place; so you must contrive to get me some gardens.

MARCUS TULLIUS CICERO

cias necesse est. Maxuma est in Scapulae celebritas, propinquitas praeterea ubi sis, ne totum diem in villa. Quare, antequam discedis, Othonem, si Romae est, convenias pervelim. Si nihil erit, etsi tu meam stultitiam consuesti ferre, eo tamen progrediar, uti stomachere. Drusus enim certe vendere vult. Si ergo aliud non erit, mea[1] erit culpa, nisi emero. Qua in re ne labar, quaeso, provide. Providendi autem una ratio est, si quid de Scapulanis possumus. Et velim me certiorem facias, quam diu in suburbano sis futurus.

Apud Terentiam tam gratia opus est nobis tua quam auctoritate. Sed facies, ut videbitur. Scio enim, si quid mea intersit, tibi maiori curae solere esse quam mihi.

XXXVIIa

CICERO ATTICO SAL.

Scr. Asturae III Non. Mai. a. 709

Hirtius ad me scripsit Sex. Pompeium Corduba exisse et fugisse in Hispaniam citeriorem, Gnaeum fugisse nescio quo; neque enim curo. Nihil praeterea novi. Litteras Narbone dedit XIIII Kal. Maias. Tu mihi de Canini naufragio quasi dubia misisti. Scribes igitur, si quid erit certius. Quod me a maestitia avocas, multum levaris, si locum fano dederis. Multa mihi εἰς ἀποθέωσιν in mentem veniunt, sed loco valde opus est. Quare etiam Othonem vide.

---

[1] non erit, mea *Graevius*: erit, non mea *M*.

LETTERS TO ATTICUS XII. 37–37a

Scapula's are the most public, and besides they are near and one can be there without spending the whole day in the country. So before you go away, I should very much like you to see Otho, if he is in town. If it comes to nothing, I shall go to such lengths as to rouse your wrath, accustomed though you are to my folly. For Drusus certainly is willing to sell. So, if there is nothing else, it will be my fault if I don't buy. Pray see that I don't make any mistake about it. The only way of making sure against that is to get some of Scapula's land, if possible. Please let me know, too, how long you are going to be in your suburban estate.

With Terentia her liking for you may help as much as your influence. But do as you think fit. For I know that you are generally more solicitous about my interests than I am myself.

XXXVIIa

CICERO TO ATTICUS, GREETING.

Hirtius tells me Sextus Pompeius has quitted *Astura,* Cordova and fled into Northern Spain, while Gnaeus *May 5,* B.C. has fled, I know not whither, nor do I care. No *45* other news. His letter was posted from Narbo on the 18th of April. You mentioned Caninius' shipwreck as though it was doubtful; so let me know, if there is any certain information. You bid me cast off melancholy; very well, you will take a great load off my mind, if you give me a site for the shrine. Many points occur to me in favour of deification; but I badly want a place. So see Otho too.

MARCUS TULLIUS CICERO

XXXVIII

CICERO ATTICO SAL.

Scr. Asturae prid. Non. Mai. a. 709

Non dubito, quin occupatissimus fueris, qui ad me nihil litterarum; sed homo nequam, qui tuum commodum non exspectarit, cum ob eam unam causam missus esset. Nunc quidem, nisi quid te tenuit, suspicor te esse in suburbano. At ego hic scribendo dies totos nihil equidem levor, sed tamen aberro.

Asinius Pollio ad me scripsit de impuro nostro cognato. Quod Balbus minor nuper satis plane, Dolabella obscure, hic apertissime. Ferrem graviter, si novae aegrimoniae locus esset. Sed tamen ecquid impurius? O hominem cavendum! Quamquam mihi quidem—sed tenendus dolor est. Tu, quoniam necesse nihil est, sic scribes aliquid, si vacabis.

XXXVIIIa

CICERO ATTICO SAL.

Scr. Asturae Non. Mai. a. 709

Quod putas oportere pervideri iam animi mei firmitatem graviusque quosdam scribis de me loqui quam aut te scribere aut Brutum, si, qui me fractum esse animo et debilitatum putant, sciant, quid litterarum et cuius generis conficiam, credo, si modo homines sint, existiment me, sive ita levatus sim, ut animum vacuum ad res difficiles scribendas adferam,

---

[1] His nephew, who had joined Caesar and was traducing him.

XXXVIII

CICERO TO ATTICUS, GREETING.

I have no doubt you are overwhelmingly busy, as you send me no letter. But what a scoundrel not to wait for your convenience when I sent him for that very reason! Now I suppose you are in your suburban estate, unless anything kept you. I sit here writing all day long, and get no relief, though it does distract my thoughts.

Asinius Pollio has written about my blackguardly kinsman.[1] Balbus the younger gave me a clear enough hint lately, Dolabella a vague one, and Pollio states it quite openly. I should be annoyed, if there were any room left for a new sorrow. But could anything be more blackguardly? What a dangerous fellow! Though to me—— But I must restrain my feelings. There is no necessity for you to write, only write, if you have time.

Astura, May 6, B.C. 45

XXXVIIIa

CICERO TO ATTICUS, GREETING.

You think there ought to be outward and visible signs of my composure of spirit by this time, and you say some speak more severely of me than either you or Brutus mention in your letters. If those who think my spirit is crushed and broken knew the amount and the nature of the literary work I am doing, I fancy, if they are human, they would hold me guiltless. There is nothing to blame me for, if I have so far recovered as to have my mind free to engage in difficult writing, and even some-

Astura, May 7, B.C. 45

reprehendendum non esse, sive hanc aberrationem a
dolore delegerim, quae maxime liberalis sit doctoque
homine dignissima, laudari me etiam oportere.   Sed,
cum ego faciam omnia, quae facere possim ad me
adlevandum, tu effice id, quod video te non minus
quam me laborare.   Hoc mihi debere videor, neque
levari posse, nisi solvero aut videro me posse solvere,
id est locum, qualem velim, invenero.   Heredes
Scapulae si istos hortos, ut scribis tibi Othonem
dixisse, partibus quattuor factis liceri cogitant, nihil
est scilicet emptori loci; sin venibunt, quid fieri
possit, videbimus.   Nam ille locus Publicianus, qui
est Treboni et Cusini, erat ad me allatus.   Sed scis
aream esse. Nullo pacto probo. Clodiae sane placent,
sed non puto esse venales.   De Drusi hortis, quamvis ab iis abhorreas, ut scribis, tamen eo confugiam,
nisi quid inveneris.   Aedificatio me non movet.
Nihil enim aliud aedificabo nisi id, quod etiam, si
illos non habuero. Κῦρος δ', ε' mihi sic placuit ut
cetera Antisthenis, hominis acuti magis quam eruditi.

XXXIX

CICERO ATTICO SAL.

Scr. Asturae   Tabellarius ad me cum sine litteris tuis venisset,
VIII Id.   existimavi tibi eam causam non scribendi fuisse, quod
Mai. a. 709   pridie scripsisses ea ipsa, ad quam rescripsi, epistula.

LETTERS TO ATTICUS XII. 38a–39

thing to praise me for, if I have chosen this mode of diverting my thoughts as the most cultivated and the one most worthy of a man of learning. But, when I am doing everything I can to cast off my sorrow, do you make an end of what I see you are as much concerned about as myself. I regard it as a debt and I cannot lay aside my care, till I have paid it or see that I can pay it, that is, till I have found a suitable place. If Scapula's heirs are thinking of dividing his garden into four parts and bidding for them among themselves, as you say Otho has told you, then there is no chance for a purchaser; but, if they put them up for sale, we will see what we can do. For that place of Publicius', which now belongs to Trebonius and Cusinius, has been offered to me; but you know it is a mere building plot. I can't put up with it at any price. Clodia's gardens I like, but I don't think they are for sale. Though you dislike Drusus' gardens, I shall have to come back to them, unless you find something. The building does not bother me. I shall only be building what I shall build in any case, even if I don't have the gardens. I am as pleased with "*Cyrus*, Books IV. and V." as with the rest of Antisthenes' works, though he is ingenious rather than learned.[1]

XXXIX

CICERO TO ATTICUS, GREETING.

As a postman arrived without any letter from you, I inferred the reason was what you mentioned yesterday in the letter I am answering. Still I

Astura, May 8, B.C. *45*

---

[1] Antisthenes was the founder of the Cynic School at Athens. He wrote a work in ten volumes, of which two, books 4 and 5, were called *Cyrus*.

Exspectaram tamen aliquid de litteris Asini Pollionis. Sed nimium ex meo otio tuum specto. Quamquam tibi remitto, nisi quid necesse erit, necesse ne habeas scribere, nisi eris valde otiosus.

De tabellariis facerem, quod suades, si essent ullae necessariae litterae, ut erant olim, cum tamen brevioribus diebus cotidie respondebant tempori tabellarii, et erat aliquid, Silius, Drusus, alia quaedam. Nunc, nisi Otho exstitisset, quod scriberemus, non erat; id ipsum dilatum est. Tamen adlevor, cum loquor tecum absens, multo etiam magis, cum tuas litteras lego. Sed, quoniam et abes (sic enim arbitror), et scribendi necessitas nulla est, conquiescent litterae, nisi quid novi exstiterit.

XL

CICERO ATTICO SAL.

Scr. Asturae VII Id. Mai. a. 709

Qualis futura sit Caesaris vituperatio contra laudationem meam, perspexi ex eo libro, quem Hirtius ad me misit; in quo colligit vitia Catonis, sed cum maximis laudibus meis. Itaque misi librum ad Muscam, ut tuis librariis daret. Volo enim eum divulgari; quod quo facilius fiat, imperabis tuis.

Συμβουλευτικὸν saepe conor. Nihil reperio et quidem mecum habeo et Ἀριστοτέλους et Θεοπόμπου libros πρὸς Ἀλέξανδρον. Sed quid simile? Illi, et quae ipsis honesta essent, scribebant et grata Alexandro. Ecquid tu eius modi reperis? Mihi quidem

expected something about Asinius Pollio's letter. But I am too apt to reckon your leisure by my own. However I give you leave not to think yourself bound to write, except in case of necessity, unless you have plenty of leisure.

About the letter carriers I would do as you suggest, if there were any pressing letters, as there were lately. Then, however, the carriers kept up to their time every day, though the days were shorter, and we had something to write about, Silius, Drusus, and other things. Now, if Otho had not cropped up, there is nothing; and even that nothing has been deferred. However it cheers me to talk with you when we are not together, and still more to read your letters. But, as you are not at home (for I think you are not), and there is no necessity to write, let there be a truce to writing, unless some new point arises.

XL

CICERO TO ATTICUS, GREETING.

What sort of thing Caesar's invective against my panegyric will be, I have seen from the book, which Hirtius has sent me. He has collected in it all Cato's faults, but given me high praise. So I have sent the book to Musca to pass on to your copyists; for I want it to be published. To facilitate that, please give your men orders.

I try my hand often at an essay of advice. I can't find anything to say; and yet I have by me Aristotle's and Theopompus' books to Alexander. But what analogy is there? They could write what was honourable to themselves and acceptable to Alexander. Can you find anything of that sort?

Astura, May 9, B.C. 45.

nihil in mentem venit. Quod scribis te vereri, ne et gratia et auctoritas nostra hoc meo maerore minuatur, ego, quid homines aut reprehendant aut postulent, nescio. Ne doleam? Qui potest? Ne iaceam? Quis umquam minus? Dum tua me domus levabat, quis a me exclusus? quis venit, qui offenderet? Asturam sum a te profectus. Legere isti laeti, qui me reprehendunt, tam multa non possunt, quam ego scripsi. Quam bene, nihil ad rem, sed genus scribendi id fuit, quod nemo abiecto animo facere posset. Triginta dies in hortis fui. Quis aut congressum meum aut facilitatem sermonis desideravit? Nunc ipsum ea lego, ea scribo, ut ii, qui mecum sunt, difficilius otium ferant quam ego laborem. Si quis requirit, cur Romae non sim: quia discessus est; cur non sim in eis meis praediolis, quae sunt huius temporis; quia frequentiam illam non facile ferrem. Ibi sum igitur, ubi is, qui optimas Baias habebat, quotannis hoc tempus consumere solebat. Cum Romam venero, nec vultu nec oratione reprehendar. Hilaritatem illam, qua hanc tristitiam temporum condiebamus, in perpetuum amisi, constantia et firmitas nec animi nec orationis requiretur.

De hortis Scapulanis hoc videtur effici posse, aliud tua gratia, aliud nostra, ut praeconi subiciantur. Id nisi fit, excludemur. Sin ad tabulam venimus, vince-

Nothing occurs to me. You say you are afraid my influence and my authority will be lessened by this sorrow of mine. For my part I don't see what people are complaining of or what they expect of me. Not to grieve? How is that possible! Not to be prostrated? No one was ever less prostrated. While I sought relief in your house, I was at home to every caller; and no one, who came, felt in the way. I came to Astura straight from you. Those cheerful friends of yours who blame me cannot read as much as I have written. How well it is written is not to the point, but it was a kind of writing that no one whose spirit was broken could do. I have been thirty days in these gardens. Who has failed to get access to me or conversation with me? At this very moment I am writing and reading so much that the people with me find the holiday harder work than I find working. If anyone asks why I am not in town, "because it is the vacation": why I am not at one of my little places, where it is now the season, "because I could not put up with the crowd of visitors." So I am staying where the man, who prized Baiae more than anyone, always used to spend this part of the year. When I come to Rome, they shall have nothing to find fault with in my looks or my conversation. The cheerfulness with which I used to temper the sadness of the times, I have lost for ever: but there shall be no lack of courage and firmness in my bearing or my words.

As to Scapula's gardens, it seems possible that, as a favour, partly to you and partly to me, they may be put up at auction. If not, we are cut out. But, if it comes to an auction, my eagerness

mus facultates Othonis nostra cupiditate. Nam, quod
ad me de Lentulo scribis, non est in eo. Faberiana
modo res certa sit, tuque enitare, quod facis, quod
volumus, consequemur.

Quod quaeris, quam diu hic: paucos dies. Sed
certum non habeo. Simul ac constituero, ad te
scribam, et tu ad me, quam diu in suburbano sis
futurus. Quo die ego ad te haec misi, de Pilia et
Attica mihi quoque eadem, quae scribis, et scribuntur
et nuntiantur.

XLI

CICERO ATTICO SAL.

*Scr. Asturae
V Id. Mai.
a. 709*

Nihil erat, quod scriberem. Scire tamen volebam,
ubi esses; si abes aut afuturus es, quando rediturus
esses. Facies igitur certiorem. Et, quod tu scire
volebas, ego quando ex hoc loco, postridie Idus Lanuvi constitui manere, inde postridie in Tusculano
aut Romae. Utrum sim facturus, eo ipso die scies.

Scis, quam sit φιλαίτιον συμφορά, minime in te
quidem, sed tamen avide sum adfectus de fano, quod
nisi non dico effectum erit, sed fieri videro (audebo
hoc dicere, et tu, ut soles, accipies), incursabit in te
dolor meus, non iure ille quidem, sed tamen feres
hoc ipsum, quod scribo, ut omnia mea fers ac tulisti.
Omnes tuas consolationes unam hanc in rem velim
conferas. Si quaeris, quid optem, primum Scapulae,

LETTERS TO ATTICUS XII. 40–41

for them will conquer Otho's purse. For as to what you say about Lentulus, he can't run to it. If only the business with Faberius is settled and you make an effort, as you are doing, we shall get what we want.

You ask how long I am staying here: only a few days. But I am not certain. As soon as I have made up my mind, I will write to you: and do you write to me how long you are staying in your estate. On the day on which I am sending this I too got the news you send me about Pilia and Attica by letter and by word of mouth.

XLI

CICERO TO ATTICUS, GREETING.

I have nothing to write. But I want to know, *Astura, May* where you are; and, if you are away or are going *11,* B.C. *45* away, when you will return. So please send me word. You wanted to know, when I am leaving here: I have made up my mind to stay at Lanuvium on the 16th, and then at Tusculum or at Rome on the next day. Which I am going to do, you shall know on the day itself.

You know how full of grievances misfortune makes one. I have none against you; but still I have a hungry longing for the shrine. I will venture to say so much, and you must take it as you usually do, that unless I see it being built, I don't say finished, my resentment will redound on you, quite unjustly, but you will put up with what I am saying, as you put up with all my moods and always have put up with them. I wish you would confine your attempts at consolation to that one point. If you want to know my wishes, I choose Scapula's place

deinde Clodiae, postea, si Silius nolet, Drusus aget iniuste, Cusini et Treboni. Puto tertium esse dominum, Rebilum fuisse certo scio. Sin autem tibi Tusculanum placet, ut significasti quibusdam litteris, tibi adsentiar. Hoc quidem utique perficies, si me levari vis, quem iam etiam gravius accusas, quam patitur tua consuetudo, sed facis summo amore et victus fortasse vitio meo. Sed tamen, si me levari vis, haec est summa levatio vel, si verum scire vis, una.

Hirti epistulam si legeris, quae mihi quasi πρόπλασμα videtur eius vituperationis, quam Caesar scripsit de Catone, facies me, quid tibi visum sit, si tibi erit commodum, certiorem. Redeo ad fanum. Nisi hac aestate absolutum erit, quam vides integram restare, scelere me liberatum non putabo.

XLII

CICERO ATTICO SAL.

Scr. Asturae VI Id. Mai. a. 709

Nullum a te desideravi diem litterarum; videbam enim, quae scribis, et tamen suspicabar vel potius intellegebam nihil fuisse, quod scriberes; a. d. vi Idus vero et abesse te putabam et plane videbam nihil te habere. Ego tamen ad te fere cotidie mittam; malo enim frustra, quam te non habere, cui des, si quid forte sit, quod putes me scire oportere. Itaque accepi vi Idus litteras tuas inanes. Quid enim habebas, quod scriberes? Mi tamen illud, quicquid erat,

first, and then Clodia's: after them, if Silius won't agree and Drusus acts unfairly, Cusinius' and Trebonius' property. I think there is a third owner: I know for certain that Rebilus was one. If however you prefer my place at Tusculum, as you hinted in a letter, I will agree. Get the thing finished somehow, if you want to see me consoled. You are blaming me already more severely than is your wont, but you do it most affectionately, and I suppose it is my fault for making you do so. However, if you wish to see me consoled, this is the best consolation, or rather, to tell the truth, the only one.

If you have read Hirtius' letter, which seems to me a sort of first sketch of the tirade Caesar has written against Cato, let me know what you think of it, if you can. I return to the shrine. If it is not finished this summer (and we have the whole summer before us), I shall not think myself free from guilt.

XLII

CICERO TO ATTICUS, GREETING.

Astura, May 10, B.C. 45

I have never asked you to fix a regular day for your letters: for I see the point you mention, and yet I suspect or rather I know there was nothing for you to write. On the 10th indeed I think you were away, and I am quite aware you have no news. However I shall write to you nearly every day: for I prefer to send letters to no purpose rather than for you to have no messenger to give one to, if there should be anything you think I ought to know. So on the 10th I got your letter with nothing in it. For what was there for you to put in it? However, the little

non molestum fuit, ut nihil aliud, scire me novi te nihil habere.

Scripsisti tamen nescio quid de Clodia. Ubi ergo ea est aut quando ventura? Placet mihi res sic, ut secundum Othonem nihil magis. Sed neque hanc vendituram puto (delectatur enim et copiosa est), et, illud alterum quam sit difficile, te non fugit. Sed, obsecro, enitamur, ut aliquid ad id, quod cupio, excogitemus.

Ego me hinc postridie Id. exiturum puto, sed aut in Tusculanum aut domum, inde fortasse Arpinum. Cum certum sciero, scribam ad te.

XLIII

CICERO ATTICO SAL.

Scr. Asturae Venerat mihi in mentem monere te, ut id ipsum,
IV Id. Mai. quod facis, faceres. Putabam enim commodius te
a. 709 idem istud domi agere posse interpellatione sublata.

Ego postridie Idus, ut scripsi ad te ante, Lanuvi manere constitui, inde aut Romae aut in Tusculano; scies ante, utrum. Quod scribis[1] recte illam rem fore levamento, bene facis, tamen id est[2] mihi crede perinde, ut existimare tu non potes. Res indicat quanto opere id cupiam, cum tibi audeam confiteri, quem id non ita valde probare arbitrer. Sed ferendus tibi in hoc meus error. Ferendus? immo vero etiam adiuvandus. De Othone diffido, fortasse quia cupio. Sed tamen maior etiam res est quam facultates nos-

[1] scribis *Boot* : scies *M*.
[2] tamen id est *Wesenberg* : cum id esse *M*.

LETTERS TO ATTICUS XII. 42-43

there was, was pleasant to me: if nothing else, it taught me you had no news.

But you say something or other about Clodia. Where is she then or when is she coming? I prefer her grounds to anyone's except Otho's. But I don't think she will sell: she likes the place and has plenty of money: and how difficult the other thing is, you are well aware. But pray let us make an effort to think out some way of getting what I want.

I think of leaving here on the 16th; but either for Tusculum or for Rome, and then on perhaps to Arpinum. When I know for certain, I will write.

XLIII

CICERO TO ATTICUS, GREETING.

It had occurred to me to advise you to do exactly what you are doing. For I thought you could get that particular business over more conveniently at home without any fear of interruption.

Astura, May 12, B.C. 45

As I said before, I intend to stop at Lanuvium on the 16th, and then either at Rome or Tusculum. You shall know in advance which. You are right in saying that will lighten my sorrow, but believe me it will do so to an extent which you cannot imagine. How eagerly I desire it you can judge from my daring to confess it to you, though I think you do not very much approve of it. But you must bear with my aberration. Bear with it? Nay you must help me in it. I have doubts about Otho, perhaps because I am eager for his place. But anyhow the property is beyond my means, espe-

trae, praesertim adversario et cupido et locuplete et herede. Proximum est, ut velim Clodiae. Sed, si ista minus confici possunt, effice quidvis. Ego me maiore religione, quam quisquam fuit ullius voti, obstrictum puto. Videbis etiam Trebonianos, etsi absunt domini. Sed, ut ad te heri scripsi, considerabis etiam de Tusculano, ne aestas effluat; quod certe non est committendum.

XLIV

CICERO ATTICO SAL.

Scr. Asturae III Id. Mai. a. 709

Et Hirtium aliquid ad te συμπαθῶς de me scripsisse facile patior (fecit enim humane) et te eius epistulam ad me non misisse multo facilius; tu enim etiam humanius. Illius librum, quem ad me misit de Catone, propterea volo divulgari a tuis, ut ex istorum vituperatione sit illius maior laudatio.

Quod per Mustelam agis, habes hominem valde idoneum meique sane studiosum iam inde a Pontiano. Perfice igitur aliquid. Quid autem aliud, nisi ut aditus sit emptori? quod per quemvis heredem potest effici. Sed Mustelam id perfecturum, si rogaris, puto. Mihi vero et locum, quem opto, ad id, quod volumus, dederis et praeterea ἐγγήραμα. Nam illa Sili et Drusi non satis οἰκοδεσποτικὰ mihi videntur. Quid enim? sedere totos dies in villa? Ista igitur malim, primum Othonis, deinde Clodiae. Si nihil fiet, aut Druso ludus est suggerendus aut utendum Tusculano.

cially when we have to bid against a man, who is eager for the place, wealthy and one of his heirs. What I should prefer after that is Clodia's. But, if nothing can be done about those, do anything. I count myself more bound by sacred obligation than anyone ever was by any vow. Look into Trebonius' place too, though the owners are away. But, as I wrote yesterday, consider my Tusculan place too, that the summer may not slip away. That certainly must not happen.

XLIV

CICERO TO ATTICUS, GREETING.

Astura, May 13, B.C. 45

I am not at all annoyed that Hirtius wrote to you about me in a sympathetic tone (he was acting kindly), and still less at your not sending his letter to me, in which you were acting even more kindly. The reason why I want your copyists to circulate the book he sent me about Cato, is that their abuse may enhance Cato's reputation.

You say you are negotiating through Mustela. He is a very suitable person and very devoted to me since the affair of Pontianus. So get something settled. But what is wanted except an opening for a purchaser? And that could be got through any of the heirs. But I think Mustela will manage that, if you ask him. You will have provided me not only with the very place I want for my purpose, but a place to grow old in besides. For Silius' and Drusus' places don't seem to me quite fit for a paterfamilias. Why, I should have to spend whole days in the country house. So I prefer the others, Otho's first and then Clodia's. If nothing comes of it, then we must play a trick on Drusus or fall back on the place at Tusculum.

MARCUS TULLIUS CICERO

Quod domi te inclusisti, ratione fecisti; sed, quaeso, confice et te vacuum redde nobis. Ego hinc, ut scripsi antea, postridie Idus Lanuvi, deinde postridie in Tusculano. Contudi enim animum et fortasse vici, si modo permansero. Scies igitur fortasse cras, summum perendie.

Sed quid est, quaeso? Philotimus nec Carteiae Pompeium teneri (qua de re litterarum ad Clodium Patavinum missarum exemplum mihi Oppius et Balbus miserant, se id factum arbitrari) bellumque narrat reliquum satis magnum. Solet omnino esse Fulviniaster. Sed tamen, si quid habes. Volo etiam de naufragio Caniniano scire quid sit.

Ego hic duo magna συντάγματα absolvi; nullo enim alio modo a miseria quasi aberrare possum. Tu mihi, etiamsi nihil erit, quod scribas, quod fore ita video, tamen id ipsum scribas velim, te nihil habuisse, quod scriberes, dum modo ne his verbis.

XLV

CICERO ATTICO SAL.

Scr. in Tusculano XVI K. Iun. a. 709

De Attica optime. Ἀκηδία tua me movet, etsi scribis nihil esse. In Tusculano eo commodius ero, quod et crebrius tuas litteras accipiam et te ipsum non numquam videbo; nam ceteroqui ἀνεκτότερα erant Asturae. Nec haec, quae refricant, hic me magis

---

[1] Of Fulvinius nothing is known, save what is inferred from this passage, that he was a person given to spreading false reports.

LETTERS TO ATTICUS XII. 44-45

You have done wisely in shutting yourself up at home. But please get your business over and let me find you with some leisure again. As I said before, I am going from here to Lanuvium on the 16th, then on the 17th to Tusculum. For I have crushed down my feelings and perhaps have conquered them, if only it will last. So you shall hear to-morrow perhaps, at the latest the day after.

But what is this, pray? Philotimus says Pompey is not shut in at Carteia (about that Oppius and Balbus had sent me a copy of a letter to Clodius of Patavium, saying they thought he was) and that there is quite an important war yet to come. Of course he always is a parody of Fulvinius.[1] However have you any news? I want to know the facts about the shipwreck of Caninius too.

I have finished two large treatises [2] here. It was the only way I could get away from my misery. As for you, even if you have nothing to write, which I think will be the case, write and tell me that you have nothing to say, provided you don't use those very words.

XLV

CICERO TO ATTICUS, GREETING.

That's good news about Attica. I am worried *Tusculum*, about your listlessness, though you say it is nothing. *May 17,* B.C. I shall find Tusculum more convenient, as I shall *45* get letters from you more frequently and see you yourself at times: for in other respects things were more endurable at Astura. My feelings are not

---

[2] The *Academica* and *De Finibus*, unless, as Reid suggests, the *Academica* alone is meant, as that was originally divided into two books.

MARCUS TULLIUS CICERO

angunt; etsi tamen, ubicumque sum, illa sunt mecum.
De Caesare vicino scripseram ad te, quia cognoram
ex tuis litteris. Eum σύνναον Quirini malo quam
Salutis. Tu vero pervulga Hirtium. Id enim ipsum
putaram, quod scribis, ut, cum ingenium amici
nostri probaretur, ὑπόθεσις vituperandi Catonis irrideretur.

XLVI

CICERO ATTICO SAL.

Scr. Asturae Id. Mai. a. 709

Vincam, opinor, animum et Lanuvio pergam in
Tusculanum. Aut enim mihi in perpetuum fundo
illo carendum est (nam dolor idem manebit, tantum
modo occultius), aut nescio, quid intersit, utrum illuc
nunc veniam an ad decem annos. Neque enim ista
maior admonitio, quam quibus adsidue conficior et
dies et noctes. "Quid ergo?" inquies, "nihil litterae?"
In hac quidem re vereor ne etiam contra;
nam essem fortasse durior. Exculto enim animo
nihil agreste, nihil inhumanum est.

Tu igitur, ut scripsisti, nec id incommodo tuo.
Vel binae enim poterunt litterae. Occurram etiam,
si necesse erit. Ergo id quidem, ut poteris.

---

[1] A statue of Caesar with the inscription *Deo Invicto* had been put recently in the temple of Quirinus on the Quirinal

LETTERS TO ATTICUS XII. 45-46

more harrowed by galling memories here than there; though to be sure, wherever I am, they are with me. I wrote to you about your "neighbour" Caesar, because I learned about it from your letters. I would rather see him sharing the temple of Quirinus than of Safety.[1] Yes, publish Hirtius' book. I thought the same as you say, that our friend's talent was shown by it, while its object, blackening Cato's character, only looked ridiculous.

XLVI

CICERO TO ATTICUS, GREETING.

I think I shall conquer my feelings and go from Lanuvium to Tusculum. For I must either give up that estate for ever (for my grief will remain the same, only less visible), or it does not matter a straw whether I go there now or ten years hence. The place will not remind me of her any more than the thoughts that harass me day and night. "Oh!" you will say, "so books do not help." In this respect I am afraid they make it worse: perhaps I should have been braver without. For in a cultivated mind there is no roughness and no insensibility.

So you will come to me as you said, and only that if convenient. A letter apiece will be enough. I will even come to meet you, if necessary. So that shall be as you find possible.

Astura, May 15, B.C. 45

hill, which he had restored after its destruction by fire in 49 B.C. Atticus' house and the temple of Salus were also on the Quirinal.

MARCUS TULLIUS CICERO

XLVII

CICERO ATTICO SAL.

Scr. Lanuvi XVII K. Iun. a. 709

De Mustela, ut scribis, etsi magnum opus est. Eo magis delabor ad Clodiam. Quamquam in utroque Faberianum nomen explorandum est. De quo nihil nocuerit si aliquid cum Balbo eris locutus, et quidem, ut res est, emere nos velle, nec posse sine isto nomine, nec audere re incerta. Sed quando Clodia Romae futura est, et quanti rem aestimas? Eo prorsus specto, non quin illud malim, sed et magna res est et difficile certamen cum cupido, cum locuplete, cum herede. Etsi de cupiditate nemini concedam; ceteris rebus inferiores sumus. Sed haec coram.

XLVIII

CICERO ATTICO SAL.

Scr. Lanuvi XVI K. Iun. mane a. 709

Hirti librum, ut facis, divulga. De Philotimo idem et ego arbitrabar. Domum tuam pluris video futuram vicino Caesare. Tabellarium meum hodie exspectamus. Nos de Pilia et Attica certiores faciet.

Domi te libenter esse facile credo. Sed velim scire, quid tibi restet, aut iamne confeceris. Ego te in Tusculano exspecto, eoque magis, quod Tironi statim te venturum scripsisti et addidisti te putare opus esse.

LETTERS TO ATTICUS XII. 47–48

XLVII

CICERO TO ATTICUS, GREETING.

About Mustela, do as you say, though it will be a big business. For that reason I incline more to Clodia; though in both cases we must find out about Faberius' debt. There will be no harm in your speaking to Balbus about it and telling him, what is the truth, that we want to buy, but cannot without getting in that debt, and dare not, until something is settled. But when is Clodia going to be in Rome, and how much do you think it will cost? Why I turn my thoughts to it is not that I should not prefer the other, but it is a big venture and it is difficult to contend with one who is eager for it, rich and one of the heirs. As far as eagerness goes, I yield to no one, but in the other respects we are worse off. However of this when we meet.

Lanuvium, May 16, B.C. 45

XLVIII

CICERO TO ATTICUS, GREETING.

Go on publishing Hirtius' book. About Philotimus I agree with you. I see your house will go up in value now you have Caesar for a neighbour. I am expecting my messenger to-day. He will tell me about Pilia and Attica.

I can easily believe you are glad to be at home: but I should like to know what business you still have or if you have finished now. I am expecting you at Tusculum, especially as you told Tiro you were coming at once, adding that you thought it necessary.

Lanuvium, May 17, B.C. 45

MARCUS TULLIUS CICERO

XLIX

CICERO ATTICO SAL.

Scr. in Tusculano XIV K. Iun. a. 709

Sentiebam omnino, quantum mihi praesens prodesses, sed multo magis post discessum tuum sentio. Quam ob rem, ut ante ad te scripsi, aut ego ad te totus aut tu ad me, quod licebit.

Heri non multo post, quam tu a me discessisti, puto, quidam urbani, ut videbantur, ad me mandata et litteras attulerunt a. C. Mario C. f. C. n. multis verbis "agere mecum per cognationem, quae mihi secum esset, per eum Marium, quem scripsissem, per eloquentiam L. Crassi, avi sui, ut se defenderem," causamque suam mihi perscripsit. Rescripsi patrono illi nihil opus esse, quoniam Caesaris, propinqui eius, omnis potestas esset, viri optimi et hominis liberalissimi; me tamen ei fauturum. O tempora! fore, cum dubitet Curtius consulatum petere! Sed haec hactenus.

De Tirone mihi curae est. Sed iam sciam, quid agat. Heri enim misi, qui videret; cui etiam ad te litteras dedi. Epistulam ad Ciceronem tibi misi. Horti quam in diem proscripti sint, velim ad me scribas.

---

[1] An impostor named Amatias or Herophilus. He was a veterinary surgeon, and was put to death by Antony after he had set up a column in the forum in Caesar's memory.

XLIX

CICERO TO ATTICUS, GREETING.

Tusculum, May 19, B.C. 45

I felt all the time how much good your presence was doing me: but I feel it still more since you have gone. So, as I wrote to you before, either I must come to you entirely or you to me, according as it can be managed.

Yesterday, soon after your departure, I think, some people, who looked like city men, brought me a message and a letter from Gaius Marius, son and grandson of Gaius.[1] He begged me in the name of our relationship, in the name of Marius, on whom I had written, and by the eloquence of his grandfather, L. Crassus, to defend him: and he stated his case in full. I wrote back that he had no need of an advocate since his relative Caesar was omnipotent, and he was the best and most liberal of men: but I would support him. What times these are! To think of Curtius wondering whether to stand for the consulship. But enough of this.

I am anxious about Tiro. But I shall know soon how he is, for yesterday I sent a man to see, and I gave him a letter to you too. I have sent you a letter for my son. Please tell me for what day the sale of the gardens is advertised.

Marius married Julia, aunt of Caesar; their son was adopted by Gratidia, grandmother of Cicero, and married a daughter of L. Crassus, the orator. Hence the claims of relationship asserted in this letter.

MARCUS TULLIUS CICERO

L

CICERO ATTICO SAL.

Scr. in Tusculano XV K. Iun. a. 709

Ut me levarat tuus adventus, sic discessus adflixit. Quare, cum poteris, id est cum Sexti auctioni operam dederis, revises nos. Vel unus dies mihi erit utilis, quid dicam "gratus"? Ipse Romam venirem, ut una essemus, si satis consultum quadam de re haberem.

LI

CICERO ATTICO SAL.

Scr. in Tusculano XIII K. Iun. a. 709

Tironem habeo citius, quam verebar. Venit etiam Nicias, et Valerium hodie audiebam esse venturum. Quamvis multi sint, magis tamen ero solus, quam si unus esses. Sed exspecto te, a Peducaeo utique, tu autem significas aliquid etiam ante. Verum id quidem, ut poteris.

De Vergilio, ut scribis. Hoc tamen velim scire, quando auctio. Epistulam ad Caesarem mitti video tibi placere. Quid quaeris? mihi quoque hoc idem maxime placuit, et eo magis, quod nihil est in ea nisi optimi civis, sed ita optimi, ut tempora; quibus parere omnes πολιτικοὶ praecipiunt. Sed scis ita nobis esse visum, ut isti ante legerent. Tu igitur id curabis. Sed, nisi plane iis intelleges placere, mittenda non est. Id autem utrum illi sentiant anne simulent, tu intelleges. Mihi simulatio pro repudiatione fuerit. Τοῦτο δὲ μηλώσῃ.

LETTERS TO ATTICUS XII. 50-51

L

CICERO TO ATTICUS, GREETING.

Your departure has depressed me as much as *Tusculum,* your arrival cheered me. So, when you can, that *May 18,* B.C. is after you have attended Sextus' auction, visit *45* me again. Even a single day will do me good, not to speak of the pleasure. I would come to Rome that we might be together, if I could make up my mind satisfactorily on a certain point.

LI

CICERO TO ATTICUS, GREETING.

I have Tiro back with me earlier than I expected. *Tusculum,* Nicias has come too and to-day I hear Valerius is *May 20,* B.C. coming. However many come, I shall be more *45* lonely than if you alone were here. But I expect you, at any rate after you've finished with Peducaeus; and you give some hint of an even earlier date. But let that be as you can.

For Vergilius, as you say. I should however like to know when the auction is. I see you think the letter ought to be sent to Caesar. Well, I thought so too very strongly, especially as there is nothing in it that the most loyal of citizens might not have written; loyal, that is to say, in the present circumstances, to which all politicians tell us we should bow. But you know I thought your Caesarian friends ought to read it first: so you must see to that. But, unless you feel sure they approve of it, it must not be sent. You will know whether they really think so or are pretending. I shall count pretence as rejection. You must probe that point.

MARCUS TULLIUS CICERO

De Caerellia quid tibi placeret, Tiro mihi narravit; debere non esse dignitatis meae, perscriptionem tibi placere:

"Hoc métuere, alterum ín metu non pónere."

Sed et haec et multa alia coram. Sustinenda tamen, si tibi videbitur, solutio est nominis Caerelliani, dum et de Metone et de Faberio sciamus.

LII

CICERO ATTICO SAL.

Scr. in Tusculano XII K. Iun. a. 709

L. Tullium Montanum nosti, qui cum Cicerone profectus est. Ab eius sororis viro litteras accepi Montanum Planco debere, quod praes pro Flaminio sit, HS $\overline{XX}$; de ea re nescio quid te a Montano rogatum. Sane velim, sive Plancus est rogandus, sive qua re potes illum iuvare, iuves. Pertinet ad nostrum officium. Si res tibi forte notior est quam mihi, aut si Plancum rogandum putas, scribas ad me velim, ut, quid rei sit et quid rogandum, sciam. De epistula ad Caesarem quid egeris, exspecto. De Silio non ita sane laboro. Tu mi aut Scapulanos aut Clodianos efficias necesse est. Sed nescio quid videris dubitare de Clodia; utrum quando veniat, an sintne venales? Sed quid est, quod audio Spintherem fecisse divortium?

LETTERS TO ATTICUS XII. 51-52

Tiro has told me what you think about Caerellia: that it ill suits my dignity to be in debt, and that I should give a note of hand,

"That you should fear the one and hold the other safe!"

But of this, and much else, when we meet. However, we must hold over the debt to Caerellia, if you agree, till we know about Meton and Faberius.

LII

CICERO TO ATTICUS, GREETING.

You know L. Tullius Montanus who has gone with *Tusculum,* my son. I have received a letter from his sister's *May 21*, B.C. husband saying that, through going bail for Flaminius, *45* Montanus owes Plancus nearly £200;[1] and that he has made some request to you about it. I should like you to assist him by speaking to Plancus or in any other way you can. I feel under an obligation to help him. If you know more about it than I do, or if you think Plancus should be spoken to, I wish you would write to me, that I may know how the matter stands, and what I ought to ask him. I am awaiting news as to what you have done about the letter to Caesar. About Silius I am not much concerned. You must get me either Scapula's or Clodia's gardens. But you seem to have some doubts about Clodia. Is it about the date of her arrival or as to whether the gardens are for sale? But what is this that I hear about Spinther divorcing his wife?

[1] 20,000 sesterces.

MARCUS TULLIUS CICERO

De lingua Latina securi es animi. Dices: "Qui talia conscribis?" Ἀπόγραφα sunt, minore labore fiunt; verba tantum adfero, quibus abundo.

LIII

CICERO ATTICO SAL.

Scr. in Tusculano XI K. Iun. a. 709

Ego, etsi nihil habeo, quod ad te scribam, scribo tamen, quia tecum loqui videor. Hic nobiscum sunt Nicias et Valerius. Hodie tuas litteras exspectabamus matutinas. Erunt fortasse alterae posmeridianae, nisi te Epiroticae litterae impedient; quas ego non interpello. Misi ad te epistulas ad Marcianum et ad Montanum. Eas in eundem fasciculum velim addas, nisi forte iam dedisti.

LETTERS TO ATTICUS XII. 52-53

Make your mind easy about the Latin language. You will say, "What, when you write on such subjects?"[1] They are copies, and don't give me much trouble. I only supply words, and of them I have plenty.

LIII

CICERO TO ATTICUS, GREETING

Tusculum, May 22, B.C. 45

Though I have nothing to say to you, I write all the same, because I feel as though I were talking to you. Nicias and Valerius are here with me. I am expecting a letter from you early to-day. Perhaps there will be another in the afternoon, unless your letter to Epirus hinders you: I don't want to interrupt that. I have sent you letters for Marcianus and for Montanus. Please put them in the same packet, unless you have sent it off already.

---

[1] Atticus had commented on the difficulty of rendering Greek philosophic terms in Latin.

M. TULLI CICERONIS
EPISTULARUM AD ATTICUM
LIBER TERTIUS DECIMUS

I

CICERO ATTICO

Scr. in Tusculano X K. Iun. a. 709

Ad Ciceronem ita scripsisti, ut neque severius neque temperatius scribi potuerit, nec magis quem ad modum ego maxime vellem; prudentissime etiam ad Tullios. Quare aut ista proficient, aut aliud agamus. De pecunia vero video a te omnem diligentiam adhiberi vel potius iam adhibitam esse. Quod si efficis, a te hortos habebo. Nec vero ullum genus possessionis est, quod malim, maxime scilicet ob eam causam, quae suscepta est; cuius festinationem mihi tollis, quoniam de aestate polliceris vel potius recipis. Deinde etiam ad καταβίωσιν maestitiamque minuendam nihil mihi reperiri potest aptius; cuius rei cupiditas impellit me interdum, ut te hortari velim. Sed me ipse revoco; non enim dubito, quin, quod me valde velle putes, in eo tu me ipsum cupiditate vincas. Itaque istuc iam pro facto habeo.

Exspecto, quid istis placeat de epistula ad Caesarem. Nicias te, ut debet, amat vehementerque tua sui memoria delectatur. Ego vero Peducaeum nos-

CICERO'S LETTERS TO ATTICUS
BOOK XIII

I

CICERO TO ATTICUS, GREETING.

You used just the right amount of severity and of moderation in your letter to my son, and it was exactly as I should have wished it to be. Your notes, too, to the Tullii[1] were full of good advice. So either those letters will set things right or we shall have to try some other means. As to the money, I see you are making every effort, or rather you have done so already. If you manage it, I shall owe the gardens to you. Indeed, there is no other kind of property I should prefer, especially for the matter I have in hand. You remove my impatience by your promise, or rather your pledge, about the summer. There is nothing either that could be found more likely to solace my declining years and my sorrow. My eagerness for it impels me at times to urge you to haste. But I restrain myself, for I have no doubt that, as you know I want it very much, your eagerness more than equals mine. So I count the matter as already settled.

I am waiting to hear what your friends decide about the letter to Caesar. Nicias is as devoted to you, as he ought to be, and is highly delighted at your remembering him. I am extremely fond of

Tusculum, May 23, B.C. 45

[1] L. Tullius Montanus and M. Tullius Marcianus, who were at Athens with Cicero's son.

trum vehementer diligo; nam et, quanti patrem feci, totum in hunc et ipsum per se aeque amo atque illum amavi, te vero plurimum, qui hoc ab utroque nostrum fieri velis. Si hortos inspexeris, et si de epistula certiorem me feceris, dederis mihi, quod ad te scribam; si minus, scribam tamen aliquid. Numquam enim derit.

II
CICERO ATTICO SAL.

Scr. in Tusculano IX K. Iun. a. 709

Gratior mihi celeritas tua quam ipsa res. Quid enim indignius? Sed iam ad ista obduruimus et humanitatem omnem exuimus. Tuas litteras hodie exspectabam, nihil equidem ut ex iis novi; quid enim? verum tamen ——.

IIa
CICERO ATTICO SAL.

Scr. in Tusculano VI K. Iun. a. 709

Oppio et Balbo epistulas deferri iubebis et tamen Pisonem sicubi de auro. Faberius si venerit, videbis, ut tantum attribuatur, si modo attribuetur, quantum debetur. Accipies ab Erote.

Ariarathes, Ariobarzani filius, Romam venit. Vult, opinor, regnum aliquod emere a Caesare; nam, quo modo nunc est, pedem ubi ponat in suo, non habet. Omnino eum Sestius noster, parochus publicus, occupavit; quod quidem facile patior. Verum tamen,

Peducaeus; for all I felt for his father I have given to him, and I love him for himself as much as I loved his father; and you most of all for trying to promote this feeling between us. If you see the gardens, and if you let me know about the letter, you will supply me with something to write about; but, anyhow, I will write something. For there will always be something to say.

II

CICERO TO ATTICUS, GREETING.

Your promptitude was more pleasing to me than the news you sent. For what could be more insulting? However I have hardened myself to insult, and put off all human feeling. I am looking forward to your letter to-day, not that I expect any news. What could there be? However ———.

Tusculum, May 24, B.C. 45

IIa

CICERO TO ATTICUS, GREETING.

Please have the letters sent to Balbus and Oppius, and anyhow speak to Piso about the gold when you can. If Faberius comes, see that the right amount of the debt is put to my credit, if anything is. Eros will tell you about it.

Tusculum, May 27, B.C. 45

Ariarathes, son of Ariobarzanes, has come to Rome, I suppose he wants to buy some kingdom from Caesar: for, as things are at present, he cannot set foot in his own. Our friend Sestius, in his character of public host, has monopolized him; and I am not sorry for it. However, as I am intimate with

quod mihi summo beneficio meo magna cum fratribus illius necessitudo est, invito eum per litteras, ut apud me deversetur. Ad eam rem cum mitterem Alexandrum, has ei dedi litteras.

IIb

CICERO ATTICO SAL.

Scr. in Tusculano IV K. Iun. a. 709

Cras igitur auctio Peducaei. Cum poteris ergo, Etsi impediet fortasse Faberius. Sed tamen cum licebit. Dionysius noster graviter queritur et tamen iure a discipulis abesse se tam diu. Multis verbis scripsit ad me, credo item ad te. Mihi quidem videtur etiam diutius afuturus. Ac nollem; valde enim hominem desidero.

A te litteras exspectabam, nondum scilicet; nam has mane rescribebam.

III

CICERO ATTICO SAL.

Scr. in Tusculano III K. Iun. a. 709

Ego vero ista nomina sic probo, ut nihil aliud me moveat, nisi quod tu videris dubitare. Illud enim non accipio in bonam partem, quod ad me refers; qui, si[1] ipse negotium meum gererem, nihil gererem[2] nisi consilio tuo. Sed tamen intellego magis te id facere diligentia, qua semper uteris, quam quod dubites de nominibus istis. Etenim Caelium non pro-

[1] qui si] quid Δ. [2] nihil gererem *omitted by* Δ.

his brothers on account of the great service I rendered them, I am sending a letter to invite him to stay at my house. As I was sending Alexander with it, I gave him this letter.

IIb

CICERO TO ATTICUS, GREETING.

So to-morrow is Peducaeus' auction. Come[1] when you can, then. But perhaps Faberius will prevent you. However, when you can manage it. Our friend Dionysius is complaining loudly at being so long away from his pupils, and there is some justice in his complaint. He has written a long letter to me, and I expect to you too. I think he will be away for some time still: and I am sorry, for I miss him very much.

I am expecting a letter from you, but not yet, as I am writing in the early morning.

Tusculum, May 29, B.C. 45

III

CICERO TO ATTICUS, GREETING.

For my part I am so satisfied with the debtors you mention, that the only thing which disquiets me is that you seem to have doubts. For I don't take it at all kindly of you to refer the matter to me. If I managed my own business, I should never manage anything without your advice. However, I know you did it more from your usual carefulness than because you had any doubts about the debtors. The fact is you don't approve of Caelius and you don't

Tusculum, May 30, B.C. 45

---

[1] Or, as Shuckburgh, "buy."

MARCUS TULLIUS CICERO

bas, plura non vis. Utrumque laudo. His igitur utendum est. Praes[1] aliquando factus esses[2] in his quidem tabulis. A me igitur omnia. Quod dies longior est, teneamus modo, quod volumus, puto fore istam etiam a praecone diem, certe ab heredibus.

De Crispo et Mustela videbis, et velim scire, quae sit pars duorum. De Bruti adventu eram factus certior. Attulerat enim ab eo Aegypta libertus litteras. Misi ad te epistulam, quia commode scripta erat.

IV

CICERO ATTICO SAL.

Scr. in Tusculano K. Iun. a. 709

Habeo munus a te elaboratum decem legatorum. Et quidem de Tuditano idem[3] puto. Nam filius anno post quaestor fuit quam consul Mummius. Sed, quoniam saepius de nominibus quaeris quid placeat, ego quoque tibi saepius respondeo placere. Si quid poteris, cum Pisone conficies; Avius enim videtur in officio futurus. Velim ante possis; si minus, utique simul simus, cum Brutus veniet in Tusculanum. Magni interest mea una nos esse. Scies autem, qui dies is futurus sit, si puero negotium dederis, ut quaerat.

---

[1] est. Praes *C* : espraes *M*.
[2] esses *Bosius* : esset *M* : es et *CZ¹*.
[3] de Tuditano idem *added by Lehmann*.

like to increase their number.[1] I agree with you in both points. So we must make the best of them as they are. Sometime you would have had to go bail for me even in this sale.[2] So now I shall pay in full myself. As to the delay in collecting the money, if only I get what I want, I think I can arrange for delay with the auctioneer or at any rate with the heirs.

See about Crispus and Mustela, and I should like to know what the share of the two is. I had heard already of Brutus' arrival, for my freedman Aegypta had brought me a letter from him. I have sent it to you, as it is obligingly written.

IV

CICERO TO ATTICUS, GREETING.

Tusculum, June 1, B.C. 45

I have received your piece of work about the ten ambassadors: and I agree with you about Tuditanus. For the son was quaestor in the year after Mummius was consul.[3] But, as you keep on asking if I am satisfied about the debtors, I too keep on answering that I am. Arrange something with Piso if you can: for I think Avius will do his duty. I wish you could come first; but, if you can't, at any rate be with me, when Brutus comes here. It is of great importance to me that we should be together. You will be able to ascertain the day, if you commission a servant to find out.

[1] Apparently Faberius had offered to make over a number of debts due to him in payment of his debt to Cicero, with an alternative of a large debt from Caelius or smaller ones from several other debtors.

[2] *i.e.* even in the purchase of the gardens for Tullia's shrine, of which Atticus disapproved. But the reading may be corrupt. [3] 145 B.C.

MARCUS TULLIUS CICERO

V

CICERO ATTICO SAL.

Scr. in Tusculano IV Non. Iun. a. 709

Sp. Mummium putaram in decem legatis fuisse, sed videlicet (etenim εὔλογον) fratri fuisse. Fuit enim ad Corinthum. Misi tibi Torquatum. Colloquere tu quidem cum Silio, ut scribis, et urge. Illam diem negabat esse mense Maio, istam non negabat. Sed tu ut omnia istuc quoque ages diligenter. De Crispo et Mustela scilicet, cum quid egeris. Quoniam ad Bruti adventum fore te nobiscum polliceris, satis est, praesertim cum hi tibi dies in magno nostro negotio consumantur.

VI

CICERO ATTICO SAL.

Scr. Asturae med. m. Mart., ut videtur, a. 709

De aquae ductu probe fecisti. Columnarium vide ne nullum debeamus; quamquam mihi videor audisse a Camillo commutatam esse legem. Pisoni quid est quod honestius respondere possimus quam solitudinem Catonis? Nec de[1] coheredibus solum Herennianis, sed etiam, ut scis (tu enim mecum egisti), de puero Lucullo, quam pecuniam tutor (nam hoc quoque ad rem pertinet) in Achaia sumpserat. Sed agit liberaliter, quoniam negat se quicquam facturum contra nostram

[1] *de added by Wesenberg.*

---

[1] At its capture in 146 B.C.
[2] *i.e.* the first book of the *De Finibus*. Cf. XIII. 32.

V

CICERO TO ATTICUS, GREETING.

I had thought Sp. Mummius was one of the ten legates: but of course, as was natural, he was private legate to his brother. For he was at Corinth.[1] I have sent *Torquatus*[2] to you. Speak with Silius as you say and urge him on. He said my receiving day would not fall in May, but he did not say the same about the other.[3] But please attend to the point carefully, as you always do. As to Crispus and Mustela, yes, when you have settled anything. As you promise to be with me when Brutus comes, I am satisfied, especially as you are spending these days on important business of mine.

Tusculum, June 2, B.C. 45

VI

CICERO TO ATTICUS, GREETING.

You have done quite right about the aqueduct. Make sure whether I owe any pillar-tax at all. However, I think I heard from Camillus that the law had been changed. What better answer can we give Piso than that Cato's guardians are away? It was not only from the heirs of Herennius that he borrowed, but, as you know (for you were acting with me), from young Lucullus: and that money was taken in Achaia by his guardian. That is another point that has to be considered. But Piso is behaving generously, as he says he will not do anything

Astura, March, B.C. 45

---

[3] *i.e.* that Cicero could not get in Faberius' debt before the end of May; but that the owners of the property he thought of buying would want payment before that date. Cf. XIII. 3.

voluntatem.   Coram igitur, ut scribis, constituemus, quem ad modum rem explicemus.   Quod reliquos coheredes convenisti, plane bene.

Quod epistulam meam ad Brutum poscis, non habeo eius exemplum; sed tamen salvum est, et ait Tiro te habere oportere, et, ut recordor, una cum illius obiurgatoria tibi meam quoque, quam ad eum rescripseram, misi.   Iudiciali molestia ut caream, videbis.

VIa

CICERO ATTICO SAL.

Scr. in Tusculano prid. Non. Iun. a. 709

Tuditanum istum, proavum Hortensi, plane non noram, et filium, qui tum non potuerat esse legatus, fuisse putaram.   Mummium fuisse ad Corinthum pro certo habeo.   Saepe enim hic Spurius, qui nuper decessit,[1] epistulas mihi pronuntiabat versiculis facetis ad familiaris missas a Corintho.   Sed non dubito, quin fratri fuerit legatus, non in decem.   Atque hoc etiam accepi, non solitos maiores nostros eos legare in decem, qui essent imperatorum necessarii, ut nos ignari pulcherrimorum institutorum aut neglegentes potius M. Lucullum et L. Murenam et ceteros ad L. Lucullum misimus.   Illudque εὐλογώτατον, illum fratri in primis eius legatis fuisse.   O operam tuam multam, qui et haec cures et mea expedias et sis in tuis non multo minus diligens quam in meis!

[1] decessit *Müller*: est *MSS*.

against our will. So, as you say, we will arrange, when we meet, how the matter is to be straightened out. It is quite as well that you have seen the other joint heirs.

You ask for my letter to Brutus. I have not a copy: but there is one in existence and Tiro says you ought to have it: and, so far as I recollect, I sent you my answer along with his letter of reproof. Please see that I am not troubled with serving on a jury.

VIa

CICERO TO ATTICUS, GREETING.

The Tuditanus you mention, great-grandfather of Hortensius, I had never heard of, and I thought it was the son who was the ambassador, though he could not have been at the time. I take it as certain that Mummius was at Corinth. For Spurius, who died lately, often used to recite to me letters Mummius wrote to his friends from Corinth in clever verse. But I have no doubt he was a special legate to his brother, not among the ten ambassadors. Here is another point too that I have been taught, that it was not the custom of our ancestors to appoint among the ten ambassadors anyone who was related to the generals, as we in ignorance of, or rather in contempt for, the soundest institutions did in sending M. Lucullus and L. Murena and others to L. Lucullus. But it was most natural that he should be among the first of his brother's legates. What a lot of work you get through, attending to points like this, managing my affairs and bestowing nearly as much care on your own affairs as on mine!

Tusculum, June 4, B.C. 45

MARCUS TULLIUS CICERO

VII

CICERO ATTICO SAL.

Scr. in Tus-culano V Id. Iun. a. 709

Sestius apud me fuit et Theopompus pridie. Venisse a Caesare narrabat litteras; hoc scribere, sibi certum esse Romae manere, causamque eam ascribere, quae erat in epistula nostra, ne se absente leges suae neglegerentur, sicut esset neglecta sumptuaria (est εὔλογον, idque eram suspicatus. Sed istis mos gerendus est, nisi placet hanc ipsam sententiam nos persequi), et Lentulum cum Metella certe fecisse divortium. Haec omnia tu melius. Rescribes igitur, quicquid voles, dum modo aliquid. Iam enim non reperio, quid te rescripturum putem, nisi forte de Mustela, aut si Silium videris.

VIIa

CICERO ATTICO SAL.

Scr. in Tus-culano IV Id. Iun. a. 709

Brutus heri venit in Tusculanum post horam decimam. Hodie igitur me videbit, ac vellem tum tu adesses. Iussi equidem ei nuntiari te, quoad potuisses, exspectasse eius adventum venturumque, si audisses, meque, ut facio, continuo te certiorem esse facturum.

VIII

CICERO ATTICO SAL.

Scr. in Tus-culano VI Id. Iun. a. 709

Plane nihil erat, quod ad te scriberem; modo enim discesseras et paulo post triplicis remiseras. Velim cures fasciculum ad Vestorium deferendum et alicui

LETTERS TO ATTICUS XIII. 7-8

VII

CICERO TO ATTICUS, GREETING.

Sestius came to see me yesterday and Theopompus *Tusculum,* too. He told me that Caesar had sent a letter *June 9,* saying he had resolved to stay at Rome and assign- B.C. *45* ing as a reason the one mentioned in my letter, fear that if he went away his laws would be disregarded, as his sumptuary law was. That is reasonable enough and is just what I suspected. But I must humour your friends, unless you think I could use that very line of argument. He tells me too that Lentulus has certainly divorced Metella. But you will know all this better than he does. So please send an answer,—anything you like provided it is something. For at the moment I cannot think of anything you will put in your answer, unless it is something about Mustela, or unless you see Silius.

VIIa

CICERO TO ATTICUS, GREETING.

Brutus came to Tusculum yesterday after four *Tusculum* o'clock. So to-day he will see me, and I wish you *June 10,* were with me. I sent him word that you had B.C. *45* waited for him as long as you could, and that you would come, if you heard; and I would let you know, as soon as I could, which I am doing.

VIII

CICERO TO ATTICUS, GREETING.

I have nothing to write; for you have only just *Tusculum,* left, and soon after you went, you sent me back my *June 8,* notebook. Please see that the packet is delivered B.C. *45* to Vestorius, and commission someone to find out if

des negotium, qui quaerat, Q. Staberi fundus num quis in Pompeiano Nolanove venalis sit. Epitomen Bruti Caelianorum velim mihi mittas et a Philoxeno Παναιτίου περὶ προνοίας. Te Idibus videbo cum tuis.

IX

CICERO ATTICO SAL.

Scr. in Tusculano XIV K. Quint. a. 709

Commodum discesseras heri, cum Trebatius venit, paulo post Curtius, hic salutandi causa, sed mansit invitatus. Trebatium nobiscum habemus. Hodie mane Dolabella. Multus sermo ad multum diem. Nihil possum dicere ἐκτενέστερον, nihil φιλοστοργότερον. Ventum est tamen ad Quintum. Multa ἄφατα, ἀδιήγητα, sed unum eius modi, quod nisi exercitus sciret, non modo Tironi dictare, sed ne ipse quidem auderem scribere. Sed hactenus.

Εὐκαίρως ad me venit, cum haberem Dolabellam, Torquatus, humanissimeque Dolabella, quibus verbis secum egissem, exposuit. Commodum enim egeram diligentissime; quae diligentia grata est visa Torquato. A te exspecto, si quid de Bruto. Quamquam Nicias confectum putabat, sed divortium non probari. Quo etiam magis laboro idem quod tu. Si quid est enim offensionis, haec res mederi potest.

Mihi Arpinum eundum est. Nam et opus est constitui a nobis illa praediola, et vereor, ne exeundi

LETTERS TO ATTICUS XIII. 8–9

any part of Q. Staberius' land at Pompeii or Nola is for sale. Please send me Brutus' *Epitome of the Annals of Caelius,* and get from Philoxenus Panaetius *On Foresight.* I shall see you and your family on the 13th.

IX

CICERO TO ATTICUS, GREETING.

Tusculum, June 18, B.C. 45

You had only just left yesterday, when Trebatius came, and then Curtius shortly afterwards. The latter only came to pay a call, but he stayed at my invitation. Trebatius is with me too, and this morning came Dolabella. We had a long talk till late in the day. I cannot exaggerate his cordiality and friendliness. However, we touched on young Quintus. Much of what he told me was unmentionable, unspeakable; but there was one thing so bad that, if the whole army did not know of it, I should not dare to dictate it to Tiro or even to write it down myself. But enough of this.

Torquatus came to me opportunely, while Dolabella was with me, and Dolabella very kindly repeated to him what I had just been saying. For I had just been pleading his cause very earnestly; and my earnestness seemed to please Torquatus greatly. I am waiting to know if you have any news about Brutus. However, Nicias thought that the matter was settled, but that the divorce was not approved. For that reason I am all the more eager about the thing, as you are too. For, if any offence has been given, this can remedy it.

I must go to Arpinum. For my little place there needs putting in order and I am afraid I may not

potestas non sit, cum Caesar venerit; de cuius adventu eam opinionem Dolabella habet, quam tu coniecturam faciebas ex litteris Messallae. Cum illuc venero intellexeroque, quid negotii sit, tum, ad quos dies rediturus sim, scribam ad te.

X

CICERO ATTICO SAL.

Scr. in Tusculano inter a. d. XIII et XI K. Quint. a. 709

Minime miror te et graviter ferre de Marcello et plura vereri periculi genera. Quis enim hoc timeret, quod neque acciderat antea nec videbatur natura ferre ut accidere posset? Omnia igitur metuenda. Sed illud παρὰ τὴν ἱστορίαν, tu praesertim, me reliquum consularem. Quid? tibi Servius quid videtur? Quamquam hoc nullam ad partem valet scilicet, mihi praesertim, qui non minus bene actum cum illis putem. Quid enim sumus aut quid esse possumus? domin an foris? Quodnisi mihi hoc venisset in mentem, scribere ista nescio quae, quo verterem me, non haberem.

Ad Dolabellam, ut scribis, ita puto faciendum, κοινότερα quaedam et πολιτικώτερα. Faciendum certe aliquid est; valde enim desiderat. Brutus si quid egerit, curabis, ut sciam; cui quidem quam primum agendum puto, praesertim si statuit. Sermunculum

have much chance of leaving Rome, when Caesar comes. About his coming Dolabella holds the same idea which you had inferred from Messalla's letter. When I get there and know how much there is to be done, then I will write and let you know, when I shall return.

X

CICERO TO ATTICUS, GREETING.

I am not at all surprised at your being upset about Marcellus[1] and fearing all sorts of new dangers. For who would have feared this? Such a thing never happened before and it did not seem as though nature could allow such things to happen. So one may fear anything. But fancy you of all people making such a historical slip as to call me the only surviving ex-consul. Why, what about Servius? However, that of course has not the slightest importance in any respect, least of all to me, who think my dead comrades' fate quite as happy as my own. For what am I or what can I be? Am I anything in private life or in public? If it had not occurred to me to write my books, such as they are, I should not know what to do with myself.

Tusculum, June 19-21, B.C. 45

I think I must follow your advice and dedicate something more general and more political to Dolabella. I must certainly do something for him, as he is very anxious for it. If Brutus makes any move, pray let me know. I think he ought to make one as soon as possible, especially if he has made up his mind.[2] That would either put an end to all chatter

[1] M. Marcellus had been murdered by P. Magius Chilo.
[2] About his marriage to Porcia.

enim omnem aut restinxerit aut sedarit. Sunt enim, qui loquantur etiam mecum. Sed haec ipse optime, praesertim si etiam tecum loquetur.

Mihi est in animo proficisci xi Kal. Hic enim nihil habeo, quod agam, ne hercule illic quidem nec usquam, sed tamen aliquid illic. Hodie Spintherem exspecto. Misit enim Brutus ad me. Per litteras purgat Caesarem de interitu Marcelli; in quem, ne si insidiis quidem ille interfectus esset, caderet ulla suspicio. Nunc vero, cum de Magio constet, nonne furor eius causam omnem sustinet? Plane, quid sit, non intellego. Explanabis igitur. Quamquam nihil habeo, quod dubitem, nisi, ipsi Magio quae fuerit causa amentiae; pro quo quidem etiam sponsor sum factus. Et nimirum id fuit. Solvendo enim non erat. Credo eum petisse a Marcello aliquid, et illum, ut erat, constantius respondisse.

XI

CICERO ATTICO SAL.

Scr. in Arpinati IX K. Quint. a. 709

"Οὐ ταὐτὸν εἶδος." Credebam esse facile; totum est aliud, posteaquam sum a te diiunctior. Sed fuit faciendum, ut et constituerem mercedulas praediorum et ne magnum onus observantiae Bruto nostro imponerem. Posthac enim poterimus commodius colere inter nos in Tusculano. Hoc autem tempore, cum

---

[1] A quotation from Euripides, *Ion*, 585:—

οὐ ταὐτὸν εἶδος φαίνεται τῶν πραγμάτων
πρόσωθεν ὄντων ἐγγυθέν θ' ὁρωμένων.

LETTERS TO ATTICUS XIII. 10-11

or at any rate lessen it. For there are people who talk even to me. But he is the best judge himself, especially if he talks it over with you too.

I am thinking of setting out on the 21st, for I have nothing to do here, and precious little to do there or anywhere else; still there is something to do there. To-day I am expecting Spinther, for Brutus has sent him to me. He writes to exculpate Caesar of Marcellus' death. But no suspicion would have fallen on Caesar, even if his death had been due to treachery; and now that Magius is known to be mad, surely that accounts for everything. I don't see his point at all. Perhaps you will explain. However, there is nothing I am in doubt about except the reason for Magius' madness; why, I had even gone security for him. That no doubt was the point; he was insolvent. I suppose he asked some favour of Marcellus, and the latter, as was his way, gave a rather decided answer.

XI

CICERO TO ATTICUS, GREETING.

"Not the same look."[1] I thought it was easy; *Arpinum,* but it is quite the reverse, now I am farther away *June 23,* from you. But it had to be done, that I might fix B.C. **45** some trifles like the rents of my farms and might not lay too great a burden of attendance on our friend Brutus. For in the future we shall find ourselves able to cultivate each other's society at Tusculum more easily. But at the present time, when he wanted

"Not the same look wear things, when seen far off and near at hand."

ille me cotidie videre vellet, ego ad illum ire non possem, privabatur omni delectatione Tusculani. Tu igitur, si Servilia venerit, si Brutus quid egerit, etiam si constituerit, quando obviam, quicquid denique erit, quod scire me oporteat, scribes. Pisonem, si poteris convenies. Vides, quam maturum sit. Sed tamen, quod commodo tuo fiat.

XII

CICERO ATTICO SAL.

Scr. in Arpinati VIII K. Quint. a. 709

Valde me momorderunt epistulae tuae de Attica nostra; eaedem tamen sanaverunt. Quod enim te ipse consolabare eisdem litteris, id mihi erat satis firmum ad leniendam aegritudinem.

Ligarianam praeclare vendidisti. Posthac, quicquid scripsero, tibi praeconium deferam. Quod ad me de Varrone scribis, scis me antea orationes aut aliquid id genus solitum scribere, ut Varronem nusquam possem intexere. Postea autem quam haec coepi φιλολογώτερα, iam Varro mihi denuntiaverat magnam sane et gravem προσφώνησιν. Biennium praeteriit, cum ille Καλλιππίδης adsiduo cursu cubitum nullum processerat, ego autem me parabam ad id, quod ille mihi misisset, ut "αὐτῷ τῷ μέτρῳ καὶ λώϊον," si modo potuissem. Nam hoc etiam Hesiodus ascribit, "αἴ κε δύνηαι."

Nunc illam περὶ τελῶν σύνταξιν sane mihi probatam

---

[1] One of the bankers from whom Cicero hoped to raise money to buy the gardens for Tullia's shrine.

LETTERS TO ATTICUS XIII. 11–12

to see me every day and I could not go to him, he got no pleasure at all out of his estate. So, if Servilia has come, if Brutus has begun to do anything, even if he has made up his mind when I am to meet Caesar, in short anything there is to tell, please write and tell me. See Piso,[1] if you can. It is high time, as you can see; however, suit your convenience.

XII

CICERO TO ATTICUS, GREETING.

Arpinum, June 24, B.C. 45

Your letter about dear Attica stung me to the quick; but it healed the wound again. For you consoled yourself in the same letter, and that I counted sufficient warrant for moderating my grief.

You have given my speech for Ligarius a magnificent start. Henceforth, when I write anything, I shall leave it to you to advertise it. As to what you say about Varro, you know formerly I have written speeches or things of such a kind, that I could not introduce him; but afterwards, when I began these more literary works, Varro had already promised to dedicate a great and important work to me. Two years have passed and that slow coach,[2] though always on the move, has not advanced an inch, while I was prepared to pay him back "full measure and more" for what he sent, if I could. For Hesiod adds "if you can."[3]

Now I have pledged my *De Finibus*, of which I

[2] It is uncertain whether the actor mentioned in Aristotle's *Poetics*, ch. 26, is referred to or someone else. Anyhow, the name seems to be used proverbially as = "a slow coach."

[3] Hesiod, *Op.* 350.

MARCUS TULLIUS CICERO

Bruto, ut tibi placuit, despondimus, idque eum non nolle mihi scripsisti. Ergo illam 'Ακαδημικήν, in qua homines nobiles illi quidem, sed nullo modo philologi nimis acute loquuntur, ad Varronem transferamus. Etenim sunt Antiochia, quae iste valde probat. Catulo et Lucullo alibi reponemus, ita tamen, si tu hoc probas; deque eo mihi rescribas velim.

De Brinniana auctione accepi a Vestorio litteras. Ait sine ulla controversia rem ad me esse conlatam. Romae videlicet aut in Tusculano me fore putaverunt a. d. VIII Kal. Quinct. Dices igitur vel amico tuo, S. Vettio, coheredi meo, vel Labeoni nostro, paulum proferant auctionem; me circiter Nonas in Tusculano fore. Cum Pisone Erotem habes. De Scapulanis hortis toto pectore cogitemus. Dies adest.

XIII, XIV

CICERO ATTICO SAL.

Scr. in Arpinati VI K. Quint. a. 709

Commotus tuis litteris, quod ad me de Varrone scripseras, totam Academiam ab hominibus nobilissimis abstuli, transtuli ad nostrum sodalem et ex duobus libris contuli in quattuor. Grandiores sunt omnino, quam erant illi, sed tamen multa detracta. Tu autem mihi pervelim scribas, qui intellexeris illum velle; illud vero utique scire cupio, quem in-

---

[1] Antiochus of Ascalon, under whom both Cicero and Varro had studied at Athens. His teaching combined the views of the Academy and Stoicism.

LETTERS TO ATTICUS XIII. 12–13, 14

think very highly, to Brutus as you advised, and you have told me he was gratified. So I must assign the *Academica* to Varro. The speakers in it are men of birth to be sure, but not scholars, and talk above their own heads. And indeed the doctrines are those of Antiochus,[1] of which Varro is a strong supporter. I will make it up to Catulus and Lucullus somewhere else[2]; that is to say, if you agree. Please write and tell me.

I have had a letter from Vestorius about the auction of Brinnius' estate. He tells me I was unanimously given the direction of it. They evidently thought I should be in town or at Tusculum on the 24th. So please tell your friend S. Vettius, my coheir, or Labienus, to put the sale off for a while; and that I shall be at Tusculum about July 7th. You have Eros to help with Piso. Let us throw ourselves heart and soul into the purchase of Scapula's gardens. The time is drawing near.

XIII, XIV

CICERO TO ATTICUS, GREETING.

Under the influence of your letters about Varro *Arpinum*, I have taken the whole of my *Academica* from its *June 26,* eminent interlocutors and transferred it to our friend: B.C. 45 and from two books I have turned it into four. They are certainly finer than the first draft though a good deal has been cut out. But I should very much like you to tell me how you knew Varro wanted it: and one thing at any rate I want to know, who

---

[2] They were the chief speakers in the first draft of the *Academica*.

tellexeris ab eo ζηλοτυπεῖσθαι nisi forte Brutum. Id
hercle restabat. Sed tamen scire pervelim. Libri
quidem ita exierunt, nisi forte me communis φιλαυτία
decipit, ut in tali genere ne apud Graecos quidem
simile quicquam. Tu illam iacturam feres aequo
animo, quod illa, quae habes de Academicis, frustra
descripta sunt. Multo tamen haec erunt splendi-
diora, breviora, meliora. Nunc autem ἀπορῶ, quo me
vertam. Volo Dolabellae valde desideranti; non reperio, quid, et simul "αἰδέομαι Τρῶας" neque, si
aliquid, potero μέμψιν effugere. Aut cessandum
igitur aut aliquid excogitandum. Sed quid haec
levia curamus?

Iliad, vi. 442;
xxii. 100

Attica mea, obsecro te, quid agit? Quae me valde
angit. Sed crebro regusto tuas litteras; in his ac-
quiesco. Tamen exspecto novas.

Brinni libertus, coheres noster, scripsit ad me velle,
si mihi placeret, coheredes, se et Sabinum Albium,
ad me venire. Id ego plane nolo. Hereditas tanti
non est. Et tamen obire auctionis diem facile poterunt (est enim III Idus), si me in Tusculano postridie
Nonas mane convenerint. Quodsi laxius volent pro-
ferre diem, poterunt vel biduum vel triduum, vel ut
videbitur; nihil enim interest. Quare, nisi iam profecti sunt, retinebis homines. De Bruto, si quid
egerit, de Caesare, si quid scies, si quid erit praeterea,
scribes.

was it of whom you noticed he was jealous: unless perhaps it was Brutus. Upon my word that is the only possible answer:[1] but still I should much like to know. Unless I am deceived like most people by egotism, the books have turned out superior to anything of the kind even in Greek. You must not be annoyed at the loss you have incurred in having the part of the *Academica* you have copied in vain. The new draft will be far finer, shorter, and better. But now I don't know where to turn. I want to do something for Dolabella, as he is very anxious for it. But I can't think of anything, and at the same time "I fear the Trojans,"[2] and even if I can think of something, I shall not escape criticism. So I must either be idle or rack my brains for something. But why do I bother about trifles like this?

Pray tell me how dear Attica is. I am very anxious about her. But I keep dipping into your letter again and again, and that solaces me. Nevertheless I am looking forward to a fresh one.

Brinnius' freedman, my co-heir, has written to me that the rest of the heirs want him and Sabinus Albius to come to me, if I am willing. I am all against that: it is more than the legacy is worth. However, they can easily manage to attend the auction, which is on the 13th, if they meet me at my place at Tusculum early on the 8th. But, if they want to put off the date still further, they can do so two or three days or as much as they like: it does not matter to me. So, unless the people have started already, stop them. If Brutus has done anything, or if you have any news about Caesar or anything else, let me know.

---

[1] Or "that is the last straw," or "the height of absurdity."
[2] *i.e.* public opinion. Cf. *Att.* II. 5.

MARCUS TULLIUS CICERO

XIV, XV

CICERO ATTICO SAL.

Scr. in Arpinati V K. Quint. a. 709

Illud etiam atque etiam consideres velim, placeatne tibi mitti ad Varronem, quod scripsimus. Etsi etiam ad te aliquid pertinet. Nam scito te ei dialogo adiunctum esse tertium. Opinor igitur, consideremus. Etsi nomina iam facta sunt; sed vel induci vel mutari possunt.

Quid agit, obsecro te, Attica nostra? Nam triduo abs te nullas acceperam; nec mirum. Nemo enim venerat, nec fortasse causa fuerat. Itaque ipse, quod scriberem, non habebam. Quo autem die has Valerio dabam, exspectabam aliquem meorum. Qui si venisset et a te quid attulisset, videbam non defuturum, quod scriberem.

XVI

CICERO ATTICO SAL.

Scr. in Arpinati IV K. Quint. a. 709

Nos, cum flumina et solitudinem sequeremur, quo facilius sustentare nos possemus, pedem e villa adhuc egressi non sumus; ita magnos et adsiduos imbres habebamus. Illam Ἀκαδημικὴν σύνταξιν totam ad Varronem traduximus. Primo fuit Catuli, Luculli, Hortensi; deinde, quia παρὰ τὸ πρέπον videbatur, quod erat hominibus nota non illa quidem ἀπαιδευσία, sed in iis rebus ἀτριψία, simul ac veni ad villam, eosdem illos sermones ad Catonem Brutumque transtuli. Ecce tuae litterae de Varrone. Nemini visa

XIV, XV

CICERO TO ATTICUS, GREETING.

Please give your earnest consideration to deciding *Arpinum,* whether what I have written ought to be sent to *June 27,* Varro: though the point has some personal interest B.C. *45* for you too: for you must know I have brought you in as a third speaker in the dialogue. So I think we must consider. The names, however, have been entered, but they can be scratched out or altered.

Pray tell me how Attica is. It is three days since I heard from you, and no wonder: for no one has come here, and perhaps there was no reason for writing. So I myself have nothing to write. However, I am expecting one of my messengers the very day I am giving this to Valerius. If he comes and brings something from you, I foresee I shall have no lack of material.

XVI

CICERO TO ATTICUS, GREETING.

Though I was looking for streams and solitude, to *Arpinum,* make life more endurable, at present I have not *June 28,* stirred a foot away from the house; we have had B.C. *45* such heavy and continuous rain. The "Academic Treatise" I have transferred entirely to Varro. At first it was assigned to Catulus, Lucullus, and Hortensius; then, as that seemed inappropriate because they were well-known not to be up in such matters, though not illiterate, as soon as I came here I transferred the conversations to Cato and Brutus. Then came your letter about Varro and he seemed the most appropriate person possible to air Antiochus'

135

est aptior Antiochia ratio. Sed tamen velim scribas ad me, primum placeatne tibi aliquid ad illum, deinde, si placebit, hocne potissimum.

Quid? Servilia iamne venit? Brutus ecquid agit et quando? De Caesare quid auditur? Ego ad Nonas, quem ad modum dixi. Tu cum Pisone, si quid poteris.

XVII, XVIII

CICERO ATTICO SAL.

Scr. in Arpinati III K. Quint. a. 709

v Kal. exspectabam Roma aliquid novi. Imperassem igitur aliquid tuis. Nunc eadem illa, quid Brutus cogitet, aut, si aliquid egit, ecquid a Caesare. Sed quid ista, quae minus curo? Attica nostra quid agat, scire cupio. Etsi tuae litterae (sed iam nimis veteres sunt) recte sperare iubent, tamen exspecto recens aliquid.

Vides, propinquitas quid habeat. Nos vero conficiamus hortos. Conloqui videbamur, in Tusculano cum essem; tanta erat crebritas litterarum. Sed id quidem iam erit. Ego interea admonitu tuo perfeci sane argutulos libros ad Varronem, sed tamen exspecto, quid ad ea, quae scripsi ad te, primum qui intellexeris eum desiderare a me, cum ipse homo πολυγραφώτατος numquam me lacessisset; deinde quem ζηλοτυπεῖν nisi forte Brutum, quem si non ζηλοτυπεῖ,[1] multo Hortensium minus aut eos, qui de re

---

[1] nisi ... ζηλοτυπεῖ *added by Bosius.*

LETTERS TO ATTICUS XIII. 16-17, 18

views. However, I should like you to write whether you approve of dedicating anything to him, and, if you do, whether you approve of this particular book.

What about Servilia? Has she come? Has Brutus done anything, and when? What news of Caesar? I shall arrive on the 7th of July, as I said. Make some arrangement with Piso, if you can.

XVII, XVIII

CICERO TO ATTICUS, GREETING.

I was expecting some news from Rome on the 27th. Then I should have given some orders to your men. Now I have only the same old questions. What is Brutus thinking of doing, or, if he has done anything, has any comment come from Caesar? But why do I ask about these things, when I care very little about them? I do want to know how our dear Attica is getting on. Though your letter (but that is quite out of date now) bids me be hopeful, still I am anxious for fresh news.

Arpinum, June 29, B.C. 45

You see the advantage of being near at hand. Certainly let us settle about the gardens. We seemed to be talking to one another, when I was at Tusculum, so frequent was the interchange of letters. But that will be the same again soon. Meantime I have taken your hint and finished off some really quite clever books for Varro. But I am waiting for your answer to my questions: first, how you knew he wanted anything from me, when in spite of his voluminous writings he has never challenged me; and next, who it was of whom he was jealous, unless it may have been Brutus. If he is not jealous of him, he certainly cannot be of Hortensius or the speakers in the *De*

publica loquuntur. Plane hoc mihi explices velim, in primis maneasne in sententia, ut mittam ad eum, quae scripsi, an nihil necesse putes. Sed haec coram.

XIX

CICERO ATTICO SAL.

Scr. in Arpinati prid. K. Quint. a. 709

Commodum discesserat Hilarus librarius IV Kal., cui dederam litteras ad te, cum venit tabellarius cum tuis litteris pridie datis; in quibus illud mihi gratissimum fuit, quod Attica nostra rogat te, ne tristis sis, quodque tu ἀκίνδυνα esse scribis.

Ligarianam, ut video, praeclare auctoritas tua commendavit. Scripsit enim ad me Balbus et[1] Oppius mirifice se probare, ob eamque causam ad Caesarem eam se oratiunculam misisse. Hoc igitur idem tu mihi antea scripseras.

In Varrone ista causa me non moveret, ne viderer φιλένδοξος (sic enim constitueram, neminem includere in dialogos eorum, qui viverent); sed, quia scribis et desiderari a Varrone et magni illum aestimare, eos confeci et absolvi, nescio quam bene, sed ita accurate, ut nihil posset supra, Academicam omnem quaestionem libris quattuor. In eis, quae erant contra ἀκαταληψίαν praeclare collecta ab Antiocho, Varroni dedi. Ad ea ipse respondeo; tu es tertius in sermone nostro. Si Cottam et Varronem fecissem inter se disputantes, ut a te proximis litteris admoneor,

[1] et *added by Vict.*

LETTERS TO ATTICUS XIII. 17, 18–19

Republica. I should like you to make this quite clear to me, especially whether you abide by your opinion that I should send him what I have written, or whether you think it is unnecessary. But of this when we meet.

XIX

CICERO TO ATTICUS, GREETING.

The copyist Hilarus had just left on the 28th, and I had given him a letter to you, when your messenger came with your letter of the day before. What I was most glad to see in it was the sentence "Our dear Attica begs you not to be anxious" and your own statement that there is no danger.

I see your influence has given my speech for Ligarius a good start. For Balbus has written to me with Oppius, saying that he is extraordinarily pleased with it; and for that reason he has sent the little thing to Caesar. So that is what you wrote to me some time ago.

In Varro's case I should not be disturbed about appearing to be tuft-hunting—for my principle has always been not to insert any living characters in my dialogues; but it was because you say Varro wants it, and appreciates the compliment, that I have finished off the work and have comprised the whole of the Academic philosophy—how well I cannot say, but with all possible care—in four books. All the fine array of arguments against the uncertainty of apperceptions collected by Antiochus I have given to Varro; I answer him myself, and you are the third speaker in our conversation. If I had made Cotta and Varro carry on the argument between them, as you suggest in your last letter, I

Arpinum, June 30, B.C. 45

meum κωφὸν πρόσωπον esset. Hoc in antiquis personis suaviter fit, ut et Heraclides in multis et nos in VI "de re publica" libris fecimus. Sunt etiam "de oratore" nostri tres mihi vehementer probati. In eis quoque eae personae sunt, ut mihi tacendum fuerit. Crassus enim loquitur, Antonius, Catulus senex, C. Iulius, frater Catuli, Cotta, Sulpicius. Puero me hic sermo inducitur, ut nullae esse possent partes meae. Quae autem his temporibus scripsi, Ἀριστοτέλειον morem habent, in quo sermo ita inducitur ceterorum, ut penes ipsum sit principatus. Ita confeci quinque libros περὶ τελῶν, ut Epicurea L. Torquato, Stoica M. Catoni, περιπατητικὰ M. Pisoni darem. Ἀζηλοτύπητον id fore putaram, quod omnes illi decesserant. Haec "Academica," ut scis, cum Catulo, Lucullo, Hortensio contuleram. Sane in personas non cadebant; erant enim λογικώτερα, quam ut illi de iis somniasse umquam viderentur. Itaque, ut legi tuas de Varrone, tamquam ἕρμαιον arripui. Aptius esse nihil potuit ad id philosophiae genus, quo ille maxime mihi delectari videtur, easque partes, ut non sim consecutus, ut superior mea causa videatur. Sunt enim vehementer πιθανὰ Antiochia; quae diligenter a me expressa acumen habent Antiochi, nitorem orationis nostrum, si modo is est aliquis in nobis. Sed tu, dandosne putes hos libros Varroni, etiam atque etiam videbis. Mihi quaedam occurrunt; sed ea coram.

should have been a mere lay figure. That suits admirably when the characters are persons of olden times; and that is what Heraclides often did in his works; and I myself did so in my six books *De Republica*. It is the same, too, in my three books *De Oratore*, of which I think very highly; in them, too, the characters were such that I could properly keep silent. For the speakers are Crassus, Antonius, old Catulus, his brother C. Julius, Cotta and Sulpicius; and the conversation is supposed to take place when I was a boy, so that I could have no part in it. But in a modern work, I follow Aristotle's practice: the conversation of the others is so put forward as to leave him the principal part. I arranged the five books *De Finibus* so as to give the Epicurean parts to L. Torquatus, the Stoic to M. Cato, and the Peripatetic to M. Piso. I thought that could not make anybody jealous, as they were all dead. This present work, the *Academica*, as you know, I had shared between Catulus, Lucullus and Hortensius. I must admit that the work did not suit the characters; for it was far too philosophical for them to have ever dreamt of such things. So, when I read your note about Varro, I jumped at it as a godsend. Nothing could have been more appropriate for expounding the system of philosophy in which he seems to be specially interested, and for introducing a part which prevents me from seeming to give my own cause the superiority. For the views of Antiochus are very persuasive, and I have put them carefully with all Antiochus' acuteness and my own polished style, if I possess one. But do you consider carefully, whether you think I ought to dedicate the books to Varro. Some objections occur to me; but of that when we meet.

MARCUS TULLIUS CICERO

XX

CICERO ATTICO SAL.

Scr. in Arpinati VI aut V Non. Quint. a. 709

A Caesare litteras accepi consolatorias datas pridie Kal. Maias Hispali. De urbe augenda quid sit promulgatum, non intellexi. Id scire sane velim. Torquato nostra officia grata esse facile patior eaque augere non desinam. Ad Ligarianam de uxore Tuberonis et privigna neque possum iam addere (est enim pervulgata) neque Tuberonem volo offendere; mirifice est enim φιλαίτιος. Theatrum quidem sane bellum habuisti. Ego, etsi hoc loco facillime sustentor, tamen te videre cupio. Itaque, ut constitui, adero. Fratrem credo a te esse conventum. Scire igitur studeo, quid egeris.

De fama nihil sane laboro; etsi scripseram ad te tunc stulte "nihil melius"; curandum enim non est. Atque hoc "in omni vita sua quemque a recta conscientia traversum unguem non oportet discedere" viden quam φιλοσόφως? An tu nos frustra existimas haec in manibus habere? Δεδῆχθαι te nollem, quod nihil erat. Redeo enim rursus eodem. Quicquamne me putas curare in toto,[1] nisi ut ei ne desim? Id ago scilicet, ut iudicia videar tenere. "Μὴ γὰρ αὐτοῖς—." Vellem tam domestica ferre possem quam ista con-

---

[1] *For* in toto *many suggestions have been made (e.g.* in Torquato *Müller:* in Bruto *Schmidt), and for* ei *Wieland suggested* mihi.

---

[1] Tubero was the prosecutor of Ligarius.

XX

CICERO TO ATTICUS, GREETING.

I have received a letter of consolation from Caesar, posted on the last of April at Hispalis. I did not understand what the proposals for improving the city are; and I should much like to know. I am not displeased that Torquatus is satisfied with my attentions, and I shall not cease to increase them. To the speech for Ligarius I cannot add anything now about Tubero's[1] wife and step-daughter, since the speech is widely circulated, and I do not wish to offend Tubero; for he is most touchy. You certainly had a good audience. Though I am happy enough here, I am longing to see you; so I shall come as arranged. I think you have met my brother; so I am anxious to know what happened.

About my reputation I don't care a straw; though I did once write to you foolishly that there was nothing better; for it is not worth bothering about. And see what deep philosophy there is in this other sentiment of mine, "In all one's life one ought not to stray a nail's breadth from the straight path of conscience." Do you think I am engaged in philosophical treatises for nothing? I should be sorry for you to distress yourself about a mere nothing. Now I come back to my point. Do you suppose I care for anything in the whole matter, except that I should not be untrue to it.[2] I am striving, it seems then, to maintain my position in the law courts. God forbid! Would I could bear my private sorrow as easily as I despise them. But do

Arpinum, July 2 or 3, B.C. 45

---

[2] The sense and the reading of this sentence are very doubtful.

temnere. Putas autem me voluisse aliquid, quod perfectum non sit? Non licet scilicet sententiam suam, sed tamen, quae tum acta sunt, non possum non probare, et tamen non curare pulchre possum, sicuti facio. Sed nimium multa de nugis.

XXI

CICERO ATTICO SAL.

Scr. Asturae IV K. Sext. a. 709

Ad Hirtium dederam epistulam sane grandem, quam scripseram proxime in Tusculano. Huic, quam tu mihi misisti, rescribam alias. Nunc alia malo. Quid possum de Torquato, nisi aliquid a Dolabella? Quod simul ac, continuo scietis. Exspectabam hodie aut summum cras ab eo tabellarios; qui simul ac venerint, mittentur ad te. A Quinto exspecto. Proficiscens enim e Tusculano VIII Kal., ut scis, misi ad eum tabellarios.

Nunc, ad rem ut redeam, "inhibere" illud tuum, quod valde mihi adriserat, vehementer displicet. Est enim verbum totum nauticum. Quamquam id quidem sciebam, sed arbitrabar sustineri remos, cum inhibere essent remiges iussi. Id non esse eius modi didici heri, cum ad villam nostram navis appelleretur. Non enim sustinent, sed alio modo remigant. Id ab ἐποχῇ remotissumum est. Quare facies, ut ita sit in libro, quem ad modum fuit. Dices hoc idem Varroni, si

you suppose there was some aspiration which was left unfulfilled? Of course one should not praise one's own principles, but I cannot help praising my past life, and yet I can well enough feel indifferent about it, as indeed I do. But that is enough and more than enough about such a trifle.

XXI

CICERO TO ATTICUS, GREETING.

I have sent a very bulky letter to Hirtius, which I *Astura,* wrote lately at Tusculum. This letter which you have *July 29,* sent, I will answer later. Just now I prefer other B.C. *45* things. What can I do for Torquatus, unless I hear from Dolabella? As soon as I hear, you shall know at once. I am expecting messengers from him to-day or to-morrow at the latest; and, as soon as they come, they shall be sent on to you. I am expecting to hear from Quintus. For when I was starting from Tusculum on the 25th, as you know, I sent messengers to him.

To return to business, the word *inhibere* suggested by you, which at first took my fancy very much, I strongly disapprove of now. For it is exclusively a nautical word. That, however, I knew before; but I thought rowers rested on their oars, when told to *inhibere*. Yesterday, when a ship put in by my house, I learned that was not so. They don't rest on their oars, they back water. That is very different to the Greek ἐποχή. So change the word back to what it was in the book[1]; and tell Varro to do

---

[1] *Academica* II. 94. Ἐποχή, of which the Latin rendering is here discussed, is the technical term in philosophy for "suspension of judgment."

forte mutavit. Nec est melius quicquam quam ut Lucilius:

" Sustineas currum ut bonus saepe agitator equosque."

Semperque Carneades προβολὴν pugilis et retentionem aurigae similem facit ἐποχῇ. Inhibitio autem remigum motum habet, et vehementiorem quidem, remigationis navem convertentis ad puppim. Vides, quanto hoc diligentius curem quam aut de rumore aut de Pollione. De Pansa etiam, si quid certius (credo enim palam factum esse), de Critonio, si quid est, sed certe [1] de Metello et Balbino.

XXIa

CICERO ATTICO SAL.

Scr. in Arpinati prid K. aut K. Quint. a. 709

Dic mihi, placetne tibi primum edere iniussu meo? Hoc ne Hermodorus quidem faciebat, is qui Platonis libros solitus est divulgare, ex quo "λόγοισιν Ἑρμόδωρος." Quid? illud rectumne existimas cuiquam ante quam [2] Bruto, cui te auctore προσφωνῶ? Scripsit enim Balbus ad me se a te quintum "de finibus" librum descripsisse; in quo non sane multa mutavi, sed tamen quaedam. Tu autem commode feceris, si reliquos continueris, ne et ἀδιόρθωτα habeat Balbus et ἕωλα Brutus. Sed haec hactenus, ne videar περὶ μικρὰ σπουδάζειν. Etsi nunc quidem maxima mihi sunt haec; quid est enim aliud?

---

[1] est, sed certe *Wesenberg*: esset certe ne *MSS.*
[2] ante quam *added by Vict.*

LETTERS TO ATTICUS XIII. 21–21a

the same, if he has altered it. One can't improve on Lucilius: "Pull up chariot and horses as a good driver oft does." And Carneades always compares the philosopher's suspension of judgment (ἐποχή) to the guard of a boxer and the pulling up of a charioteer. But the *inhibitio* of rowers implies motion, and indeed the rather violent motion of rowing to back the boat. You see how much more attention I pay to this than either to rumour or to Pollio. Let me know too about Pansa, if anything definite is known, and I suppose it has come out, about Critonius, if there is any news, and anyhow about Metellus and Balbinus.

XXIa

CICERO TO ATTICUS, GREETING.

Come now, do you really think you ought to publish without my orders? Even Hermodorus never did such a thing, though he used to circulate Plato's books, and that gave rise to the line " our Hermodorus deals in dialogues."[1] Do you really think you were justified in sending to anyone before you sent to Brutus, to whom at your advice I dedicated the work. For Balbus has written to me that you let him have a copy of the fifth book of the *De Finibus*, in which I have made a few alterations, though not many. However, I shall be obliged if you will keep back the others, so that Balbus may not get unrevised copies and Brutus what is stale. But enough of this; I don't want to seem to make a fuss about trifles. Though these are now my important things, for what else have I?

Arpinum, June 30 or July 1, B.C. 45

---

[1] The verse ends with ἐμπορεύεται.

MARCUS TULLIUS CICERO

Varroni quidem quae scripsi te auctore, ita propero mittere, ut iam Romam miserim describenda. Ea si voles, statim habebis. Scripsi enim ad librarios, ut fieret tuis, si tu velles, describendi potestas. Ea vero continebis, quoad ipse te videam; quod diligentissime facere soles, cum a me tibi dictum est. Quo modo antea fugit me tibi dicere? Mirifice Caerellia, studio videlicet philosophiae flagrans, describit a tuis; istos ipsos " de finibus " habet. Ego autem tibi confirmo (possum falli ut homo) a meis eam non habere; numquam enim ab oculis meis afuerunt. Tantum porro aberat, ut binos scriberent; vix singulos confecerunt. Tuorum tamen ego nullum delictum arbitror itemque te volo existimare; a me enim praetermissum est, ut dicerem me eos exire nondum velle. Hui, quam diu de nugis! de re enim nihil habeo quod loquar.

De Dolabella tibi adsentior. Coheredes, ut scribis, in Tusculano. De Caesaris adventu scripsit ad me Balbus non ante Kal. Sextiles. De Attica optime, quod levius ac levius, et quod fert εὐκόλως. Quod autem de illa nostra cogitatione scribis, in qua nihil tibi cedo, ea, quae novi, valde probo, hominem, domum, facultates. Quod caput est, ipsum non novi, sed audio laudabilia, de Scrofa etiam proxime. Accedit, si quid hoc ad rem, εὐγενέστερος est etiam quam pater. Coram igitur et quidem propenso animo ad probandum. Accedit enim, quod patrem, ut scire te puto, plus etiam quam non modo tu, sed quam ipse scit, amo idque et merito et iam diu.

---

[1] Or " copies."

LETTERS TO ATTICUS XIII. 21a

I am in such a hurry to send what I have written to Varro, as you suggested, that I have sent it already to Rome to be copied. If you like, you shall have it at once. For I wrote to my copyist telling them to give your people leave to copy, if you liked. Please keep it, however, till I see you. You are generally most careful to do so, when I have told you. I was nearly forgetting to say that Caerellia, inspired of course by love of philosophy, is copying from your people[1]; she has those very books *De Finibus*. I assure you, so far as it is humanly possible to affirm anything, that she did not get it from mine, for my copy was never out of my sight. So far were my people from making two copies, that they could scarcely make up one. However, I am not finding any fault in your people, and I hope you will not either, for I omitted to say that I did not want the books circulated yet. Dear me, how I do harp on trifles. The fact is I have nothing of importance to say.

I agree about Dolabella. My co-heirs I will meet at Tusculum, as you suggest. As to Caesar's arrival, Balbus has written that he won't be here till the first of August. It is good news that Attica's attack gets slighter and slighter and that she is bearing it cheerfully. As to that idea of ours, about which I am quite as eager as you are, so far as I know anything about the man, I approve of him, his family, and his fortune. What is most important is that, though I do not know him himself, I hear very well of him, even quite recently from Scrofa. If it is of any importance, one may add that he is even better bred than his father. So we will speak of it when we meet, and I am disposed to approve. For in addition, as I think you know, I am with good reason and long have been fonder of his father than either you or he himself is aware.

MARCUS TULLIUS CICERO

XXII

CICERO ATTICO SAL.

Scr. in Arpinati IV Non. Quint. a. 709

De Varrone non sine causa quid tibi placeat tam diligenter exquiro. Occurrunt mihi quaedam. Sed ea coram. Te autem ἀσμεναίτατα intexui, faciamque id crebrius. Proximis enim tuis litteris primum te id non nolle cognovi. De Marcello scripserat ad me Cassius antea, τὰ κατὰ μέρος Servius. O rem acerbam! Ad prima redeo. Scripta nostra nusquam malo esse quam apud te, sed ea tum foras dari, cum utrique nostrum videbitur. Ego et librarios tuos culpa libero, neque te accuso, et tamen aliud quiddam ad te scripseram, Caerelliam quaedam habere, quae nisi a te[1] habere non potuerit. Balbo quidem intellegebam sat faciendum fuisse, tantum nolebam aut obsoletum Bruto aut Balbo inchoatum dari. Varroni, simul ac te videro, si tibi videbitur, mittam. Quid autem dubitarim, cum videro te, scies.

Attributos quod appellas, valde probo. Te de praedio Oviae exerceri moleste fero. De Bruto nostro perodiosum, sed vita fert. Mulieres autem vix satis humane, quae iniquo animo ferant, cum utraque

---

[1] habere ... te *omitted by MSS.; added by Ascensius and old editors.*

---

[1] M. Marcellus, consul in 51 B.C. and a partisan of Pompey, had just been murdered by M. Magius Cibo at Athens out of jealousy for the favour shown him by Caesar, who had

XXII

CICERO TO ATTICUS, GREETING.

I have my reasons for asking so persistently for your opinion about Varro. Some objections occur to me; but of those when we meet. Your name I introduced with the greatest pleasure and I shall do so more frequently, for I see for the first time from your last letter that you do not disapprove. About Marcellus Cassius had already written to me, and Servius sent some details.[1] What a sad thing! I return to my former point. There are no hands in which I would rather have my writings than in yours, but I should prefer them not to leave your hands till we have agreed on it. I acquit your copyists of fault and I bring no charge against you; but there was something different that I did mention in a letter, that Caerellia had some things she could only have got from you. In Balbus' case I realize of course that you had to satisfy him; only I am sorry that Brutus should get anything stale or Balbus anything unfinished. I will send to Varro, as soon as I have seen you, if you agree. Why I have hesitated, you shall know, when I do see you.

I strongly approve of your calling in those debts which have been transferred to me. I am sorry you are being bothered about Ovia's estate. About Brutus it is a great nuisance, but such is life. The ladies, however, are not very considerate in being annoyed, though both of them observe the pro-

granted him permission to return to Rome, an event celebrated in Cicero's speech *Pro Marcello*. Servius' letter is preserved, *Ad Fam.* IV. 12, and gives full details of the murder. Cf. also *Att.* XIII. 10.

officio pareat. Tullium scribam nihil fuit quod appellares; nam tibi mandassem, si fuisset. Nihil enim est apud eum positum nomine voti, sed est quiddam apud illum meum. Id ego in hanc rem statui conferre. Itaque et ego recte tibi dixi, ubi esset, et tibi ille recte negavit. Sed hoc quoque ipsum continuo adoriamur. Lucum hominibus non sane probo, quod est desertior, sed habet εὐλογίαν. Verum hoc quoque, ut censueris, quippe qui omnia. Ego, ut constitui, adero, atque utinam tu quoque eodem die! Sin quid (multa enim), utique postridie. Etenim coheredes: a quibus sine tua opprimi malitia. Est[1] alteris iam litteris nihil ad me de Attica. Sed id quidem in optima spe pono; illud accuso non te, sed illam, ne salutem quidem. At tu et illi et Piliae plurimam, nec me tamen irasci indicaris. Epistulam Caesaris misi, si minus legisses.

XXIII

CICERO ATTICO SAL.

Scr. in Tusculano VI Id. Quint. a. 709

Antemeridianis tuis litteris heri statim rescripsi; nunc respondeo vespertinis. Brutus mallem me ar-

[1] a quis sine te opprimi militia est *MSS.: the reading I have adopted is that of Tyrrell.*

---

[1] Cato's daughter Porcia, to whom Brutus was to be married, and his mother Servilia, who being a partisan of

prieties.¹ There was no necessity for you to dun my secretary Tullius; I should have told you, if there had been. For he has nothing of mine towards carrying out my vow.² But he has some of my money, and that I am thinking of devoting to that purpose. So we were both right, I in telling you where it was, and he in denying he had it. But let us get hold of this same money also at once. I do not very much approve of a grove for mortals, as it is not much frequented; but there is something to say for it. However, let that too be as you like, since you decide everything. I shall come to town when I arranged, and I hope to goodness you will be there the same day. But, if anything prevents you, and lots of things may, the next day at any rate. For there are my co-heirs, and without your shrewdness I shall be done for. This is the second letter with no news of Attica. But that I take as a hopeful sign. There is one thing I have a grievance about, not against you, but against her, that she does not even send her regards. But pay my best respects to her and to Pilia, and don't hint that I am angry anyhow. I am sending Caesar's letter, in case you should not have read it.

XXIII

CICERO TO ATTICUS, GREETING.

The morning's letter I answered yesterday at *Tusculum*, once, now I am answering yours of the evening. I would rather Brutus had asked me to Rome. It

July 10, B.C. 45

Caesar opposed the marriage. Most editors however adopt Orelli's reading *in utraque*, in which case it would mean "though Brutus is attentive to both."

² *i.e.* no money deposited with him towards the building of the shrine.

cesseret. Et aequius erat, cum illi iter instaret et subitum et longum, et me hercule nunc, cum ita simus adfecti, ut non possimus plane simul vivere (intellegis enim profecto, in quo maxime posita sit συμβίωσις), facile patiebar nos potius Romae una esse quam in Tusculano.

Libri ad Varronem non morabantur, sunt enim detexti, ut vidisti; tantum librariorum menda tolluntur. De quibus libris scis me dubitasse, sed tu videris. Item, quos Bruto mittimus, in manibus habent librarii.

Mea mandata, ut scribis, explica. Quamquam ista retentione omnes ait uti Trebatius; quid tu istos putas? Nosti domum. Quare confice εὐαγώγως. Incredibile est, quam ego ista non curem. Omni tibi adseveratione adfirmo, quod mihi credas velim, mihi maiori offensioni esse quam delectationi possessiunculas meas. Magis enim doleo me non habere, cui tradam, quam gaudeo[1] habere, qui utar. Atque illud Trebatius se tibi dixisse narrabat; tu autem veritus es fortasse, ne ego invitus audirem. Fuit id quidem humanitatis, sed, mihi crede, iam ista non curo. Quare da te in sermonem et perseca et confice, et ita cum Polla loquere, ut te cum illo Scaeva loqui

[1] gaudeo *added by Gronovius.*

---

[1] By the Julian law of 49 B.C. debtors could make over property to their creditors on the valuation it had before the Civil war, and could deduct all interest already paid from the debt.

[2] *Domum* may refer to some house offered in payment of a debt to Cicero, or it may possibly be used in the sense I, following most editors, have given it, for which however *familia* is commoner. Reid would read *dominum*, referring it to Caesar.

would have been fairer, as he is on the point of a sudden long journey, and upon my soul I should have much preferred that we should meet in Rome rather than in my house at Tusculum, now that the state of our feelings prevents us from living together at all, for of course you understand what constitutes good company.

There is no delay about the books dedicated to Varro. They are finished, as you have seen; there is only the correction of the copyists' mistakes. About those books you know I have had some hesitation, but you must look to it. The copyists have in hand, too, those I am dedicating to Brutus.

Carry out my instructions as you say. However what about that abatement?[1] Trebatius says everybody is taking advantage of it. What do you suppose my debtors will do? You know the gang.[2] So settle the matter accommodatingly. You would never believe how little I care about such things. I give you my solemn word for it, and I hope you will believe me, that the little I have causes me more annoyance than pleasure. For I am more grieved at having no one to leave it to than pleased at having enough for my own enjoyment. Trebatius tells me he told you so; but perhaps you feared I should be sorry at the news. That was certainly kind of you; but, believe me, I don't care about such things now. So get you to your conferences, hack away at it and finish the business; and in talking with Polla consider you are talking with that fellow Scaeva,[3] and

---

[3] Caesar had a favourite centurion named Scaeva, and that may be the person here referred to. If so it means "remember they are all people who have shared Caesar's plunder." But many regard the name and the words *da* to *confice* as a quotation from some play.

putes, nec existimes eos, qui non debita consectari soleant, quod debeatur, remissuros. De die tantum videto et id ipsum bono modo.

XXIV

CICERO ATTICO SAL.

Scr. in Tusculano V Id. Quint. a. 709

Quid est, quod Hermogenes mihi Clodius Andromenem sibi dixisse se Ciceronem vidisse Corcyrae? Ego enim audita tibi putaram. Nil igitur ne ei quidem litterarum? An non vidit? Facies ergo ut sciam.

Quid tibi ego de Varrone rescribam? Quattuor διφθέραι sunt in tua potestate. Quod egeris, id probabo. Nec tamen "αἰδέομαι Τρῶας." Quid enim? Sed, ipsi quam res illa probaretur, magis verebar. Sed, quoniam tu suscipis, in alteram aurem.

De retentione rescripsi ad tuas accurate scriptas litteras. Conficies igitur, et quidem sine ulla dubitatione aut retretctatione. Hoc fieri et oportet et opus est.

XXV

CICERO ATTICO SAL.

Scr. in Tusculano IV Id. Quint. a. 709

De Andromene, ut scribis, ita putaram. Scisses enim mihique dixisses. Tu tamen ita mihi de Bruto scribis, ut de te nihil. Quando autem illum putas? Nam ego Romam pridie Idus. Bruto ita volui scri-

---

[1] Cf. *Att.* XIII. 13.

don't imagine that those who are in the habit of taking what is not owing to them, will abate anything that is. Only be careful that they pay up to time and allow some latitude there too.

XXIV

CICERO TO ATTICUS, GREETING.

What am I to make of this? Hermogenes Clodius *Tusculum,* tells me that Andromenes said he saw my son at *July 11,* Corcyra. For I supposed you had heard of it. Then B.C. *45* didn't he give any letter even to him? Or perhaps he didn't see him. You must let me know, please.

What answer am I to give you about Varro? You have the four parchment rolls: and whatever you do I shall approve. It is not that "I fear the Trojans."[1] Why should I? But I am more afraid how he may regard it. However, as you undertake the matter, I shall sleep in peace.[2]

About the abatement I have answered your careful letter. You must get the matter over, and that too without any hesitation or refusal. That ought to be and must be done.

XXV

CICERO TO ATTICUS, GREETING.

About Andromenes I thought exactly what you *Tusculum,* say, for you would have known and told me. How- *July 12,* ever, you have written such a lot about Brutus that B.C. *45* you say nothing of yourself. But when do you think he is coming? For I shall come to Rome on the 14th. What I meant to say in my letter to

[2] Lit. "on both ears." Supply *dormire licet*.

bere (sed, quoniam tu te legisse scribis, fui fortasse ἀσαφέστερος), me ex tuis litteris intellexisse nolle eum me quasi prosequendi sui causa Romam nunc venire. Sed, quoniam iam adest meus adventus, fac, quaeso, ne quid eum Idus impediant, quo minus suo commodo in Tusculano sit. Nec enim ad tabulam eum desideraturus eram (in tali enim negotio cur tu unus non satis es?), sed ad testamentum volebam, quod iam malo alio die, ne ob eam causam Romam venisse videar. Scripsi igitur ad Brutum iam illud, quod putassem, Idibus nihil opus esse. Velim ergo totum hoc ita gubernes, ut ne minima quidem re ulla Bruti commodum impediamus.

Sed quid est tandem, quod perhorrescas, quia tuo periculo iubeam libros dari Varroni? Etiam nunc si dubitas, fac, ut sciamus. Nihil est enim illis elegantius. Volo Varronem, praesertim cum ille desideret; sed est, ut scis,

Iliad, xi. 654

"δεινὸς ἀνήρ· τάχα κεν καὶ ἀναίτιον αἰτιόῳτο."

Ita mihi saepe occurrit vultus eius querentis fortasse vel hoc, meas partis in iis libris copiosius defensas esse quam suas, quod me hercule non esse intelleges, si quando in Epirum veneris. Nam nunc Alexionis epistulis cedimus. Sed tamen ego non despero probatum iri Varroni, et id, quoniam impensam fecimus in macrocolla, facile patior teneri. Sed, etiam atque etiam dico, tuo periculo fiet. Quare, si addubitas, ad Brutum transeamus; est enim is quoque Antiochius.

Brutus was that I had gathered from your note that he did not wish me to come to Rome now just to pay my respects to him—but, as you say you have read the letter, perhaps I was not quite clear. However, as I am just on the point of coming, please see that my presence on the 15th does not prevent his coming to Tusculum at his convenience. For I shall not want him at the auction—surely in such a business you alone will be enough: but I do want him when I make my will. That I would rather postpone for another day now, so as not to seem to have come to Rome expressly for that purpose. So I have written to Brutus now that I shall not want him, as I had thought, on the 15th. I should like you to look after all this and see that we don't inconvenience Brutus in the least.

But what on earth is the reason why you are so frightened at my bidding you send the books to Varro on your own responsibility? Even now, if you have any doubts, let me know. Nothing could be more finished than they are. I want Varro, especially as he desires it: but, as you know, he is "a fearsome man; the blameless he would blame." I often picture him to myself complaining of this perhaps, that my side in the books is more fully defended than his own, though I assure you, if ever you come to Epirus, I will convince you it is not. For at present I have to give way to Alexio's[1] letters. However, I don't despair of winning Varro's approval; and, as I have gone to the expense of a large paper copy, I should like to stick to my plan. But I repeat again, it must be on your responsibility. So, if you have doubts, let us change to Brutus: he is also a

---

[1] Atticus' steward.

MARCUS TULLIUS CICERO

O Academiam volaticam et sui similem! modo huc, modo illuc. Sed, quaeso, epistula mea ad Varronem valdene tibi placuit? Male mi sit, si umquam quicquam tam enitar. Ergo ne Tironi quidem dictavi, qui totas περιοχὰς persequi solet, sed Spintharo syllabatim.

XXVI

CICERO ATTICO SAL.

Scr. in Tusculano prid. Id. Mai. a. 709

De Vergili parte valde probo. Sic ages igitur. Et quidem id erit primum, proximum Clodiae. Quodsi neutrum, metuo, ne turbem et inruam in Drusum. Intemperans sum in eius rei cupiditate, quam nosti. Itaque revolvor identidem in Tusculanum. Quidvis enim potius, quam ut non hac aestate absolvatur.

Ego, ut tempus est nostrum, locum habeo nullum, ubi facilius esse possim quam Asturae. Sed, quia, qui mecum sunt, credo, quod maestitiam meam non ferunt, domum properant, etsi poteram remanere, tamen, ut scripsi tibi, proficiscar hinc, ne relictus videar. Quo autem? Lanuvio conor equidem in Tusculanum. Sed faciam te statim certiorem. Tu litteras conficies. Equidem credibile non est quantum scribam, quin etiam noctibus. Nihil enim somni. Heri etiam effeci epistulam ad Caesarem; tibi enim placebat. Quam non fuit malum scribi, si forte opus

---

[1] Like Cicero's treatise, which had already been rewritten twice: cf. XIII. 16.

follower of Antiochus. O that fickle Academy, always the same, now one thing, now another.[1] But pray tell me, were you very pleased with my letter to Varro. May I be hanged if I ever take so much trouble with anything again. So I did not even dictate it to Tiro, who can follow whole sentences as dictated, but syllable by syllable to Spintharus.

XXVI

CICERO TO ATTICUS, GREETING.

About Vergilius'[2] share I approve; so arrange it like that. And indeed it will be my first choice, next to Clodia's. If neither, I fear I shall run amuck and make a dash for Drusus. As you know, I have lost control of myself in my desire for this. So I keep coming back to the idea of my place at Tusculum. For anything is better than not getting it finished this summer.

Tusculum, May 14, B.C. 45

Under the present circumstances I am as comfortable at Astura as I could be anywhere. But as those who are with me are in a hurry to go home, I suppose because they cannot put up with my melancholy, though I might remain, I shall leave here, as I told you, so as not to seem deserted. But where am I to go? From Lanuvium I am trying to bring myself to go to Tusculum. But I will let you know soon. Please write the letters. You wouldn't believe how much writing I get done by night as well as day, for I cannot sleep. Yesterday I even composed a letter to Caesar, as you desired. There was no harm in writing it in case you thought it necessary: as

[2] Vergilius was one of the four co-heirs of Scapula. Cf. XII. 38a.

esse putares; **ut** quidem nunc est, nihil sane est necesse mittere. Sed id quidem, ut tibi videbitur. Mittam tamen ad te exemplum fortasse Lanuvio, nisi forte Romam. Sed cras scies.

XXVII

CICERO ATTICO SAL.

Scr. in Tusculano VIII K. Iun. a. 709

De epistula ad Caesarem nobis vero semper rectissime placuit, ut isti ante legerent. Aliter enim fuissemus et in hos inofficiosi, et in nosmet ipsos, si illum offensuri fuimus, paene periculosi. Isti autem ingenue; mihique gratum, quod, quid sentirent, non reticuerunt, illud vero vel optime, quod ita multa mutari volunt, ut mihi de integro scribendi causa non sit. Quamquam de Parthico bello quid spectare debui, nisi quod illum velle arbitrabar? Quod enim aliud argumentum epistulae nostrae nisi κολακεία fuit? An, si ea, quae optima putarem, suadere voluissem, oratio mihi defuisset? Totis igitur litteris nihil opus est. Ubi enim ἐπίτευγμα magnum nullum fieri possit, ἀπότευγμα vel non magnum molestum futurum sit, quid opus est παρακινδυνεύειν? praesertim cum illud occurrat, illum, cum antea nihil scripserim, existimaturum me nisi toto bello confecto nihil scripturum fuisse. Atque etiam vereor, ne putet me hoc quasi Catonis μείλιγμα esse voluisse. Quid quaeris? valde me paenitebat, nec mihi in hac quidem re quicquam magis ut vellem accidere potuit, quam quod σπουδὴ nostra non est probata. Incidissimus etiam in illos, in eis in cognatum tuum.

---

[1] Or "come into contact with." *Cognatum* refers to young Quintus.

things are, there is certainly no need to send it. But let that be as you like. However, I will send you a copy, perhaps from Lanuvium, unless I happen to come to Rome. But you shall know to-morrow.

XXVII

CICERO TO ATTICUS, GREETING.

Tusculum, May 25, B.C. 45

As for the letter to Caesar, I was always ready to let your friends read it first. If I had not been, I should not have done my duty by them, and should very nearly have imperilled myself, if I were likely to offend him. But they have acted frankly, and I am thankful to them for not concealing their feelings; but the best thing of all is that they want to make so many alterations that there is no sense in my writing it all over again. However, what view ought I to have taken of the Parthian war except what I thought he wanted? Indeed what other purpose had my letter save to kowtow to him? Do you suppose I should have been at a loss for words, if I had wanted to give him the advice which I really thought best? So the whole letter is unnecessary. For, when I cannot make a *coup*, and a fiasco, however slight, would be unpleasant, why should I run unnecessary risk? Especially as it occurs to me that, as I have not written before, he would think I should not have written until the whole war were over. Besides I am afraid he may think it is to sugar the pill of my Cato. In fact I am very sorry I wrote it, and nothing could suit my wishes better than that they do disapprove of my zeal. I should have fallen foul of [1] Caesar's party, and among them your relative.

Sed redeo ad hortos. Plane illuc te ire nisi tuo
magno commodo nolo; nihil enim urget. Quicquid
erit, operam in Faberio ponamus. De die tamen
auctionis, si quid scies. Eum, qui e Cumano venerat,
quod et plane valere Atticam nuntiabat et litteras se
habere aiebat, statim ad te misi.

XXVIII

CICERO ATTICO SAL.

Scr. in Tus-
culano VII
K. Iun. a.
709

Hortos quoniam hodie eras inspecturus, quid visum
tibi sit, cras scilicet. De Faberio autem, cum venerit.
De epistula ad Ceasaremiurato, mihi crede, non possum; nec me turpitudo deterret, etsi maxime debe-
bat. Quam enim turpis est adsentatio, cum vivere
ipsum turpe sit nobis! Sed, ut coepi, non me hoc
turpe deterret. Ac vellem quidem (essem enim, qui
esse debebam), sed in mentem nihil venit. Nam,
quae sunt ad Alexandrum hominum eloquentium et
doctorum suasiones, vides, quibus in rebus versentur.
Adulescentem incensum cupiditate verissimae gloriae,
cupientem sibi aliquid consilii dari, quod ad laudem
sempiternam valeret, cohortantur ad decus. Non
deest oratio; ego quid possum? Tamen nescio quid
e quercu exsculpseram, quod videretur simile simu-
lacri. In eo quia non nulla erant paulo meliora quam
ea, quae fiunt et facta sunt, reprehenduntur; quod
me minime paenitet. Si enim pervenissent istae
litterae, mihi crede, nos paeniteret. Quid? tu non

But to return to the gardens. I don't in the least want you to go there, unless it is quite convenient to you: for there is no hurry. Whatever happens let us direct our efforts towards Faberius. However send me the date of the auction, if you know it. I have sent this man, who came from Cumae, straight on to you, as he said Attica was quite well and he had letters.

XXVIII

CICERO TO ATTICUS, GREETING.

Tusculum, May 26, B.C. 45

As you are going to look at the garden to-day, I shall of course hear from you to-morrow what you think of it; and about Faberius, when he has come. About the letter to Caesar, I give you my word of honour I cannot; it is not the shame of the thing that prevents me, though that is just what ought. Ah, how shameful is flattery, when life alone is a disgrace! But, as I was beginning to say, it is not the shame of it that prevents me—I only wish it were, for then I should be the man I ought to be— but I cannot think of anything to write. Just consider the subjects of the letters of advice addressed to Alexander by men of eloquence and learning. Here was a youth fired by a desire for the truest glory and desiring to have some advice given him on the subject of eternal fame, and they exhort him to follow honour. There is plenty to say on that: but what can I say? However, from hard material I had rough hewn something that seemed to me to take shape. Because there were a few touches in it a little better than the actual facts past or present, fault is found with them; and I don't regret it a bit. For, if the letter had reached its destination, believe me, I should have regretted it. Why, don't you

vides ipsum illum Aristoteli discipulum summo ingenio, summa modestia, posteaquam rex appellatus sit, superbum, crudelem, immoderatum fuisse? Quid? tu hunc de pompa Quirini contubernalem his nostris moderatis epistulis laetaturum putas? Ille vero potius non scripta desideret quam scripta non probet. Postremo ut volet. Abiit illud, quod tum me stimulabat, cum tibi dabam πρόβλημα 'Αρχιμήδειον. Multo mehercule magis nunc opto casum illum, quem tum timebam, vel quem libebit.

Nisi quid te aliud impediet, mi optato veneris. Nicias a Dolabella magno opere arcessitus (legi enim litteras), etsi invito me, tamen eodem me auctore, profectus est.

Hoc manu mea. Cum quasi alias res quaererem de philologis e Nicia, incidimus in Talnam. Ille de ingenio nihil nimis, modestum et frugi. Sed hoc mihi non placuit. Se scire aiebat ab eo nuper petitam Cornificiam, Q. filiam, vetulam sane et multarum nuptiarum; non esse probatum mulieribus, quod ita reperirent, rem non maiorem $\overline{\text{DCCC}}$. Hoc putavi te scire oportere.

XXIX

CICERO ATTICO SAL.

Scr. in Tusculano VI K Iun. a. 709

De hortis ex tuis litteris cognovi et Chrysippo. In villa, cuius insulsitatem bene noram, video nihil

---

[1] Caesar. Cf. *Att.* XII. 45, 3.

LETTERS TO ATTICUS XIII. 28-29

see that even that pupil of Aristotle, in spite of his high ability and his high character, became proud, cruel, and ungovernable, after he got the title of king? How do you suppose this puppet messmate of Quirinus[1] will like my moderate letters? Let him rather look for what I do not write than disapprove of what I have written. In short let it be as he pleases. What was spurring me on when I put that insoluble problem[2] before you has all gone now. Upon my word now I should far rather welcome the misfortune I feared then or any other.

If there is nothing to prevent you, come to me and welcome. Nicias at Dolabella's urgent request (for I read the letter) has gone, against my will though not against my advice.

The rest I have written myself. When I was discussing men of learning with Nicias, we chanced to speak of Talna. He had not much to say for his intelligence, though he gave him a good and steady character. But there was one thing that seemed to me unsatisfactory. He said he knew he had lately sought in marriage Cornificia, Quintus' daughter, though quite an old woman and married more than once before; but the ladies would not agree as they found he was not worth more than 7,000 guineas.[3] I thought you ought to know this.

XXIX

CICERO TO ATTICUS, GREETING.

I have heard all about the gardens from your letter and from Chrysippus. I was well aware of the bad taste shown in the house, and I see there

Tusculum, May 27, B.C. 45

[2] What to write to Caesar. Cf. *Att.* XII. 40, 2.
[3] 800,000 sesterces.

aut pauca mutata; balnearia tamen laudat maiora, de minoribus ait hiberna effici posse. Tecta igitur ambulatiuncula addenda est; quam ut tantam faciamus, quantam in Tusculano fecimus, prope dimidio minoris constabit isto loco. Ad id autem, quod volumus, ἀφίδρυμα nihil aptius videtur quam lucus, quem ego noram; sed celebritatem nullam tum habebat, nunc audio maximam. Nihil est, quod ego malim. In hoc τὸν τυφόν μου πρὸς θεῶν τροποφόρησον. Reliquum est, si Faberius nobis nomen illud explicat, noli quaerere, quanti; Othonem vincas volo. Nec tamen insaniturum illum puto; nosse enim mihi hominem videor. Ita male autem audio ipsum esse tractatum, ut mihi ille emptor non esse videatur. Quid enim? pateretur? Sed quid argumentor? Si Faberianum explicas, emamus vel magno; si minus, ne parvo quidem possumus. Clodiam igitur. A qua ipsa ob eam causam sperare videor, quod et multo minoris sunt, et Dolabellae nomen tam expeditum videtur, ut etiam repraesentatione confidam. De hortis satis. Cras aut te aut causam; quam quidem puto[1] futuram Faberianam. Sed, si poteris.

Ciceronis epistulam tibi remisi. O te ferreum, qui illius periculis non moveris! Me quoque accusat. Eam tibi epistulam misi semissem.[2] Nam illam alte-

---

[1] puto *added by Wesenberg.*
[2] misi semissem *Purser*: misissem *MSS.*

LETTERS TO ATTICUS XIII. 29

has been little or no alteration; however, he praises the larger bath and thinks the smaller could be made into a winter snuggery. So a covered passage would have to be added, and, if I made one the same size as that at my place at Tusculum, the cost would be about half as much in that district. However, for the erection we want to make nothing could be more suitable than the grove, which I used to know well; then it was not at all frequented, now I hear it is very much so. There is nothing I should prefer. In this, humour my whim, in heaven's name. For the rest, if Faberius pays that debt, don't bother about the cost; I want you to outbid Otho: and I don't think he will bid wildly, for I fancy I know the man. Besides I hear he has had such bad luck that I doubt if he will buy. For would he put up with it, if he could help it?[1] But what is the good of talking? If you get the money from Faberius, let us buy even at a high price; if not, we cannot even at a low. So then we must fall back on Clodia. In her case I see more grounds for hope, as her property is worth much less, and Dolabella's debt seems so safe that I feel confident of being able to pay in ready money. Enough about the gardens. To-morrow I shall either see you or hear the reason why not. I expect that will be the business with Faberius. But come, if you can.

I am sending young Quintus' letter. How hardhearted of you not to tremble at his hair-breadth escapes. He complains about me too. I have sent you half the letter. The other half about his

[1] Probably, as Manutius suggests, this means "would he endure the wrong he has suffered, if he had any means left."

ram de rebus gestis eodem exemplo puto.   In Cumanum hodie misi tabellarium.   Ei dedi tuas ad Vestorium, quas Pharnaci dederas.

XXX

CICERO ATTICO SAL.

Scr. in Tusculano V K. Iun. post ep. XXXI a. 709

Commodum ad te miseram Demean, cum Eros ad me venit. Sed in eius epistula nihil erat novi nisi auctionem biduum. Ab ea igitur, ut scribis, et velim confecto negotio Faberiano; quem quidem negat Eros hodie, cras mane putat. A te colendus est; istae autem κολακεῖαι non longe absunt a scelere. Te, ut spero, perendie.

Mi, sicunde potes, erues, qui decem legati Mummio fuerint. Polybius non nominat. Ego memini Albinum consularem et Sp. Mummium; videor audisse ex Hortensio Tuditanum. Sed in Libonis annali xiiii annis post praetor est factus Tuditanus quam consul Mummius. Non sane quadrat. Volo aliquem Olympiae aut ubi visum πολιτικὸν σύλλογον more Dicaearchi, familiaris tui.

XXXI

CICERO ATTICO SAL.

Scr. in Tusculano eodem die quo ep. XXX paulo ante a. 709

v Kal. mane accepi a Demea litteras pridie datas, ex quibus aut hodie aut cras exspectare te deberem. Sed, ut opinor, idem ego, qui exspecto tuum adventum, morabor te. Non enim puto tam expeditum Faberianum negotium futurum, etiamsi est futurum,

170

adventures I think you have in duplicate. I have sent a messenger to-day to Cumae. I have given him your letter to Vestorius, which you had given to Pharnaces.

XXX

CICERO TO ATTICUS, GREETING.

I had just sent Demeas to you, when Eros arrived. But in his letter there was no news except that the auction lasts two days. So you will come after it, as you say, and I hope the business with Faberius will be settled. Eros thinks he will not settle up to-day, but will to-morrow morning. You must be polite to him; though such kowtowing is almost criminal. I hope you will come the day after to-morrow.

Tusculum, May 28, B.C. 45

Dig out for me from somewhere, if you can, the names of Mummius' ten legates. Polybius does not give them. I remember Albinus the ex-consul and Sp. Mummius; and I think Hortensius told me Tuditanus. But in Libo's annals Tuditanus was praetor fourteen years after Mummius' consulship. That does not square at all. I am thinking of writing a kind of political conference, held at Olympia or wherever you like, like that of your friend Dicaearchus.

XXXI

CICERO TO ATTICUS, GREETING.

On the 28th in the morning Demeas delivered a letter dated the day before, from which I ought to expect you either to-day or to-morrow. But, I suppose, I who am looking forward to your coming, shall be the very person who will delay it. For I don't expect the business with Faberius will be so far settled, even if it is to be settled, that it will not

Tusculum, May 28, B.C. 45

ut non habeat aliquid morae. Cum poteris igitur. Quoniam etiamnum abes, Dicaearchi, quos scribis, libros sane velim mi mittas, addas etiam καταβάσεως.

De epistula ad Caesarem κέκρικα; atqui[1] id ipsum, quod isti aiunt illum scribere, se nisi constitutis rebus non iturum in Parthos, idem ego suadebam in illa epistula. Utrum liberet, facere posse auctore me. Hoc enim ille exspectat videlicet neque est facturus quicquam nisi de meo consilio. Obsecro, abiciamus ista et semiliberi saltem simus; quod adsequemur et tacendo et latendo.

Sed adgredere Othonem, ut scribis. Confice, mi Attice, istam rem. Nihil enim aliud reperio, ubi et in foro non sim et tecum esse possim. Quanti autem, hoc mihi venit in mentem. C. Albanius proximus est vicinus. Is ↀ iugerum de M. Pilio emit, ut mea memoria est, HS $\overline{\text{cxv}}$. Omnia scilicet nunc minoris. Sed accedit cupiditas, in qua praeter Othonem non puto nos ullum adversarium habituros. Sed eum ipsum tu poteris movere, facilius etiam, si Canum haberes. O gulam insulsam! Pudet me patris. Rescribes, si quid voles.

[1] atqui *Wesenberg*: atque *MSS*.

cause some delay. So come when you can. Since you are still away, I should like you to send me the books of Dicaearchus, which you mention, with the *Descent*.[1]

As for the letter to Caesar I have made up my mind; and yet precisely what they say he says in his letter, that he will not go against the Parthians until affairs are arranged here, is what I advised in my letter. I told him he could do whichever he chose with my full leave. For of course he wants that and won't do anything without my advice. For heaven's sake let us give up flattery and be at least half-free; and that we can manage by keeping quiet and out of sight.

But approach Otho, as you say, and finish that business, my dear Atticus. For I don't see any other way of keeping away from the forum and yet being with you. As to the price, this has just occurred to me. The nearest neighbour is C. Albanius. He bought some 600 acres[2] of M. Pilius, so far as I can recollect for £110,000.[3] Of course everything has gone down in value now. But on the other side counts our eagerness to purchase, though I don't suppose we shall have anyone bidding against us except Otho. Him however you can influence personally, and could still more easily, if you had Canus with you. What senseless gluttony![4] Shame on his father! Answer, if you want to say anything.

[1] So called because it described a visit to the cave of Trophonius in Arcadia.
[2] 1,000 *jugera*.   [3] 11,500,000 sesterces.
[4] Probably this refers to some act of young Quintus Cicero.

MARCUS TULLIUS CICERO

XXXII

CICERO ATTICO SAL.

Scr. in Tusculano IV K. Iun. a. 709

Alteram a te epistulam cum hodie accepissem, nolui te una mea contentum. Tu vero age, quod scribis, de Faberio. In eo enim totum est positum id, quod cogitamus; quae cogitatio si non incidisset, mihi crede, istuc ut cetera non laborarem. Quam ob rem, ut facis (istuc enim addi nihil potest), urge, insta, perfice.

Dicaearchi περὶ ψυχῆς utrosque velim mittas et καταβάσεως. Τριπολιτικὸν non invenio et epistulam eius, quam ad Aristoxenum misit. Tres eos libros maxime nunc vellem; apti essent ad id, quod cogito. Torquatus Romae est. Misi, ut tibi daretur. Catulum et Lucullum, ut opinor, antea. His libris nova prohoemia sunt addita, quibus eorum uterque laudatur. Eas litteras volo habeas, et sunt quaedam alia. Et, quod ad te de decem legatis scripsi, parum intellexisti, credo, quia διὰ σημείων scripseram. De C. Tuditano enim quaerebam, quem ex Hortensio audieram fuisse in decem. Eum video in Libonis praetorem P. Popilio, P. Rupilio coss. Annis XIIII ante, quam praetor factus est, legatus esse potuisset, nisi admodum sero quaestor esset factus? quod non arbitror. Video

---

[1] So called because it represented the ideal State as a mixture of monarchy, aristocracy, and democracy.

LETTERS TO ATTICUS XIII. 32

XXXII

CICERO TO ATTICUS, GREETING.

Tusculum, May 29, B.C. 45

As I have received two letters from you to-day, I did not think it right that you should content yourself with only one of mine. Pray do as you say about Faberius. For on that depends entirely what I am thinking of. And, if that idea had never occurred to me, believe me I should not bother about that any more than anything else. So continue your energy—for you cannot add to it—and push on and finish the matter.

Please send me Dicaearchus' two books *About the Soul* and the *Descent*. I can't find the *Mixed Constitution*[1] and the letter he sent to Aristoxenus. I should much like to have those three books now; they would bear on what I am planning. Torquatus[2] is in Rome. I have sent orders for it to be given to you. Catulus and Lucullus I believe you have already. I have added new prefaces to the books, in which each of them is mentioned with honour. Those compositions I should like you to have, and there are some others too. What I said about the ten legates, you did not fully understand, I suppose because I wrote it in shorthand.[3] I was asking about C. Tuditanus, who Hortensius told me was one of them. I see in Libo that he was praetor in the consulship of P. Popilius and P. Rupilius. Could he have been legate fourteen years before he was praetor, unless he was very late in getting the quaestorship? I don't think that was the case; for

---

[2] *i.e. De Finibus*, Bk. I., in which Torquatus is the chief speaker. Similarly, Catulus and Lucullus are the first two books of the *Academica* in its first form.

[3] Or *demi-mots*, as Tyrrell renders it.

enim curules magistratus eum legitimis annis perfacile cepisse. Postumium autem, cuius statuam in Isthmo meminisse te dicis, nesciebam fuisse. Is autem est, qui cos. cum L. Lucullo fuit; quem tu mihi addidisti sane ad illum σύλλογον personam idoneam. Videbis igitur, si poteris, ceteros, ut possimus πομπεῦσαι καὶ τοῖς προσώποις.

XXXIII

CICERO ATTICO SAL.

Scr. in Tusculano III Non. Iun. a. 709

O neglegentiam miram! Semelne putas mihi dixisse Balbum et Faberium professionem relatam? qui etiam eorum iussu miserim, qui profiteretur. Ita enim oportere dicebant. Professus est Philotimus libertus. Nosti, credo, librarium. Sed scribes et quidem confectum. Ad Faberium, ut tibi placet, litteras misi, cum Balbo autem puto te aliquid fecisse hodie[1] in Capitolio. In Vergilio mihi nulla est δυσωπία. Nec enim eius causa sane debeo, et, si emero, quid erit, quod postulet? Sed videbis, ne is tum sit in Africa ut Caelius.

De nomine tu videbis cum Cispio; sed, si Plancus destinat, tum habet res difficultatem. Te ad me venire uterque nostrum cupit; sed ista res nullo modo relinquenda est. Othonem quod speras posse

[1] hodie *Bosius*: H. *MSS.*: fuisti enim *Elmore*.

---

[1] Possibly a statement of income before the next census; or perhaps some formality in the transference of a debt due from Faberius to Cicero.

I see he won the curule offices quite easily in the proper years. But I did not know that Postumius, whose statue you say you remember in the Isthmus, was one of them. He was the man who was consul with L. Lucullus; and it is a very suitable person you have added to my conference. So please look up the others too, if you can, that I may make a show with my *dramatis personæ*, as well as my subject.

XXXIII

CICERO TO ATTICUS, GREETING.

What extraordinary carelessness! Do you think it was only once that I have been told by Balbus and Faberius that the return[1] had been made? Why it was at their suggestion that I sent a man to make it, for they said I ought to do so. It was my freedman Philotimus who made the return; you know him, I think, a copyist. But you must write and let me know it is finished. I have sent a letter as you advise to Faberius. With Balbus I think you have made some arrangement in the Capitol to-day. About Vergilius I have no scruples; for there is no reason why I should have in his case; and, if I buy, what claim will he have? But see that he may not be in Africa then like Caelius.[2]

The debt you must look into with Cispius; but, if Plancus intends to bid, there will be difficulties. That you should come to me would suit us both, but that business cannot possibly be thrown up. It is

---

[2] Vergilius had sided with Pompey in Spain, and Cicero apparently is afraid that, like Caelius, for whom cf. XIII. 3, he may not be in Italy when applied to for payment. But the reading and the sense are uncertain.

MARCUS TULLIUS CICERO

vinci, sane bene narras. De aestimatione, ut scribis, cum agere coeperimus; etsi nihil scripsit nisi de modo agri. Cum Pisone, si quid poterit. Dicaearchi librum accepi et καταβάσεως exspecto.

. . . negotium dederis, reperiet ex eo libro, in quo sunt senatus consulta Cn. Cornelio, L. Mummio coss. De Tuditano autem quod putas, εὔλογον est tum illum, quoniam fuit ad Corinthum (non enim temere dixit Hortensius), aut quaestorem aut tribunum mil. fuisse, idque potius credo. Tu de[1] Antiocho scire poteris videlicet[2] etiam, quo anno quaestor aut tribunus mil. fuerit; si neutrum, saltem,[3] in praefectis an in contubernalibus fuerit, modo fuerit in eo bello.

XXXIIIa
CICERO ATTICO SAL.

Scr. in Tusculano VII Id. Quint. a. 709

De Varrone loquebamur: lupus in fabula. Venit enim ad me et quidem id temporis, ut retinendus esset. Sed ego ita egi, ut non "scinderem paenulam" (memini enim tuum): et multi erant nosque imparati. Quid refert? Paulo post C. Capito cum T. Carrinate. Horum ego vix attigi paenulam. Tamen remanserunt, ceciditque belle. Sed casu

---

[1] fuisse . . . de *as Ernesti*: idque potius fuisse. sed credo te de *M*. [2] videlicet *Schmidt*: vide *MSS*.

[3] saltem *Gurlitt*: ea de *M*: cadet (et) *ZO*¹, *L* (*marg*.): eadem *O*².

welcome news that you think we can beat Otho. As you say about the assignment, when we begin to negotiate; though he has not mentioned anything except the extent of the ground. Discuss it with Piso in case he can do anything. I have received Dicaearchus' book and am expecting his *Descent*.

(If you) will commission someone, he will find out ... from the book containing the decrees passed in the consulship of Cn. Cornelius and L. Mummius. Your idea about Tuditanus is reasonable enough, he was either quaestor or military tribune, since he was at Corinth at the time and Hortensius was not speaking at random; and I think you are right. You will be able to find out from Antiochus of course in what year he was quaestor or military tribune. If he was neither, then he would at least have been among the prefects or on the staff, provided he was in the war at all.

XXXIIIa

CICERO TO ATTICUS, GREETING.

We were talking of Varro: talk of the devil, you know, for here he came and at such an hour that I had to ask him to stop. But I did not cling so closely to him as to "tear his cloak" (for I remember that phrase [1] of yours), and there were a lot of them and I was unprepared. But what does that matter? Just afterwards came C. Capito and T. Carrinas. Their cloaks I hardly touched; but they stayed and it turned out all right. By chance Capito began

Tusculum, July 9, B.C. 45

---

[1] I follow Reid and Shuckburgh in referring this to the preceding phrase and not to the following.

sermo a Capitone de urbe augenda, a ponte Mulvio
Tiberim duci secundum montes Vaticanos, campum
Martium coaedificari, illum autem campum Vaticanum
fieri quasi Martium campum. "Quid ais?" inquam;
"at ego ad tabulam, ut, si recte possem, Scapulanos
hortos." "Cave facias," inquit; "nam ista lex per-
feretur; vult enim Caesar." Audire me facile passus
sum, fieri autem moleste fero. Sed tu quid ais?
Quamquam quid quaero? Nosti diligentiam Capi-
tonis in rebus novis perquirendis. Non concedit
Camillo. Facies me igitur certiorem de Idibus. Ista
enim me res adducebat. Eo adiunxeram ceteras,
quas consequi tamen biduo aut triduo post facile
potero. Te tamen in via confici minime volo; quin
etiam Dionysio ignosco. De Bruto quod scribis, feci,
ut ei liberum esset, quod ad me attineret. Scripsi
enim ad eum heri Idibus eius opera mihi nihil opus
esse.

XXXIV

CICERO ATTICO SAL.

Scr. Asturae　Asturam veni VIII Kal. vesperi.[1] Vitandi enim
VI K. Sext. caloris causa Lanuvi tris horas acquieveram. Tu
a. 709　velim, si grave non erit, efficias, ne ante Nonas mihi
illuc veniendum sit (id potes per Egnatium Maxi-

---

[1] vesperi *Schmidt* : iul. *M*.

LETTERS TO ATTICUS XIII. 33a–34

talking of the improvements of the city: the course of the Tiber is to be diverted from the Mulvian bridge along the Vatican hills; the Campus Martius to be built over, and the Vatican plain to be a sort of Campus Martius. "What's that?" I said. "Why, I was going to the sale to buy Scapula's gardens, if I could safely." "Don't you do it," he told me; "for the law will be passed: Caesar wants it." I was not disturbed at hearing it: but I should be annoyed, if they do it. What have you got to say about it? However I need not ask. You know how eager a news-monger Capito is: not even Camillus can beat him at that. So you must let me know about the auction on the 15th: for that is what is bringing me to town. I have combined some other things with it: but those I can easily do two or three days later. However I don't want you to be tired out with travelling: nay, I even excuse Dionysius. As to what you say about Brutus, I have left it open so far as I am concerned: for yesterday I wrote and told him that I should have no need of his help on the 15th.

XXXIV

CICERO TO ATTICUS, GREETING.

I reached Astura on the evening of the 25th: for *Astura,* to avoid the heat of the day I rested three hours *July 27,* B.C. at Lanuvium. I should like you, if it is no trouble, *45* to contrive that I need not come to Rome before the 5th of next-month. You can manage it through Egnatius Maximus. The chief point is that you

mum), illud in primis, cum Publilio me apsente[1] conficias. De quo quae fama sit, scribes.

Terence, Andr. 185
"Id populus curat scilicet!"

Non mehercule arbitror; etenim haec decantata erat fabula. Sed complere paginam volui. Quid plura? ipse enim adsum, nisi quid tu prorogas. Scripsi enim ad te de hortis.

XXXV, XXXVI

CICERO ATTICO SAL.

Scr. in Tusculano III Id. Quint. a. 709

O rem indignam! Gentilis tuus urbem auget, quam hoc biennio primum vidit, et ei parum magna visa est, quae etiam ipsum capere potuerit. Hac de re igitur exspecto litteras tuas. Varroni scribis te, simul ac venerit. Dati igitur iam sunt, nec tibi integrum est, hui, si scias, quanto periculo tuo! Aut fortasse litterae meae te retardarunt; nisi eas nondum legeras, cum has proximas scripsisti. Scire igitur aveo, quo modo res se habeat.

De Bruti amore vestraque ambulatione etsi mihi nihil novi adfers, sed idem quod saepe, tamen hoc audio libentius quo saepius, eoque mihi iucundius est, quod tu eo laetaris, certiusque eo est, quod a te dicitur.

[1] me apsente *Müller* : mea pene absente *M*.

should settle with Publilius in my absence:[1] and about that you will let me know what people say. "Of course the world is all agog with that!" On my honour I don't think so; for the nine days' wonder is over. But I wanted to fill the page. What need of more: for I am almost with you, unless you put me off for a bit. For I have written to you about the gardens.

XXXV, XXXVI

CICERO TO ATTICUS, GREETING.

What a shame! A countryman of yours[2] is *Tusculum,* enlarging the city, which he had never seen two *July 13,* years ago, and he thinks it too small to hold the B.C. *45* great man alone. On that point then I am expecting a letter from you. You say you will present my book to Varro, as soon as he arrives. So they are already given and you have no choice left. Ah, if you but knew what a risk you are running! Or perhaps my letter stopped you, unless you had not read it, when you wrote your last letter. So I am eager to know how the matter stands.

As to Brutus' affection and your walk, though you give me no actual news, but only a repetition of what has often happened, yet the more often I hear it, the gladder I am; and I find it the more gratifying, because you enjoy it, and the more certain, because you tell me of it.

[1] About Cicero's divorce from Publilia.
[2] *i.e.* an Athenian.

MARCUS TULLIUS CICERO

XXXVII

CICERO ATTICO SAL.

Scr. in Tusculano IV Non. Sext. a. 709

Has alteras hodie litteras. De Xenonis nomine et de Epiroticis $\overline{xxxx}$ nihil potest fieri nec commodius nec aptius, quam ut scribis. Id erat locutus mecum eodem modo Balbus minor. Nihil novi sane nisi Hirtium cum Quinto acerrime pro me litigasse; omnibus eum locis furere maximeque in conviviis cum multa de me tum redire ad patrem; nihil autem ab eo tam ἀξιοπίστως dici quam alienissimos nos esse a Caesare; fidem nobis habendam non esse, me vero etiam cavendum (φοβερὸν ἂν ἦν, nisi viderem scire regem me animi nihil habere), Ciceronem vero meum vexari; sed id quidem arbitratu suo. Laudationem Porciae gaudeo me ante dedisse Leptae tabellario, quam tuas acceperim litteras. Eam tu igitur, si me amas, curabis, si modo mittetur, isto modo mittendam Domitio et Bruto.

De gladiatoribus, de ceteris, quae scribis ἀνεμοφόρητα, facies me cotidie certiorem. Velim, si tibi videtur, appelles Balbum et Offilium. De auctione proscribenda equidem locutus sum cum Balbo. Placebat (puto conscripta habere Offilium omnia; habet et Balbus) sed Balbo placebat propinquum diem et Romae; si Caesar moraretur, posse diem differri. Sed is quidem adesse videtur. Totum igitur considera; placet enim Vestorio.

---

[1] 4,000 sesterces.

XXXVII

CICERO TO ATTICUS, GREETING.

This is the second letter to-day. About Xeno's *Tusculum*, debt and the £40[1] owing to you in Epirus, things could not happen more conveniently than you say they are happening in your letter. Balbus the younger suggested the same to me the other day. I have no news except that Hirtius has been taking my part most valiantly in arguments against young Quintus. The latter is raving about me everywhere, especially at dinner-parties, and then he falls back on his father: nothing he says is so likely to be believed as that we are utterly irreconcilable to Caesar; that we are not to be trusted, and that I ought to be held in suspicion, which would have been terrifying, if were I not aware that the king knows I have no spirit left. He says too that my son is being bullied by me: but that he may say as much as he likes. I am glad I sent the funeral oration of Porcia to Lepta the messenger before I got your letter. So, as you love me, have it sent to Domitius and Brutus in the form you suggest, if it is to be sent at all.

Aug. 2, B.C. *45*

About the gladiatorial games and the things which you call airy nothings send me news day by day. I should like you to apply to Balbus and Offilius, if you think fit. About giving notice of the auction I have spoken with Balbus. He agreed—I imagine Offilius has a complete list, and so has Balbus—well Balbus agreed for a day near at hand and for Rome as the place: if Caesar puts off coming, the day might be deferred. But he seems to be close at hand. So think it all over; for Vestorius is content.

MARCUS TULLIUS CICERO

XXXVIII

CICERO ATTICO SAL.

Scr. in Tus-   Ante lucem cum scriberem contra Epicureos, de
culano circ.   eodem oleo et opera exaravi nescio quid ad te et
prid. Non.
Sext. a. 709   ante lucem dedi. Deinde, cum somno repetito simul
cum sole experrectus essem, datur mi epistula a
sororis tuae filio, quam ipsam tibi misi; cuius est
principium non sine maxima contumelia. Sed fortasse οὐκ ἐπέστησεν. Est autem sic: "Ego enim,
quicquid non belle in te dici potest—." Posse vult
in me multa dici non belle, sed ea se negat approbare.
Hoc quicquam pote inpurius? Iam cetera leges
(misi enim ad te) iudicabisque. Bruti nostri cotidianis adsiduisque laudibus, quas ab eo de nobis
haberi permulti mihi renuntiaverunt, commotum
istum aliquando scripsisse aliquid ad me credo et ad
te, idque ut sciam facies. Nam ad patrem de me
quid scripserit, nescio, de matre quam pie! "Volueram," inquit, "ut quam plurimum tecum essem,
conduci mihi domum et id ad te scripseram. Neglexisti. Ita minus multum una erimus. Nam ego
istam domum videre non possum; qua de causa,
scis." Hanc autem causam pater odium matris esse
dicebat. Nunc me iuva, mi Attice, consilio, "πότερον
δίκᾳ τεῖχος ὕψιον," id est utrum aperte hominem as-

XXXVIII

CICERO TO ATTICUS, GREETING.

As I was writing against the Epicureans before daybreak, I scribbled something or other to you by the same lamp and at the same sitting and despatched it before daybreak. Then as I was getting up with the sun after another sleep, I get a letter from your sister's son, which I enclose. The beginning of it is most insulting: but perhaps he did not stop to think. This is how it runs: "For, whatever there is to be said to your discredit, I . . ." He wants me to understand there is plenty to be said to my discredit, but he does not agree with it. Could anything be more disgusting? You may read the rest (for I have sent it on) and judge for yourself. I fancy it is the daily and continual complimentary remarks which, as I hear from many, our friend Brutus is making about us, which have provoked him into writing something to me and to you—let me know if he has written to you. For what he has written to his father about me I don't know: about his mother how affectionately! "I should have liked," he says, "to be with you as much as possible and to have a house taken for me somewhere: and so I told you. You took no notice: so we shall not be together much: for I cannot bear the sight of your house: you know why." His father tells me the reason is his hatred of his mother. Now, Atticus, help me with your advice. "By honest means shall I the high wall climb?"[1] that is to say shall I openly renounce and

Tusculum, circa Aug. 4, B.C. 45

[1] From a fragment of Pindar, as also the following Greek words.

perner et respuam "ἢ σκολιαῖς ἀπάταις." Ut enim
Pindaro sic "δίχα μοι νόος, ἀτρέκειαν εἰπεῖν." Omnino
moribus meis illud aptius, sed hoc fortasse temporibus. Tu autem, quod ipse tibi suaseris, idem mihi
persuasum putato. Equidem vereor maxime, ne in
Tusculano opprimar. In turba haec essent faciliora.
Utrum igitur Asturae? Quid, si Caesar subito? Iuva
me, quaeso, consilio. Utar eo, quod tu decreveris.

XXXIX

CICERO ATTICO SAL.

Scr. in Tusculano Non. Sext. a. 709

O incredibilem vanitatem! ad patrem "domo sibi
carendum propter matrem," ad matrem plenam pietatis. Hic autem iam languescit et ait sibi illum iure
iratum. Sed utar tuo consilio; "σκολιὰ" enim tibi
video placere. Romam, ut censes, veniam, sed invitus; valde enim in scribendo haereo. "Brutum,"
inquis, "eadem." Scilicet; sed, nisi hoc esset, res
me ista non cogeret. Nec enim inde venit, unde
mallem, neque diu afuit neque ullam litteram ad me.
Sed tamen scire aveo, qualis ei totius itineris summa
fuerit. Libros mihi, de quibus ad te antea scripsi,
velim mittas et maxime Φαίδρου περὶ θεῶν et περὶ
Παλλάδος.[1]

[1] περὶ Παλλάδος *Orelli*: ΠΛΛΙΔΟΣ *MSS.*: παντός *Gurlitt*: Ἀπολλοδώρου *Hirzel*.

abjure the fellow, or shall I act " with wiles"? For, like Pindar's, "my mind divided cannot truly tell." The first would suit my character best, of course, but the second perhaps the times. But take it I have made up my mind to do whatever you have made up your mind to do. I am horribly afraid of being caught at Tusculum. It would be more comfortable in company. At Astura then? What if Caesar arrives unexpectedly? Please assist me with advice. I will do what you decide.

XXXIX

CICERO TO ATTICUS, GREETING.

What incredible hypocrisy! To write to his father that "he had no home owing to his mother," and to his mother a letter full of affection. His father however is already cooling down and says the son has a right to be angry with him. But I will follow your advice; for I see "crooked ways" are what you favour. I will come to Rome, as you think I ought, though against my will; for I cannot tear myself from my writing. You say I shall find Brutus on the way: of course, but without this other reason that would not be strong enough to move me. For he has not come from the place I should wish, nor has he been long away or sent me any letter. Still I should like to know the result of his whole journey. Please send me the books I asked for before, especially Phaedrus *On the Gods* and *On Pallas*.

Tusculum, Aug. 5, B.C. 45

MARCUS TULLIUS CICERO

XL

CICERO ATTICO SAL.

Scr. in Tusculano VII aut VI Id. Sext. a. 709

Itane? nuntiat Brutus illum ad bonos viros? Εὐαγγέλια. Sed ubi eos? nisi forte se suspendit. Hic autem, ut stultum[1] est. Ubi igitur φιλοτέχνημα illud tuum, quod vidi in Parthenone, Ahalam et Brutum? Sed quid faciat? Illud optime: "Sed ne is quidem, qui omnium flagitiorum auctor, bene de nostro." At ego verebar, ne etiam Brutus eum diligeret; ita enim significarat iis litteris, quas ad me: "Ast vellem aliquid degustasses de fabulis." Sed coram, ut scribis.

Etsi quid mi auctor es? advolone an maneo? Equidem et in libris haereo et illum hic excipere nolo; ad quem, ut audio, pater hodie ad Saxa summa[2] acrimonia. Mirum quam inimicus ibat, ut ego obiurgarem. Sed ego ipse κεκέπφωμαι. Itaque posthac. Tu tamen vide, quid de adventu meo censeas, et τὰ ὅλα, cras si perspici potuerint, mane statim ut sciam.

[1] stultum *Tunstall*: fultum *MSS.*: futilum *Schmidt*.
[2] summa *inserted by Schmidt*.

[1] The "Parthenon" was probably the name of the library in Brutus' house. According to Nepos (*Att.* 18), Atticus

XL

CICERO TO ATTICUS, GREETING.

Tusculum, Aug. 7 or 8, B.C. 45

Is that so? Does Brutus really say Caesar is going over to the right party? That is good news. But where will he find them, unless, perhaps, he hangs himself? But how foolish it is of Brutus! Where, then, does that masterpiece of yours, which I saw in the Parthenon, the tree of Brutus' family from Ahala and Brutus, come in?[1] But what can he do? It is excellent to hear that not even the man who began the whole criminal business has a good word to say for young Quintus. Indeed, I was beginning to be afraid that even Brutus was fond of him; for in his letter to me he said, "But I wish you could have had a taste of his tales." But when we meet, as you say.

However, what do you advise? Shall I fly to meet him or stay where I am? For my part I am glued to my books, and I don't want to receive him here. I hear his father has gone to-day to Saxa Rubra[2] to meet him in a fury. He was so extraordinarily enraged against him that I remonstrated with him. But I am capable of acting the "giddy goat" too. So it rests with the future. Do you please see what you think about my movements and everything else. If you can see the way to-morrow, let me know early.

---

compiled a pedigree of the Junian family from its origin for Brutus.

[2] About ten miles from Rome on the Via Flaminia.

MARCUS TULLIUS CICERO

XLI

CICERO ATTICO SAL.

Scr. in Tus-culano VI aut V Id. Sext. a. 709

Ego vero Quinto epistulam ad sororem misi. Cum ille quereretur filio cum matre bellum et se ob eam causam domo cessurum filio diceret, dixi illum commodas ad matrem litteras, ad te nullas. Ille alterum mirabatur, de te autem suam culpam, quod saepe graviter ad filium scripsisset de tua in illum iniuria. Quod autem relanguisse se dicit, ego ei tuis litteris lectis σκολιαῖς ἀπάταις significavi me non fore iratum.[1] Tum enim mentio Canae. Omnino, si id consilium placeret, esset necesse; sed, ut scribis, ratio est habenda gravitatis, et utriusque nostrum idem consilium esse debet, etsi in me graviores iniuriae et certe notiores. Si vero etiam Brutus aliquid adferet, nulla dubitatio est. Sed coram. Magna enim res et multae cautionis. Cras igitur, nisi quid a te commeatus.

XLII

CICERO ATTICO SAL.

Scr. in Tus-culano ex. m. Dec. a. 709

Venit ille ad me καὶ μάλα κατηφής. Et ego: "Σὺ δὲ δὴ τί σύννους;" "Rogas?" inquit, "cui iter instet et iter ad bellum idque cum periculosum tum etiam

[1] iratum *inserted by Lambinus.*

LETTERS TO ATTICUS XIII. 41-42

XLI

CICERO TO ATTICUS, GREETING.

I sent Quintus your letter for your sister. When he complained that his son was at daggers drawn with his mother and said he should give up the house to his son on that account, I said young Quintus had sent an amiable letter to his mother and none to you. He was surprised at the first, but said it was his fault about you, as he had often written in anger to his son about your unfairness to him. However, he said his anger had abated, so I read your letter, and "by crooked ways" hinted that I should not bear malice. For then he began to mention Cana.[1] To be sure, if that plan found favour, we should have to make it up; but, as you say, we must consider our dignity, and we ought to concert our plans together, though his attacks on me were the worst and certainly the most public. If Brutus, too, should come to our aid, we need not hesitate. But we must discuss it together; for it is an important matter and requires great caution. So to-morrow, unless you give me furlough.

Tusculum, Aug. 8 or 9, B.C. 45

XLII

CICERO TO ATTICUS, GREETING.

Young Quintus has come to me very down in the mouth. So I asked, why he had the blues. "Need you ask," said he, "when I have a journey before me, a journey to a war, and one that is both

Tusculum, Dec. B.C. 45

---

[1] Daughter of Q. Gellius Canus. Negotiations for her marriage with young Quintus were going on.

turpe!" "Quae vis igitur?" inquam. "Aes," inquit, "alienum et tamen ne viaticum quidem." Hoc loco ego sumpsi quiddam de tua eloquentia; nam tacui. At ille: "Sed me maxime angit avunculus." "Quidnam?" inquam. "Quod mihi," inquit, "iratus est." "Cur pateris?" inquam, "malo enim ita dicere quam cur committis?" "Non patiar," inquit, "causam enim tollam." Et ego: "Rectissume quidem; sed, si grave non est, velim scire, quid sit causae." "Quia, dum dubitabam, quam ducerem, non satis faciebam matri; ita ne illi quidem. Nunc nihil mihi tanti est. Faciam, quod volunt." "Feliciter velim," inquam, "teque laudo. Sed quando?" "Nihil ad me," inquit, "de tempore, quoniam rem probo." "At ego," inquam, "censeo, priusquam proficiscaris. Ita patri quoque morem gesseris." "Faciam," inquit, "ut censes." Hic dialogus sic conclusus est.

Sed heus tu, diem meum scis esse III Nonas Ianuarias; aderis igitur. Scripseram iam: ecce tibi orat Lepidus, ut veniam. Opinor augures velle habere ad templum effandum. Eatur; μὴ σκόρδου.[1] Videbimus te igitur.

[1] μὴ σκόρδου *Tyrrell*: ΜΙΑΣΚΟΡΔΟΥ *M*: μίασμα δρύος *Gronovius*.

dangerous and even disgraceful." "What is there to compel you then?" I said. "Debt," said he, "and yet not enough money for the journey." At that point I borrowed something from your style of eloquence: I held my tongue. Well, he went on. "But what worries me most is my uncle." "Why?" said I. "Because he is angry with me," he answered. "Why do you let him be so?" I said, "for I would rather put it that way than say, Why do you make him angry?" "I will not let him," he said, "for I will remove the reason." I replied, "Very right of you, too; but, if it is not a serious matter, I should like to know what the reason is." "Because my hesitation which wife I should take annoyed my mother, and consequently him, too. Now nothing is worth that, and I will do anything they like." "I hope you will have luck," I said, "and I approve of your resolution. But when are you going to do it?" "The time doesn't matter to me," said he, "since I have made up my mind to it." "Well, I think you ought to do it before you go," I said. "You would oblige your father, too, by doing so." "I will do as you advise," he said; and there the conversation ended.

But, look here, you know it is my birthday on the 3rd of January. So you must come. I was just writing, and here is a request from Lepidus for me to come to town. I suppose the augurs want me for consecrating a temple. I must go; anything for a quiet life.[1] So you will see me.

---

[1] Tyrrell explains this as an allusion to the proverb ἵνα μὴ σκόροδα μηδὲ κυάμους (φάγω) (that I may not eat garlic or beans), which was applied to persons wishing for a quiet life.

MARCUS TULLIUS CICERO

XLIII

CICERO ATTICO SAL.

Scr. in Tus-
culano prid.
Id. Quint. a.
709

Ego vero utar prorogatione diei, tuque humanissime fecisti, qui me certiorem feceris, atque ita, ut eo tempore acciperem litteras, quo non exspectarem, tuque ut ab ludis scriberes. Sunt omnino mihi quaedam agenda Romae, sed consequemur biduo post.

XLIV

CICERO ATTICO SAL.

Scr. in Tus-
culano XIII
aut XII K.
Sext. a. 709

O suavis tuas litteras! (etsi acerba pompa. Verum tamen scire omnia non acerbum est, vel de Cotta) populum vero praeclarum, quod propter malum vicinum ne Victoriae quidem ploditur! Brutus apud me fuit; cui quidem valde placebat me aliquid ad Caesarem. Adnueram; sed pompa deterret. Tu tamen ausus es Varroni dare! Exspecto, quid iudicet. Quando autem pelleget? De Attica probo. Est quiddam etiam animum levari cum spectatione tum etiam religionis opinione et fama. Cottam mi velim mittas; Libonem mecum habeo et habueram ante Cascam. Brutus mihi T. Ligari verbis nuntiavit, quod appelletur L. Corfidius in oratione Ligariana, erratum esse meum. Sed, ut aiunt, μνημονικὸν ἁμάρτημα. Sciebam Corfidium pernecessarium Ligari-

---

[1] A procession at the Ludi Circenses, in which Caesar's image was carried among the gods, next to Victory.

XLIII

CICERO TO ATTICUS, GREETING.

Tusculum, July 14, B.C. 45

Yes, I will take advantage of the postponement of the day of sale; and it was very kind of you to inform me of it, especially to let me have a letter, when I did not expect one, and to write it at the games. There are, to be sure, some things I have to do at Rome; but I will attend to them two days later.

XLIV

CICERO TO ATTICUS, GREETING.

Tusculum, July 20 or 21, B.C. 45

What a delightful letter yours was! Though the procession[1] was unpleasant news; still it is not unpleasant to know everything, even about Cotta.[2] The people were splendid not even to clap Victory because of her bad neighbour. Brutus was staying with me and highly approved of my writing something to Caesar. I assented; but the procession puts me off. Have you really dared to send my book to Varro! I am eager for his opinion. But when will he finish reading it? I agree about Attica. It is something that the spirits are relieved by the spectacle and by the general feeling of religious associations. I wish you would send me Cotta; I have Libo and before that I had Casca. Brutus brought me a message from T. Ligarius that the mention of L. Corfidius in my speech for Ligarius is a mistake. But it is a *lapsus memoriae*, as they say. I knew that Corfidius was extremely

---

[2] Cotta had suggested that Caesar should adopt the title of king, stating that the Sibylline books said Parthia could only be conquered by a king.

orum; sed eum video ante esse mortuum. Da igitur, quaeso, negotium Pharnaci, Antaeo, Salvio, ut id nomen ex omnibus libris tollatur.

XLV

CICERO ATTICO SAL.

Scr. in Tus-culano III Id. Sext. a. 709

Fuit apud me Lamia post discessum tuum epistulamque ad me attulit missam sibi a Caesare. Quae quamquam ante data erat quam illae Diocharinae, tamen plane declarabat illum ante ludos Romanos esse venturum. In qua extrema scriptum erat, ut ad ludos omnia pararet, neve committeret, ut frustra ipse properasset. Prorsus ex his litteris non videbatur esse dubium, quin ante eam diem venturus esset, itemque Balbo, cum eam epistulam legisset, videri Lamia dicebat.

Dies feriarum mihi additos video, sed quam multos, fac, si me amas, sciam. De Baebio poteris et de altero vicino Egnatio.

Quod me hortaris, ut eos dies consumam in philosophia explicanda, currentem tu quidem; sed cum Dolabella vivendum esse istis diebus vides. Quodnisi me Torquati causa teneret, satis erat dierum, ut Puteolos excurrere possem et ad tempus redire. Lamia quidem a Balbo, ut videbatur, audiverat multos nummos domi esse numeratos, quos oporteret quam primum dividi, magnum pondus argenti; auctionem praeter praedia primo quoque tempore fieri oportere. Scribas ad me velim, quid tibi placeat.

friendly with the Ligarii; but I see he was dead before the trial. So please get Pharnaces, Antaeus and Salvius to erase the name from all copies.

XLV

CICERO TO ATTICUS, GREETING.

Tusculum, Aug. 11, B.C. 45

Lamia was with me after you left, and brought me a letter Caesar had sent to him. Though it was despatched earlier than those of Diochares, still it asserted plainly that he would come before the Roman games.[1] At the end he told him to make all preparations for the games and not let him hurry back for nothing. From this letter there certainly seemed no doubt that he would come before that date; and Lamia said that Balbus thought so too, when he read the letter.

I see I have some additional days' holiday, but please let me know how many. You can find out from Baebius or your other neighbour Egnatius.

In exhorting me to spend the days in an exposition of philosophy, you are only spurring a willing horse; but note that I have to spend those days with Dolabella. Now, if I had not been detained on Torquatus' business, there would have been time enough to make an excursion to Puteoli and return in time. Lamia has heard from Balbus, it appears, that there is a good deal of ready money in the house, which ought to be divided as soon as possible, and a considerable amount of silver plate, and that the auction of all but the real property ought to take place at the earliest opportunity. Please write and tell me what you think. Upon

[1] September 15–19.

MARCUS TULLIUS CICERO

Equidem, si ex omnibus esset eligendum, nec diligentiorem nec officiosiorem nec mehercule nostri studiosiorem facile delegissem Vestorio; ad quem accuratissimas litteras dedi; quod idem te fecisse arbitror. Mihi quidem hoc satis videtur. Tu quid dicis? Unum enim pungit, ne neglegentiores esse videamur. Exspectabo igitur tuas litteras.

XLVI

CICERO ATTICO SAL.

Scr. in Tusculano prid. Id. Sext. a. 709

Pollex quidem, ut dixerat ad Idus Sextiles, ita mihi Lanuvi pridie Idus praesto fuit, sed plane pollex, non index. Cognosces igitur ex ipso. Balbum conveni. Lepta enim de sua munerum[1] curatione laborans me ad eum perduxerat. In eo autem Lanuvino, quod Lepido tradidit. Ex eo hoc primum: " Paulo ante acceperam eas litteras, in quibus magno opere confirmat ante ludos Romanos." Legi epistulam. Multa de meo Catone, quo saepissime legendo se dicit copiosiorem factum, Bruti Catone lecto se sibi visum disertum. Ex eo cognovi cretionem Cluvi (o Vestorium neglegentem!) liberam cretionem testibus praesentibus sexaginta diebus. Metuebam, ne ille arcessendus esset. Nunc mittendum est, ut meo

---

[1] munerum *Schmidt, coll. Fam.* vi. 19. 2 : viin *M* : vini *vulg.*

my word, if I had had the whole world to select from, I could hardly have chosen a man more painstaking, more obliging, nor, I am sure, more devoted to my interests than Vestorius. I have sent him an extremely carefully worded letter; and I think you have done the same. I think that is sufficient. What do you say? The one thing that bothers me is that we may seem too careless. So I will wait for your letter.

XLVI

CICERO TO ATTICUS, GREETING.

Pollex, having arranged to meet me on the 13th of August, has done so at Lanuvium on the 12th: but he is a mere thumb, and not a pointing finger.[1] So you must get your news from him himself. I have met Balbus: for Lepta, being anxious about the contract for the shows, took me to him. Well, he was in the place at Lanuvium, which he made over to Lepidus: and the first thing he said to me was, "I have just had a letter in which Caesar definitely asserts that he will be here before the Roman games." I read the letter. It dilated on my *Cato*, and he said that by reading it frequently he had increased his flow of language, and, when he read Brutus' *Cato*, he began to think himself eloquent. I learned from him that the formal acceptance of Cluvius' legacy was an unconditional acceptance within sixty days before witnesses. How careless of Vestorius not to tell me! I was afraid I should have to send for him: but now I must

Tusculum, Aug. 12, B.C. 45

[1] In the Latin there is a play on the proper name, which I am unable to reproduce in English.

iussu cernat. Idem igitur Pollex. Etiam de hortis Cluvianis egi cum Balbo. Nil liberalius. Se enim statim ad Caesarem scripturum, Cluvium autem a T. Hordeonio legare et Terentiae HS ↃↃↃ[1] et sepulcro multisque rebus, nihil a nobis. Subaccusa, quaeso, Vestorium. Quid minus probandum quam Plotium unguentarium per suos pueros omnia tanto ante Balbo, illum mi ne per meos quidem? De Cossinio doleo; dilexi hominem.

Quinto delegabo, si quid aeri meo alieno superabit et emptionibus, ex quibus mi etiam aes alienum faciendum puto. De domo Arpini nil scio.

Vestorium nil est quod accuses. Iam enim obsignata hac epistula noctu tabellarius noster venit, et ab eo litteras diligenter scriptas attulit et exemplum testamenti.

XLVII

CICERO ATTICO SAL.

Scr. in Tusculano Id. Sext. a. 709

"Posteaquam abs te, Agamemno," non "ut venirem" (nam id quoque fecissem, nisi Torquatus esset), sed ut scriberem, "tetigit aures nuntius, extemplo" instituta omisi; ea, quae in manibus habebam, abieci, quod iusseras, edolavi. Tu velim e Pollice cognoscas

---

[1] 50,000 sesterces.

commission him to accept at my orders. So this same Pollex can take the message. I discussed Cluvius' gardens with Balbus too, and he was most obliging. For he said he would write to Caesar at once, but that Cluvius had subtracted from Hordeonius' legacy some £500[1] for Terentia, the cost of his tomb and a lot of other things, but nothing from my share. Please remonstrate with Vestorius. It is surely most out of place for Plotius the perfumer to send his own special messengers with full particulars to Balbus so long in advance, while Vestorius does not send me news even by my messengers. I am sorry about Cossinius; I was fond of him.

I will make over to Quintus anything that may be left after paying my debts and making purchases, for which I am afraid I shall incur more debt. About the house at Arpinum I know nothing.

There is no necessity to grumble at Vestorius, for to-night, after I had sealed this letter, my messenger came bringing a letter full of details and a copy of the will.

XLVII

CICERO TO ATTICUS, GREETING.

Tusculum, Aug. 13, B.C. 45

"When from thee, Agamemnon, the message reached my ears," not "that I should come" (though I should have done that too, if it had not been for Torquatus), "straightway" I gave up what I had begun, threw down what I had in hand and made a rough sketch of what you ordered.[2] I should like you to find out from

[2] *i.e.* he gave up working at the *De Natura Deorum*, and set about writing a letter to Caesar.

rationes nostras sumptuarias. Turpe est enim nobis illum, qualiscumque est, hoc primo anno egere. Post moderabimur diligentius. Idem Pollex remittendus est, ut ille cernat. Plane Puteolos non fuit eundum, cum ob ea, quae ad te scripsi, tum quod Caesar adest. Dolabella scribit se ad me postridie Idus. O magistrum molestum!

XLVIIa

CICERO ATTICO SAL.

Scr. Asturae III K. Sext. a. 709

Lepidus ad me heri vesperi litteras misit Antio. Nam ibi erat. Habet enim domum, quam nos vendidimus. Rogat magno opere, ut sim Kal. in senatu; me et sibi et Caesari vehementer gratum esse facturum. Puto equidem nihil esse. Dixisset enim tibi fortasse aliquid Oppius, quoniam Balbus est aeger. Sed tamen malui venire frustra quam desiderari, si opus esset. Moleste ferrem postea. Itaque hodie Anti, cras ante meridiem domi. Tu velim, nisi te impedivisti, apud nos pr. Kal. cum Pilia.

Te spero cum Publilio confecisse. Equidem Kal. in Tusculanum recurram; me enim absente omnia cum illis transigi malo. Quinti fratris epistulam ad te misi, non satis humane illam quidem respondentem meis litteris, sed tamen quod tibi satis sit, ut equidem existimo. Tu videbis.

Pollex the state of my exchequer. It would be a disgrace to me that my son should run short of money in his first year, whatever he may deserve. Afterwards we will restrict him more carefully. Pollex also must be sent back, that Vestorius may accept the inheritance. Clearly I ought not to have gone to Puteoli, both on account of what you say, and because Caesar is getting near. Dolabella tells me he is coming to me on the 14th. What a tiresome school-master!

XLVIIa

CICERO TO ATTICUS, GREETING.

Yesterday evening I had a letter from Lepidus *Astura,* at Antium. That is where he is, for he has the *July 30,* B.C. house I sold. He implores me to be in the Senate 45 on the 1st, saying that both he and Caesar would take it as a great favour. I don't think it is of any importance; for Oppius would probably have said something to you, as Balbus is ill. However I would rather come for nothing, if necessary, than have my absence noticed. I should regret it afterwards. So to-day I go to Antium, tomorrow home by midday. I should like you and Pilia to come to dinner on the last of the month, if you are not engaged.

I hope you have settled with Publilius. I shall rush back to Tusculum on the 1st; for I prefer all the transactions with them to take place in my absence. I am sending my brother's letter to you: it is not a very kind answer to mine, but I think it should satisfy you. You will see for yourself.

MARCUS TULLIUS CICERO

XLVIII

CICERO ATTICO SAL.

Scr. in Tus-culano IV Non. Sext. a. 709

Heri nescio quid in strepitu videor exaudisse, cum diceres te in Tusculanum venturum. Quod utinam! iterum utinam! tuo tamen commodo.

Lepta me rogat, ut, si quid sibi opus sit, accurram; mortuus enim Babullius. Caesar, opinor, ex uncia, etsi nihil adhuc; sed Lepta ex triente. Veretur autem, ne non liceat tenere hereditatem, ἀλόγως omnino, sed veretur tamen. Is igitur si accierit, accurram; si minus, non antequam necesse erit. Tu Pollicem, cum poteris.

Laudationem Porciae tibi misi correctam. Adeo properavi, ut, si forte aut Domitio filio aut Bruto mitteretur, haec mitteretur. Id, si tibi erit commodum, magno opere cures velim et velim M. Varronis et Olli mittas laudationem, Olli utique. Nam illam legi, volo tamen regustare. Quaedam enim vix mihi credo legisse me.

XLIX

CICERO ATTICO SAL.

Scr. in Tus-culano circ. XI K. Sept. a. 709

Atticae primum salutem (quam equidem ruri esse arbitror; multam igitur salutem) et Piliae. De Tigellio, si quid novi. Qui quidem, ut mihi Gallus Fadius scripsit, μέμψιν ἀναφέρει mihi quandam iniquissimam, me Phameae defuisse, cum eius causam re-

XLVIII

CICERO TO ATTICUS, GREETING.

Yesterday in the midst of all the noise I think *Tusculum,* I caught some remark of yours about coming to *Aug. 2,* B.C. Tuscalum. I wish you would. I wish to goodness *45* you would: but at your convenience.

Lepta asks me to go to him, if there is any necessity: for Babullius is dead. Caesar, I fancy, is heir to one-twelfth of his estate—though I know nothing yet: but Lepta to a third. He is afraid he may not be allowed to take the inheritance. It is absurd of course, but still he is afraid. So, if he sends for me, I shall go at once: if not, not till it is necessary. Send back Pollex, when you can.

I am sending you the funeral oration of Porcia corrected. I have hurried about it, so that, if it should be sent to young Domitius or to Brutus, this edition should be sent. If it is convenient, I should much like you to see about it, and please send me the orations of M. Varro, and Ollius, at any rate that of Ollius. I have read it, but I want to dip into it again: for there are things in it that I can hardly believe I read.

XLIX

CICERO TO ATTICUS, GREETING.

First health to Attica (who I suppose is now in *Tusculum,* the country, so I wish her a full return to health) *circa Aug.* and to Pilia too. Let me know about Tigellius, *22,* B.C. *45* if there is any news. According to a letter of Fadius Gallus, he is very down on me most unjustly for deserting Phamea, when I had undertaken his

cepissem. Quam quidem receperam contra pueros Octavios Cn. filios non libenter; sed Phameae causa volebam. Erat enim, si meministi, in consulatus petitione per te mihi pollicitus, si quid opus esset; quod ego perinde tuebar, ac si usus essem. Is ad me venit dixitque iudicem operam dare sibi constituisse eo die ipso, quo de Sestio nostro lege Pompeia in consilium iri necesse erat. Scis enim dies illorum iudiciorum praestitutos fuisse. Respondi non ignorare eum, quid ego deberem Sestio. Quem vellet alium diem si sumpsisset, me ei non defuturum. Ita tum ille discessit iratus. Puto me tibi narrasse. Non laboravi scilicet nec hominis alieni iniustissimam iracundiam mihi curandam putavi. Gallo autem narravi, cum proxime Romae fui, quid audissem, neque nominavi Balbum minorem. Habuit suum negotium Gallus, ut scribit. Ait illum me animi conscientia, quod Phameam destituissem, de se suspicari. Quare tibi hactenus mando, de illo nostro, si quid poteris, exquiras, de me ne quid labores. Est bellum aliquem libenter odisse et, quem ad modum non omnibus dormire, ita[1] non omnibus servire. Etsi mehercule, ut tu intellegis, magis mihi isti serviunt, si observare servire est.

[1] non omnibus dormire, ita *added by Lambinus.*

---

[1] Or "I did also wish well to Phamea," as Shuckburgh.
[2] In a letter of about the same date to Gallus (*Ad Fam.* VII. 24) Cicero says, *Cipius, opinor, olim "non omnibus dormio"; sic ego non omnibus, mi Galle, servio.* It is explained

LETTERS TO ATTICUS XIII. 49

case. It went against the grain with me to take it at all against the sons of Cn. Octavius; but for Phamea's sake I agreed.[1] For, if you remember, when I was standing for the consulship, he sent a promise of any assistance he could render through you; and I appreciated it as much as if I had used it. He came to me and said the judge had undertaken to hear his case on the very same day that the jury were bound by the Pompeian law to settle that of our friend Sestius. For you know the days of those cases have been fixed by law. I answered that he could not but be aware of my obligations to Sestius. If he would choose any other day, I would not fail him. So then he left me in a temper. I think I told you about it. I did not bother myself about it of course, not thinking that a perfectly unwarrantable fit of anger of a stranger concerned me. However I told Gallus the next time I was in town what I had heard, without mentioning young Balbus. Gallus took the matter up, as he tells me. He says Tigellius asserts that I suspect him because of my bad conscience about my desertion of Phamea. Accordingly I commission you to find out what you can from young Balbus, but not to bother your head about me. It is quite a good thing to have somebody to hate with a will, and not to pander to everybody any more than to be asleep for everybody.[2] Though upon my word, as you know, Caesar's party are obsequious to me more than I to them, if attention is obsequiousness.

that Cipius used to shut his eyes to his wife's barefaced amours in his presence; but when a servant, thinking him asleep, stole a cup before his eyes, he woke up with this remark.

MARCUS TULLIUS CICERO

L

CICERO ATTICO SAL.

Scr. in Tus-culano circ. IX K. Sept. a. 709

Admonitus quibusdam tuis litteris, ut ad Caesarem uberiores litteras mittere instituerem, cum mihi Balbus nuper in Lanuvino dixisset se et Oppium scripsisse ad Caesarem me legisse libros contra Catonem et vehementer probasse, conscripsi de iis ipsis libris epistulam Caesari, quae deferretur ad Dolabellam; sed eius exemplum misi ad Oppium et Balbum, scripsique ad eos, ut tum deferri ad Dolabellam iuberent meas litteras, si ipsi exemplum probassent. Ita mihi rescripserunt, nihil umquam se legisse melius, epistulamque meam iusserunt dari Dolabellae.

Vestorius ad me scripsit, ut iuberem mancipio dari servo suo pro mea parte Hetereio cuidam fundum Brinnianum, ut ipse ei Puteolis recte mancipio dare posset. Eum servum, si tibi videbitur, ad me mittes; opinor enim ad te etiam scripsisse Vestorium.

De adventu Caesaris idem quod a te mihi scriptum est ab Oppio et Balbo. Miror te nihildum cum Tigellio. Velut hoc ipsum, quantum acceperit, prorsus aveo scire, nec tamen flocci facio. Quaeris, quid cogitem de obviam itione. Quid censes nisi Alsium? Et quidem ad Murenam de hospitio scripseram, sed opinor cum Matio profectum. Sallustius igitur urgebitur.

Scripto iam superiore versiculo Eros mihi dixit sibi Murenam liberalissime respondisse. Eo igitur utamur. Nam Silius culcitas non habet. Dida autem, opinor, hospitibus totam villam concessit.

L

CICERO TO ATTICUS, GREETING.

You suggested in one of your letters that I should set about composing a longer letter to send to Caesar, and Balbus told me lately at Lanuvium that he and Oppius had written to Caesar telling him I had read his books against Cato and strongly approved of them: so I wrote a letter to Caesar about those books to be sent to Dolabella. But I sent a copy to Oppius and Balbus, asking them to send on my letter to Dolabella, if they themselves approved of the copy. So they have answered that they never read anything better and have had my letter forwarded to Dolabella.

Vestorius has written asking me to make over my share in the property of Brinnius to a slave of his on behalf of one Hetereius, so that he can complete the transfer at Puteoli according to law. If you think it right, send the slave to me; for I suppose Vestorius has written to you too.

About Caesar's coming Oppius and Balbus tell me the same as you. I am surprised that you have not yet had a talk with Tigellius. For instance, I should much like to know just how much he got; however I don't really care a straw. You ask what I think about going to meet Caesar. Where are you thinking of, unless it is Alsium? Indeed I have written to Murena asking him to take me in; but I suppose he has gone with Matius. So I shall inflict myself on Sallustius.

When I had written the last line, Eros told me Murena gave him the kindest of answers: so let me make use of him. For Silius has no cushions, while Dida, I believe, has given up his whole villa to guests.

Tusculum, circa Aug. 24, B.C. 45

MARCUS TULLIUS CICERO

LI

CICERO ATTICO SAL.

Scr. in Tus-
culano VII
K. Sept. a.
709

Ad Caesarem quam misi epistulam, eius exemplum fugit me tum tibi mittere. Nec id fuit, quod suspicaris, ut me puderet tui, ne ridicule Μίκυλλος,[1] nec mehercule scripsi aliter, ac si πρὸς ἴσον ὁμοιόνque scriberem. Bene enim existimo de illis libris, ut tibi coram. Itaque scripsi et ἀκολακεύτως et tamen sic, ut nihil eum existimem lecturum libentius.

De Attica nunc demum mihi est exploratum; itaque ei de integro gratulare. Tigellium totum mihi, et quidem quam primum; nam pendeo animi. Narrabo tibi, Quintus cras; sed, ad me an ad te, nescio. Mi scripsit Romam VIII Kal. Sed misi, qui invitaret. Etsi hercle iam Romam veniendum est, ne ille ante advolet.

LII

CICERO ATTICO SAL.

Scr. in Puteo-
lano XII K.
Ian. a. 709

O hospitem mihi tam gravem ἀμεταμέλητον! Fuit enim periucunde. Sed, cum secundis Saturnalibus ad Philippum vesperi venisset, villa ita completa a militibus est, ut vix triclinium, ubi cenaturus ipse Caesar esset, vacaret, quippe hominum CIƆ CIƆ. Sane sum commotus, quid futurum esset postridie; ac mihi Barba Cassius subvenit, custodes dedit. Castra in

[1] Μίκυλλος *Schmidt, comparing Lucian Gall. I, Tyrann.* 14: micillus *MSS.*

LI

CICERO TO ATTICUS, GREETING.

It escaped my memory to send you a copy of the letter I sent to Caesar at the time. It was not, as you suspect, that I was ashamed of showing it to you, for fear I should seem too much of a flatterer; nor, I assure you, did I write otherwise than I should to an equal. For I have got a high opinion of those books of his, as I told you when we met. So I wrote without flattery, and yet I think he will read it with great pleasure.

Tusculum, Aug. 26, B.C. 45

At last I have full news of Attica; so please congratulate her again. Tell me all about Tigellius and that too as soon as possible; for I am feeling anxious. There is one thing I must mention. Young Quintus is coming to-morrow; but, whether to me or to you, I don't know. He wrote to me he was coming to Rome on the 25th. I have sent someone to invite him here. Though to be sure I must go to Rome now, for fear Caesar may forestall me.

LII

CICERO TO ATTICUS, GREETING.

To think that my formidable guest leaves no regret behind! For indeed it passed off splendidly. However, when he reached Philippus on the evening of the 18th, the house was so full of soldiers that there was hardly a room left for Caesar himself to dine in. Two thousand men if you please! I was much disturbed as to what was going to happen the next day; and Cassius Barba came to the rescue and gave me guards. A camp was pitched in the fields,

Puteoli, Dec 21, B.C. 45

agro, villa defensa est. Ille tertiis Saturnalibus apud Philippum ad h. vii nec quemquam admisit ; rationes, opinor, cum Balbo. Inde ambulavit in litore. Post h. viii in balneum. Tum audivit de Mamurra, vultum non mutavit. Unctus est, accubuit. Ἐμετικὴν agebat. Itaque et edit et bibit ἀδεῶς et iucunde, opipare sane et apparate nec id solum, sed

"bene cocto et
condito sermone bono et, si quaeris, libenter."

Praeterea tribus tricliniis accepti οἱ περὶ αὐτὸν valde copiose. Libertis minus lautis servisque nihil defuit. Nam lautiores eleganter accepi. Quid multa? homines visi sumus. Hospes tamen non is, cui diceres: "Amabo te, eodem ad me, cum revertere." Semel satis est. Σπουδαῖον οὐδὲν in sermone, φιλόλογα multa. Quid quaeris? delectatus est et libenter fuit. Puteolis se aiebat unum diem fore, alterum ad Baias.

Habes hospitium sive ἐπισταθμείαν odiosam mihi, dixi, non molestam. Ego paulisper hic, deinde in Tusculanum. Dolabellae villam cum praeteriret, omnis armatorum copia dextra, sinistra ad equum nec usquam alibi. Hoc ex Nicia.

---

[1] A quotation from Lucilius.

LETTERS TO ATTICUS XIII. 52

and the house put under guard. On the 19th he stayed with Philippus till one o'clock and admitted no one: at his accounts, I believe, with Balbus. Then he walked on the shore. After two he took his bath. Then he heard about Mamurra without changing countenance. He was anointed and sat down to dinner. He was undergoing a course of emetics, so he ate and drank at his pleasure without fear. It was a lordly dinner and well-served, and not only that, but

" Well cooked, and seasoned, and, the truth to tell,
   With pleasant discourse all went very well." [1]

Besides his chosen circle were entertained very liberally in three rooms: and freedmen of lower degree and slaves could not complain of stint. The upper sort were entertained in style. In fact, I was somebody.[2] Still he was not the sort of guest to whom one would say: "Be sure to look me up on the way back." Once is enough. There was no serious talk, but plenty of literary. In a word he was pleased and enjoyed himself. He said he would spend one day at Puteoli and another near Baiae.

There you have all about my entertainment, or billeting you might say, objectionable, as I have said, but not uncomfortable. I am staying here a while and then go to Tusculum. As he passed Dolabella's house and nowhere else the whole troop formed up on the right and left of him. So Nicias tells me.

[2] Or, as Tyrrell suggests, "we were quite friendly together," *i.e.* Caesar did not "assume the god"; or possibly even "we all felt we were in civilised society."

M. TULLI CICERONIS
EPISTULARUM AD ATTICUM
LIBER QUARTUS DECIMUS

I

CICERO ATTICO SAL.

Scr. in suburbano Mati VII Id. Apr. a. 710

Deverti ad illum, de quo tecum mane. Nihil perditius; explicari rem non posse. "Etenim, si ille tali ingenio exitum non reperiebat, quis nunc reperiet?" Quid quaeris? perisse omnia aiebat (quod haud scio an ita sit; verum ille gaudens) adfirmabatque minus diebus xx tumultum Gallicum. In sermonem se post Idus Martias praeterquam Lepido venisse nemini. Ad summam non posse istaec sic abire. O prudentem Oppium! qui nihilo minus illum desiderat, sed loquitur nihil, quod quemquam bonum offendat. Sed haec hactenus.

Tu, quaeso, quicquid novi (multa autem exspecto), scribere ne pigrere, in his, de Sexto satisne certum, maxime autem de Bruto nostro. De quo quidem ille, ad quem deverti, Caesarem solitum dicere: "Magni refert, hic quid velit, sed, quicquid volt, valde volt"; idque eum animadvertisse, cum pro Deiotaro Nicaeae dixerit; valde vehementer eum visum et libere dicere;

CICERO'S LETTERS
TO ATTICUS
BOOK XIV

I

CICERO TO ATTICUS, GREETING.

I have stopped for a visit with the man we were speaking of in the morning.[1] His view is that nothing could be more disgraceful and the thing was quite hopeless. "For, if Caesar with his genius could not find a solution, who will find it now?" In a word he said the end had come (which may be true, but he was pleased about it), and assured me that in less than twenty days there would be a rising in Gaul. He has not discussed the matter with anyone except Lepidus since the 15th of March: and, in fine, things cannot pass off like this. What a wise man is Oppius! He regrets Caesar quite as much, but says nothing that can offend any of the loyal party. So much for that.

Pray do not delay in sending me any news—and I expect there is plenty: among other things whether we may be sure of Sextus, but especially about our friend Brutus. About him the man I am staying with says Caesar used to say: "What he wants is of great importance, but whatever he wants, he wants it badly"; and that he noticed it, when he pleaded for Deiotarus at Nicaea, for he seemed to speak with emphasis and with boldness. Again—I like to write

At Matius' villa, April 7, B.C. 44

[1] C. Matius Calvena.

atque etiam (ut enim quicque succurrit, libet scribere) proxime, cum Sesti rogatu apud eum fuissem exspectaremque sedens, quoad vocarer, dixisse eum: "Ego dubitem, quin summo in odio sim, cum M. Cicero sedeat nec suo commodo me convenire possit? Atqui, si quisquam est facilis, hic est. Tamen non dubito, quin me male oderit." Haec et eius modi multa. Sed ad propositum. Quicquid erit non modo magnum, sed etiam parvum, scribes. Equidem nihil intermittam.

II

CICERO ATTICO SAL.

Scr. in suburbano Mati VI Id. Apr. a. 710

Duas a te accepi epistulas heri. Ex priore theatrum Publiliumque cognovi, bona signa consentientis multitudinis. Plausus vero L. Cassio datus etiam facetus mihi quidem visus est. Altera epistula de Madaro scripta, apud quem nullum φαλάκρωμα, ut putas. Processit enim, sed minus. Diutius sermone eius sum retentus. Quod autem ad te scripseram obscure fortasse, id eius modi est. Aiebat Caesarem secum, quo tempore Sesti rogatu veni ad eum, cum exspectarem sedens, dixisse: "Ego nunc tam sim stultus, ut hunc ipsum facilem hominem putem mihi esse amicum, cum tam diu sedens meum commodum exspectet?" Habes igitur φαλάκρωμα inimicissimum otii, id est Bruti.

---

[1] *i.e.* the production of a mime by Publilius Syra.

the first thing that comes into my head—recently, when at Sestius' request I paid Caesar a visit and was sitting waiting to be called in, he remarked: "Can I doubt that I am heartily detested, when Cicero sits waiting and cannot visit me at his convenience? Yet, if ever there was a good-natured man, he is one. However, I have no doubt that he detests me." That and more to the same effect. But to return to the point. Write me anything there is to write, not only important matters, but even petty details. I shall not let anything escape me.

II

CICERO TO ATTICUS, GREETING.

I had two letters from you yesterday. From the first I learned about the theatre and Publilius,[1] good signs of the unanimous feeling of the people. The applause given to Cassius I thought even overdone. The other letter was about Bald-pate,[2] though he is not so bald as you think. For he has advanced, though not very far. I have been detained too long by his talk. What I mentioned to you, perhaps a little obscurely, was like this. He said Caesar remarked to him, when I went to see him at Sestius' request and was sitting waiting: "Can I be foolish enough to think that this man, good-natured though he is, is friendly to me, when he has to sit and wait for my convenience so long." So you have in Baldpate a bitter enemy of peace, that is to say, of Brutus.

From Matius' villa. April 8, B.C. 44

---

[2] *Madaro* = μαδαρῷ, "bald-pate," a pun on Calvena, Matius' *agnomen*. The reading and rendering of the rest of the sentence is doubtful.

MARCUS TULLIUS CICERO

In Tusculanum hodie, Lanuvi cras, inde Asturae cogitabam. Piliae paratum est hospitium, sed vellem Atticam. Verum tibi ignosco. Quarum utrique salutem.

III

CICERO ATTICO SAL.

Scr. in Tusculano V Id. Apr. a. 710

Tranquillae tuae quidem litterae. Quod utinam diutius! nam Matius posse negabat. Ecce autem structores nostri ad frumentum profecti, cum inanes redissent, rumorem adferunt magnum Romae domum ad Antonium frumentum omne portari. Πανικὸν certe; scripsisses enim. Corumbus Balbi nullus adhuc. Est mihi notum nomen; bellus enim esse dicitur architectus.

Ad obsignandum tu adhibitus non sine causa videris. Volunt enim nos ita putare; nescio, cur non animo quoque sentiant. Sed quid haec ad nos? Odorare tamen Antoni διάθεσιν; quem quidem ego epularum magis arbitror rationem habere quam quicquam mali cogitare.

Tu, si quid pragmaticum habebis, scribes; sin minus, populi ἐπισημασίαν et mimorum dicta perscribito. Piliae et Atticae salutem.

IV

CICERO ATTICO SAL.

Scr. Lanuvii IV Id. Apr. a. 710

Numquid putas me Lanuvi? At ego te istic cotidie aliquid novi suspicor. Tument negotia. Nam, cum Matius, quid censes ceteros? Equidem doleo, quod

I am thinking of going to Tusculum to-day, to Lanuvium to-morrow, and then to Astura. I am ready to entertain Pilia, though I should like Attica. However, I forgive you. So greet me to them both.

III

CICERO TO ATTICUS, GREETING.

Tusculum, April 9, B.C. 44

Your letter is full of peace, and I only hope peace may last some time. Matius does not think it can. Here are my builders, who had gone off harvesting, returning empty-handed and bringing a strong report that all the corn is being taken to Antony's house at Rome. Of course it is a false alarm, or I should have heard it from you. Not a sign as yet of Balbus' man Corumbus. I know the name; he is said to be a good architect.

It appears to me there was reason in their asking you to be present at the sealing of that will: for they want us to think them friendly, and I don't see why that should not be their real feeling. But what does it matter to us? However, scent out Antony's intentions; I fancy he is more concerned about his banquets than about plotting any harm.

If you have any news of practical importance, let me hear it; if not, give me full details as to who were cheered by the people at the mimes, and the epigrams of the actors. My love to Pilia and Attica.

IV

CICERO TO ATTICUS, GREETING.

Lanuvium, April 10, B.C. 44

Do you suppose I get any news at Lanuvium? But I suspect you hear something fresh every day in town. The trouble is coming to a head: for when Matius thinks so, what do you suppose others think? What

numquam in ulla civitate accidit, non una cum libertate rem publicam recuperatam. Horribile est, quae loquantur, quae minitentur. Ac vereor Gallica etiam bella, ipse Sextus quo evadat. Sed omnia licet concurrant, Idus Martiae consolantur. Nostri autem ἥρωες, quod per ipsos confici potuit, gloriosissime et magnificentissime confecerunt; reliquae res opes et copias desiderant, quas nullas habemus. Haec ego ad te. Tu, si quid novi (nam cotidie aliquid exspecto), confestim ad me, et, si novi nihil, nostro more tamen ne patiamur intermitti litterulas. Equidem non committam.

V

CICERO ATTICO S. D.

Scr. Asturae III Id. Apr. a. 710

Spero tibi iam esse, ut volumus, quoniam quidem ᾐσίτησας, cum leviter commotus esses; sed tamen velim scire, quid agas. Signa bella, quod Calvena moleste fert se suspectum esse Bruto; illa signa non bona, si cum signis legiones veniunt e Gallia. Quid tu illas putas, quae fuerunt in Hispania? nonne idem postulaturas? quid, quas Annius transportavit? C. Asinium volui, sed μνημονικὸν ἁμάρτημα. Ab aleatore[1] φυρμὸς πολύς. Nam ista quidem Caesaris libertorum coniuratio facile opprimeretur, si recta saperet Antonius. O meam stultam verecundiam! qui legari noluerim ante res prolatas, ne deserere viderer hunc

[1] a balneatore *some MSS. and editors: in which case it refers to the Pseudo-Marius.*

worries me is what never happened in any other state, that the constitution has not been recovered when freedom has. It is frightful to listen to the rumours and the threats: and I am afraid of a war in Gaul and of what side Sextus will take. But though all the world conspire against us, the Ides of March console me. Our heroes accomplished most gloriously and magnificently all that they could accomplish by themselves; the other matters require money and forces, and we have neither. That is all I have to say to you. If you have any news (for I expect something every day), let me know quickly, and, even if there is no news, don't let us break our custom and not exchange notes. I will see that I don't.

V

CICERO TO ATTICUS, GREETING.

I hope you are as well as I wish you to be by now, as you were fasting owing to a slight indisposition: but I should like to know how you are. It is a good sign that Calvena is annoyed at Brutus' suspicions; but it will be by no means a good sign if the legions come from Gaul with their ensigns. What do you think about those that were in Spain? Won't they make the same demands? And what of those that Annius took across? I meant to say C. Asinius, but I had a *lapsus memoriae*. A fine mess the gambler [1] is making. For that conspiracy of Caesar's freedmen might have been put down easily, if Antony had his wits about him. How foolish were my scruples in refusing a free legation before the vacation for fear of appearing to shirk this turmoil. Of course, if I could

Astura,
April 11,
B.C. *44*

[1] Antony.

MARCUS TULLIUS CICERO

rerum tumorem; cui certe si possem mederi, desse non deberem. Sed vides magistratus, si quidem illi magistratus, vides tamen tyranni satellites in imperiis, vides eiusdem exercitus, vides in latere veteranos, quae sunt εὑρίπιστα omnia, eos autem, qui orbis terrae custodiis non modo saepti, verum etiam magni[1] esse debebant, tantum modo laudari atque amari, sed parietibus contineri. Atque illi quoquo modo beati, civitas misera. Sed velim scire, qui adventus Octavi, num qui concursus ad eum, num quae νεωτερισμοῦ suspicio. Non puto equidem, sed tamen, quicquid est, scire cupio. Haec scripsi ad te proficiscens Astura III Idus.

VI

CICERO ATTICO S. D.

Scr. Fundis prid. Id. Apr. a. 710

Pridie Idus Fundis accepi tuas litteras cenans. Primum igitur melius esse, deinde meliora te nuntiare. Odiosa illa enim fuerant, legiones venire. Nam de Octavio susque deque. Exspecto, quid de Mario; quem quidem ego sublatum rebar a Caesare. Antoni conloquium cum heroibus nostris pro re nata non incommodum. Sed tamen adhuc me nihil delectat praeter Idus Martias. Nam, quoniam Fundis sum cum Ligure nostro, discrucior Sextili fundum a verberone Curtilio possideri. Quod cum dico, de toto genere dico. Quid enim miserius quam ea nos

[1] *For* magni *Manutius proposed* vagi, *Orelli* ἅγιοι, *and Reid* muniti.

LETTERS TO ATTICUS XIV. 5-6

have helped to remedy it, I had no right to fail in my duty. But you see the magistrates, if they can be called magistrates; you see, in spite of all, the tyrant's satellites in authority; you see his army, you see his veterans on our flank. All these can easily be fanned into flame. But those who ought to be hedged about and even honoured by the watchful care of the whole world, are only praised and admired —and confined to their houses. And they, be that as it may, are happy, while the state is in misery. But I should like to know about Octavius' arrival, whether there was a rush to meet him and whether there was any suspicion of a *coup d'état*. I don't suppose there was, but still I should like to know, whatever happened. I am writing this as I leave Astura on the 11th of April.

VI

CICERO TO ATTICUS, GREETING.

On the 12th I received your letter at Fundi during dinner. First you are better, and secondly you send better news. For the news about the coming of the legions was annoying. That about Octavius is neither here nor there. I am anxious to hear about Marius. I thought Caesar had got rid of him. Antony's conversation with our heroes is not unsatisfactory under the circumstances. However, nothing at present gives me any pleasure except the Ides of March. For now that I am at Fundi with our friend Ligur, I am annoyed at an estate of a Sextilius being in the hands of a knave like Curtilius. In mentioning this instance I am speaking of a whole class. For can there be a more wretched state of affairs than

Fundi, April 12, B.C. 44

tueri, propter quae illum oderamus? etiamne consules et tribunos pl. in biennium, quos ille voluit? Nullo modo reperio, quem ad modum possim πολιτεύεσθαι. Nihil enim tam σόλοικον quam tyrannoctonos in caelo esse, tyranni facta defendi. Sed vides consules, vides reliquos magistratus, si isti magistratus, vides languorem bonorum. Exsultant laetitia in municipiis. Dici enim non potest, quanto opere gaudeant, ut ad me concurrant, ut audire cupiant mea verba de re p. Nec ulla interea decreta. Sic enim πεπολιτεύμεθα, ut victos metueremus.

Haec ad te scripsi apposita secunda mensa; plura et πολιτικώτερα postea, et tu, quid agas, quidque agatur.

VII

CICERO ATTICO SAL.

Scr. in Formiano XVII K. Mai. a. 710

Postridie Idus Paulum Caietae vidi. Is mihi de Mario et de re publica quaedam sane pessima. A te scilicet nihil; nemo enim meorum. Sed Brutum nostrum audio visum sub Lanuvio. Ubi tandem est futurus? Nam cum reliqua tum de hoc scire aveo omnia. Ego e Formiano exiens XVII Kal., ut inde altero die in Puteolanum, scripsi haec.

A Cicerone mihi litterae sane πεπινωμέναι et bene longae. Cetera autem vel fingi possunt, πίνος litterarum significat doctiorem. Nunc magno opere a te

that we should keep up the things for which we detested him? Are we to have consuls and tribunes, too, for the next two years selected by him? I don't see how I can possibly take part in politics. For nothing could be more topsy-turvy than to belaud the slayers of the tyrant to the skies and to defend the tyrant's acts. But you see the consuls; you see the other magistrates, if they can be called magistrates; you see the indifference of the loyalists. In the country towns they are jumping for joy. I cannot describe their rejoicing, how they flock round me, how they want to hear what I have to say about the state. And in the meantime no senatorial decrees. For our policy is this, that we are afraid of the conquered party.

This I have written during dessert. I will write fuller and more about politics later, and do you write what you are doing and what is being done.

VII

CICERO TO ATTICUS, GREETING.

I saw Paulus at Caieta on the 14th. He told me *Formiae,* about Marius and gave me very bad news about the *April 15,* State. From you, of course, I have nothing, as none B.C. *44* of my men have arrived. But I hear our friend Brutus has been seen near Lanuvium. Where on earth is he going to be? For I want to know among other things everything about him. I am writing this as I leave Formiae on the 15th, and I hope to reach Puteoli on the next day.

I have had a letter from my son in quite the best style, and fairly long. Other things may be put on, but the style of the letter shows that he is learning something. Now I appeal to you earnestly to see

peto, de quo sum nuper tecum locutus, ut videas, ne quid ei desit. Id cum ad officium nostrum pertinet tum ad existimationem et dignitatem; quod idem intellexi tibi videri. Omnino, si ego, ut volo, mense Quinctili in Graeciam, sunt omnia faciliora; sed, cum sint ea tempora, ut certi nihil esse possit, quid honestum mihi sit, quid liceat, quid expediat, quaeso, da operam, ut illum quam honestissime copiosissimeque tueamur.

Haec et cetera, quae ad nos pertinebunt, ut soles, cogitabis ad meque aut, quod ad rem pertineat, aut, si nihil erit, quod in buccam venerit, scribes.

VIII

CICERO ATTICO SAL.

Scr. in Sinuessano XVII K. Mai. a. 710

Tu me iam rebare, cum scribebas, in actis esse nostris, et ego accepi XVII Kal. in deversoriolo Sinuessano tuas litteras. De Mario probe, etsi doleo L. Crassi nepotem. Optime iam etiam Bruto nostro probari Antonium. Nam, quod Iuniam scribis moderate et amice scriptas litteras attulisse, mihi Paulus dicit ad se a fratre missas; quibus in extremis erat sibi insidias fieri; se id certis auctoribus comperisse. Hoc nec mihi placebat et multo illi minus. Reginae fuga mihi non molesta est. Clodia quid egerit, scribas ad me velim. De Byzantiis curabis ut cetera et Pelopem ad te arcesses. Ego, ut postulas, Baiana

---

[1] The Pseudo-Marius had just been put to death by Antony.
[2] Both letters came from M. Lepidus, husband of Junia—the one to Brutus, the other to L. Aemilius (Lepidus) Paulus.

LETTERS TO ATTICUS XIV. 7-8

that he wants for nothing: I had already mentioned the point to you. It is a point that concerns my duty and my reputation and dignity as well; and I see you take that view yourself. Of course, if I go to Greece, as I should like, in July, everything will be easier: but, as the times are such that I cannot be sure what will be honourable, possible, or expedient for me, I beg you to be careful that we supply him with a reasonable and liberal income.

As usual you will consider these points and others that concern me, and will write and tell me the pertinent facts or, if there are none, whatever comes into your head.

VIII

CICERO TO ATTICUS, GREETING.

When you wrote, you thought I was already in one *Sinuessa*, of my seaside houses, and I have received your letter *April 15*, on the 15th in my lodge at Sinuessa. I am glad about B.C. *44* Marius, though I am sorry for the grandson of L. Crassus.[1] It is a very good thing that Antony is so approved of even by our friend Brutus. You say Junia brought a letter written in a moderately friendly tone: Paulus tells me it was sent to him by his brother,[2] and that at the end of it there was a statement that there was a plot against him, which he had ascertained on excellent authority. I was annoyed about that and he was still more annoyed. I see nothing to object to in Cleopatra's flight. I should like you to tell me what Clodia has done. You must look after the people of Byzantium like everything else, and get Pelops[3] to call on you. I

---

[3] Possibly the Pelops mentioned by Plutarch (*Cic.* 25), to whom Cicero wrote about some honours the Byzantines proposed to confer on him.

negotia chorumque illum, de quo scire vis, cum perspexero, tum scribam, ne quid ignores. Quid Galli, quid Hispani, quid Sextus agat, vehementer exspecto. Ea scilicet tu declarabis, qui cetera. Nauseolam tibi tuam causam otii dedisse facile patiebar. Videbare enim mihi legenti tuas litteras requiesse paulisper. De Bruto semper ad me omnia perscribito, ubi sit, quid cogitet. Quem quidem ego spero iam tuto vel solum tota urbe vagari posse. Verum tamen ——.

IX

CICERO ATTICO SAL.

Scr. Puteolis
XV K. Mai.
a. 710

De re publica multa cognovi ex tuis litteris; quas quidem multiiuges accepi uno tempore a Vestori liberto. Ad ea autem, quae requiris, brevi respondebo. Primum vehementer me Cluviana delectant. Sed quod quaeris, quid arcessierim Chrysippum, tabernae mihi duae corruerunt, reliquaeque rimas agunt, itaque non solum inquilini, sed mures etiam migraverunt. Hanc ceteri calamitatem vocant, ego ne incommodum quidem. O Socrates et Socratici viri! numquam vobis gratiam referam. Di immortales, quam mihi ista pro nihilo! Sed tamen ea ratio aedificandi initur consiliario quidem et auctore Vestorio, ut hoc damnum quaestuosum sit.

Hic turba magna est eritque, ut audio, maior.

will look into all that lot of fellows[1] at Baiae, about whom you wish to know, as you ask me, and will let you know all about them. I am very anxious to hear what the Gauls, and the Spaniards, and Sextus are doing. You will, of course, inform me of that as of other things. I am not sorry your slight attack of sickness gave you an excuse for rest, for, judging by your letters, you seem to have taken a little holiday. Always give me full news about Brutus, his movements and his intentions. I hope he will soon be able to walk about the whole city safely even by himself. However ——.

IX

CICERO TO ATTICUS, GREETING.

From your letters I have learned much about politics. I had a whole batch of them at the same time from the freedman of Vestorius. However, I will answer your questions shortly. Firstly, I am delighted about the Cluvian property. You ask why I sent for Chrysippus. Two of my shops have fallen down and the rest are cracking: so not only the tenants, but even the mice, have migrated. Other people call it a calamity, but I don't count it even a nuisance. O Socrates and followers of Socrates, I can never thank you sufficiently. Ye gods! how insignificant I count all such things. However, at the advice and on the suggestion of Vestorius I have adopted a plan of rebuilding which will make my loss a profit.

There are lots of people here, and I hear there

Puteoli, April 17, B.C. 44

[1] *negotium* here seems to be used as a contemptuous term in the sense of "fellow," for which cf. *Att.* I. 12 and v. 18; and to refer to Hirtius, Pansa, and Balbus who were idling at Baiae.

Duo quidem quasi designati consules. O di boni! vivit tyrannis, tyrannus occidit! Eius interfecti morte laetamur, cuius facta defendimus! Itaque quam severe nos M. Curtius accusat, ut pudeat vivere, neque iniuria. Nam mori miliens praestitit quam haec pati; quae mihi videntur habitura etiam vetustatem.

Et Balbus hic est multumque mecum. Ad quem a Vetere litterae datae pridie Kal. Ianuar., cum a se Caecilius circumsederetur et iam teneretur, venisse cum maximis copiis Pacorum Parthum; ita sibi esse eum ereptum multis suis amissis. In qua re accusat Volcacium. Ita mihi videtur bellum illud instare. Sed Dolabella et Nicias viderint. Idem Balbus meliora de Gallia. xxi die litteras habebat Germanos illasque nationes re audita de Caesare legatos misisse ad Aurelium, qui est praepositus ab Hirtio, se, quod imperatum esset, esse facturos. Quid quaeris? omnia plena pacis, aliter ac mihi Calvena dixerat.

X

CICERO ATTICO SAL.

Scr. in Puteolano XIII K. Mai. a. 710

Itane vero? hoc meus et tuus Brutus egit, ut Lanuvi esset, ut Trebonius itineribus deviis proficisceretur in provinciam, ut omnia facta, scripta, dicta, promissa, cogitata Caesaris plus valerent, quam si ipse viveret? Meministine me clamare illo ipso primo

will be more. Two of them are the so-called consuls designate. Good God, the tyranny lives though the tyrant is dead! We rejoice at his assassination and defend his actions. So see how severely M. Curtius criticizes us! We feel ashamed to live, and he is perfectly right. For to die is a thousand times better than to suffer such things, which seem to me to be likely to continue for some considerable time.

Balbus, too, is here, and is often with me. He has had a letter from Vetus, dated the last of December, saying that when Caecilius was besieged and already within his grasp, the Parthian Pacorus came with a large force, and so Caecilius was snatched from his hands and he lost many men. For that he blames Volcacius. So I suppose there is a war imminent there. But that is Dolabella's and Nicias' look out. Balbus also has better news about Gaul. Twenty-one days ago he had a letter that the Germans and the tribes there, on hearing about Caesar, sent ambassadors to Aurelius, who was appointed by Hirtius, saying that they would do as they were bidden. In fact everything seems peaceable there, contrary to what Calvena said.

X

CICERO TO ATTICUS, GREETING.

Is this what it comes to? Is this what our hero Brutus, my hero and yours, has achieved, that he should have to stay at Lanuvium, that Trebonius must make his way to his province by roundabout routes; that all the acts, notes, words, promises, and projects of Caesar should have more validity than if he were alive? Do you remember that I cried aloud

Puteoli, April 19, B.C. 44

MARCUS TULLIUS CICERO

Capitolino die debere senatum in Capitolium a praetoribus vocari? Di immortales, quae tum opera effici potuerunt laetantibus omnibus bonis, etiam sat bonis, fractis latronibus! Liberalia tu accusas. Quid fieri tum potuit? iam pridem perieramus. Meministine te clamare causam perisse, si funere elatus esset? At ille etiam in foro combustus, laudatusque miserabiliter, servique et egentes in tecta nostra cum facibus immissi. Quae deinde? ut audeant dicere: "Tune contra Caesaris nutum?" Haec et alia ferre non possum. Itaque "γῆν πρὸ γῆς" cogito; tua tamen ὑπηνέμιος.

Nausea iamne plane abiit? Mihi quidem ex tuis litteris coniectanti ita videbatur. Redeo ad Tebassos, Scaevas, Frangones. Hos tu existimas confidere se illa habituros stantibus nobis? in quibus plus virtutis putarunt, quam experti sunt. Pacis isti scilicet amatores et non latrocinii auctores. At ego, cum tibi de Curtilio scripsi Sextilianoque fundo, scripsi de Censorino, de Messalla, de Planco, de Postumo, de genere toto. Melius fuit perisse illo interfecto, quod numquam accidisset, quam haec videre.

Octavius Neapolim venit XIIII Kal. Ibi eum Balbus

---

[1] The murderers of Caesar barricaded themselves on the Capitol after the murder, and were visited by Cicero and others.

[2] At a meeting of the Senate on March 17 it was decreed that Caesar's *acta* should be confirmed, that he should have a public funeral, and that his will should be read.

[3] Greece.

on that first day on the Capitol[1] that the Senate should be summoned thither by the praetors? Ye gods! what might we not have accomplished then, when all the loyalists were rejoicing, and even the half loyal, while the knaves were crushed. You blame the Liberalia.[2] What could have been done then? We were done for already. Do you remember you exclaimed our cause was lost if the funeral took place? But he was even burned in the forum and a moving oration was delivered in his praise, and slaves and paupers were incited to attack our houses with torches. And the end of it all is that they dare to say: "Are you going to oppose Caesar's will?" Such things as these I cannot bear. So I am thinking of shifting from land to land. But your land[3] is too exposed.

Has your sickness left you entirely now? So far as I can guess from your letters it has. I return to the Tebassi, Scaevae, and Frangones.[4] Do you suppose they will have any confidence in their homesteads, while we have any power? They have found us to have less courage than they expected. I suppose we must hold them lovers of peace and not a gang of brigands. But, when I wrote to you of Curtilius and Sextilianus' farm, I wrote of Censorinus, Messalla, Plancus, Postumus, and all such cases It were better to have perished when he was slain— though it would never have come to that[5]—than to see such things.

Octavius came to Naples on the 18th of April. There Balbus met him the next morning, and the

---

[4] Veterans of Caesar's army, who had had lands of the Pompeian party given to them.
[5] Cicero implies that the republican party would have prevailed, if they had been bolder after Caesar's death.

MARCUS TULLIUS CICERO

mane postridie, eodemque die mecum in Cumano, illum hereditatem aditurum. Sed, ut scribis, ῥιζόθεμιν magnam cum Antonio. Buthrotia mihi tua res est, ut debet, eritque curae. Quod quaeris, iamne ad centena Cluvianum, adventare videtur. Scilicet primo anno $\overline{\text{LXXX}}$ detersimus.

Quintus pater ad me gravia de filio, maxime quod matri nunc indulgeat, cui antea bene merenti fuerit inimicus. Ardentes in eum litteras ad me misit. Ille autem quid agat, si scis, nequedum Roma es profectus, scribas ad me velim, et hercule si quid aliud. Vehementer delector tuis litteris.

XI

CICERO ATTICO SAL.

Scr. in Cumano XI K. Mai. a. 710

Nudius tertius dedi ad te epistulam longiorem; nunc ad ea, quae proxime. Velim mehercule, Asturae Brutus. Ἀκολασίαν istorum scribis. An censebas aliter? Equidem etiam maiora exspecto. Cum contionem lego " de tanto viro, de clarissimo civi," ferre non queo. Etsi ista iam ad risum. Sed memento, sic alitur consuetudo perditarum contionum, ut nostri illi non heroes, sed di futuri quidem in gloria sempiterna sint, sed non sine invidia, ne sine periculo quidem. Verum illis magna consolatio conscientia

[1] Left in Caesar's will.
[2] Saving the people of Buthrotum from confiscation of their land for distribution among Caesar's veterans.

same day he was with me at Cumae and said Octavius would accept that inheritance.[1] But as you say, there will be a crow to pick with Antony. I am attending to your business at Buthrotum,[2] as I ought, and I will continue to do so. You ask if Cluvius' legacy amounts to £1,000 yet. Well, in the first year I cleared about £800.[3]

Quintus is grumbling to me about his son, chiefly because he is now making much of his mother, while formerly he disliked her in spite of all she did for him. The letter against him he sent me was written in a blazing fury. If you know what the youth is doing, and have not left Rome yet, I should be glad to hear from you, and uncommonly glad for any other news too. Your letters give me so much pleasure.

XI

CICERO TO ATTICUS, GREETING.

Two days ago I sent you a fairly long letter: now I answer your last. I wish to heaven Brutus were at Astura. You speak of the wild conduct of the Caesareans. Did you expect anything else? For my part I look for worse. When I read a speech about "so great a man, so illustrious a citizen," it is more than I can bear, though now such talk is an absurdity. But take note, the habit of wild public speaking is so fostered nowadays, that, though eternal glory will be the portion of those friends of ours, who will be held not merely heroes but gods, they will not escape dislike or even danger. However, they have the great consolation of being

Cumae, April 21, B.C. 44

---

[3] 100,000 and 80,000 sesterces respectively.

maximi et clarissimi facti, nobis quae, qui interfecto rege liberi non sumus? Sed haec fortuna viderit, quoniam ratio non gubernat.

De Cicerone quae scribis, iucunda mihi sunt; velim sint prospera. Quod curae tibi est, ut ei suppeditetur ad usum et cultum copiose, per mihi gratum est, idque ut facias, te etiam atque etiam rogo. De Buthrotiis et tu recte cogitas, et ego non dimitto istam curam. Suscipiam omnem etiam actionem, quam video cotidie faciliorem. De Cluviano, quoniam in re mea me ipsum diligentia vincis, res ad centena perducitur. Ruina rem non fecit deteriorem, haud scio an etiam fructuosiorem.

Hic mecum Balbus, Hirtius, Pansa. Modo venit Octavius et quidem in proximam villam Philippi mihi totus deditus. Lentulus Spinther hodie apud me. Cras mane vadit.

XII

CICERO ATTICO SAL.

Scr. Puteolis X K. Mai. a. 710

O mi Attice, vereor, ne nobis Idus Martiae nihil dederint praeter laetitiam et odii poenam ac doloris. Quae mihi istim adferuntur! quae hic video!

"'Ὦ πράξεως καλῆς μέν, ἀτελοῦς δέ.''

Scis, quam diligam Siculos et quam illam clientelam honestam iudicem. Multa illis Caesar neque me

conscious of a heroic and magnificent deed, but what have we, who have killed a king and yet are not free? However, this lies in fortune's hands, since reason no longer rules.

What you tell me of my son is welcome news; I hope all will go well. I am exceedingly grateful to you for arranging that he shall be supplied with sufficient for luxury as well as necessities, and I beg you again and again to continue to do so. You are right about the people of Buthrotum, and I am not remitting my attention. I will undertake their whole case, which is daily looking simpler. As for Cluvius' inheritance, since you are more anxious about my affairs than I am myself, it is approaching £1,000.[1] The fall of some houses did not depreciate it; indeed, I am not sure it did not make it better.

Balbus, Hirtius, and Pansa are here with me. Octavius has just come to stay, and that, too, in the very next house, Philippus' place, and he is devoted to me. Lentulus Spinther is staying with me to-day. To-morrow early he is going.

XII

CICERO TO ATTICUS, GREETING.

Puteoli, April 22, B.C. 44

My dear Atticus, I fear the Ides of March may have given us nothing but our joy and satisfaction of our hatred and resentment. What news I get from Rome! What things I see here! "The deed was fair but its result is naught."

You know how fond I am of the Sicilians, and how great an honour I count it to be their patron. Caesar granted them many privileges, and I was pleased at

[1] 100,000 sesterces.

invito, etsi Latinitas erat non ferenda. Verum tamen. Ecce autem Antonius accepta grandi pecunia fixit legem a dictatore comitiis latam, qua Siculi cives Romani; cuius rei vivo illo mentio nulla. Quid? Deiotari nostri causa non similis? Dignus ille quidem omni regno, sed non per Fulviam. Sescenta similia. Verum illuc redeo. Tam claram tamque testatam rem tamque iustam Buthrotiam non tenebimus aliqua ex parte? et eo quidem magis, quo iste plura?

Nobiscum hic perhonorifice et peramice Octavius. Quem quidem sui Caesarem salutabant, Philippus non, itaque ne nos quidem; quem nego posse bonum civem. Ita multi circumstant, qui quidem nostris mortem minitantur. Negant haec ferri posse. Quid censes, cum Romam puer venerit, ubi nostri liberatores tuti esse non possunt? Qui quidem semper erunt clari, conscientia vero facti sui etiam beati. Sed nos, nisi me fallit, iacebimus. Itaque exire aveo, " ubi nec Pelopidarum," inquit. Haud amo vel hos designatos, qui etiam declamare me coëgerunt, ut ne apud aquas quidem acquiescere liceret. Sed hoc meae nimiae facilitatis. Nam id erat quondam quasi necesse, nunc, quoquo modo se res habet, non est item.

Quam dudum nihil habeo, quod ad te scribam! Scribo tamen, non ut delectem his litteris, sed ut

---

[1] The full quotation, which comes from the *Pelops* of Accius, runs:
" evolem,
ubi nec Pelopidarum nomen nec facta aut famam audiam."

it, though to give them the Latin rights was intolerable. However———. But here is Antony taking a huge bribe and posting up a law said to have been carried by the dictator in the Comitia, which gives the Sicilians the citizenship, though there was no mention of such a thing when Caesar was alive. Again, is not our friend Deiotarus' case just the same? He is certainly worthy of any kingdom, but not of one bought through Fulvia. There are thousands of other cases. However, to return to my point. Shall I not be able to maintain to some extent my case for the people of Buthrotum, since it is so well supported by witnesses and so just, especially as he is free with his grants?

Octavius is here with us on terms of respect and friendship. His people address him as Caesar, but Philippus does not, and so I do not either. I hold that it is impossible for a loyal citizen to do so. We are surrounded by so many who threaten death to our friends, and declare they cannot bear the present state of affairs. What do you think will happen, when this boy comes to Rome, where those who have set us free cannot live in safety. They, indeed, will ever be famous, and even happy in the consciousness of their deed. But we, unless I am much mistaken, shall be crushed. So I long to go " where no bruit of the sons of Pelops may reach my ears,"[1] as the saying is. I have no love even for these consuls designate, who have forced me to declaim to them, so that I can't have peace even by the sea. But that is due to my excess of good nature. For at one time declamation was more or less a necessity; now, however things turn out, it is not.

How long it is since I have had anything to write to you! However, I write, not to charm you with

eliciam tuas. Tu, si quid erit de ceteris, de Bruto utique, quicquid. Haec conscripsi x Kal. accubans apud Vestorium, hominem remotum **a dialecticis, in arithmeticis satis exercitatum.**

XIII

CICERO ATTICO SAL.

Scr. Puteolis Septimo denique die litterae mihi redditae sunt,
VI K. Mai. quae erant a te xiii Kal. datae; quibus quaeris atque
a. 710 etiam me ipsum nescire arbitraris, utrum magis tumulis prospectuque an ambulatione ἁλιτενεῖ delecter. Est mehercule, ut dicis, utriusque loci tanta amoenitas, ut dubitem, **utra anteponenda sit.**

Iliad x. 228
— — "Ἀλλ' οὐ δαιτὸς ἐπηράτου ἔργα μέμηλεν,
ἀλλὰ λίην μέγα πῆμα, διοτρεφές, εἰσορόωντες
δείδιμεν· ἐν δοιῇ δὲ σαωσέμεν ἢ ἀπολέσθαι."

Quamvis enim tu magna et mihi iucunda scripseris de D. Bruti adventu ad suas legiones, in quo spem maximam video, tamen, si est bellum civile futurum, quod certe erit, si Sextus in armis permanebit, quem permansurum esse certo scio, quid nobis faciendum sit, ignoro. Neque enim iam licebit, quod Caesaris bello licuit, neque huc neque illuc. Quemcumque enim haec pars perditorum laetatum Caesaris morte putabit (laetitiam autem apertissime tulimus omnes), hunc in hostium numero habebit; quae res ad caedem maximam spectat. Restat, ut in castra Sexti aut, **si forte,** Bruti nos conferamus. Res odiosa et

my letter, but to draw your answers. Do you send me any news you have, especially about Brutus, but about anything else too. I write this on the 22nd while at dinner with Vestorius,[1] a man who has no idea of philosophy, but who is well up in arithmetic.

XIII

CICERO TO ATTICUS, GREETING.

The letter you sent on the 19th did not reach me for seven days. In it you ask whether I take more pleasure in hills and a view or a walk by the silver sea, and you seem to think I may not know myself. Upon my word, both are so beautiful, as you say, that I doubt which to prefer. "But no thought have we of the service of a dainty meal; nay, seeing a woeful heavy bane sent on us by heaven, we shudder in doubt whether we shall be saved or perish." For although you have sent me great and welcome news about D. Brutus having joined his troops, in which I see great hopes, still, if there is going to be civil war—and that there must be, if Sextus stays under arms, as I know for certain he will—I don't know what we are to do. For now there will be no chance of sitting on the fence, as there was in Caesar's war. For, if this gang of ruffians thinks anyone was rejoiced at the death of Caesar—and we all of us showed our joy quite openly—they will count him an enemy; and that looks like a considerable massacre. Our alternative is to take refuge in Sextus' camp, or join ourselves to Brutus if we can. That is a hateful business and unsuitable for our age,

Puteoli, April 26, B.C. 44

[1] A banker at Puteoli.

MARCUS TULLIUS CICERO

aliena nostris aetatibus et incerto exitu belli, et nescio quo pacto tibi ego possum, mihi tu dicere.

Iliad v. 428

"Τέκνον ἐμόν, οὔ τοι δέδοται πολεμήια ἔργα,
ἀλλὰ σύγ' ἱμερόεντα μετέρχεο ἔργα λόγοιο."

Sed haec fors viderit, ea quae talibus in rebus plus quam ratio potest. Nos autem id videamus, quod in nobis ipsis esse debet, ut, quicquid acciderit, fortiter et sapienter feramus, et accidisse hominibus meminerimus, nosque cum multum litterae tum non minimum Idus quoque Martiae consolentur. Suscipe nunc meam deliberationem, qua sollicitor. Ita multa veniunt in mentem in utramque partem. Proficiscor, ut constitueram, legatus in Graeciam : caedis inpendentis periculum non nihil vitare videor, sed casurus in aliquam vituperationem, quod rei publicae defuerim tam gravi tempore. Sin autem mansero, fore me quidem video in discrimine, sed accidere posse suspicor, ut prodesse possim rei publicae. Iam illa consilia privata sunt, quod sentio valde esse utile ad confirmationem Ciceronis me illuc venire; nec alia causa profectionis mihi ulla fuit tum, cum consilium cepi legari a Caesare. Tota igitur hac de re, ut soles, si quid ad me pertinere putas, cogitabis.

Redeo nunc ad epistulam tuam. Scribis enim esse rumores me, ad lacum quod habeo, venditurum, minusculam vero villam Quinto traditurum vel impenso pretio, quo introducatur, ut tibi Quintus filius dixerit, dotata Aquilia. Ego vero de venditione nihil cogito,

especially considering the uncertainty of war; and somehow or other it seems to me that I can say to you and you to me: "My son, to thee are not given the arts of war; nay, do thou rather compass the witching arts of speech."[1] But that we must leave to chance, which is of more importance in such matters than reason. For ourselves, let us look to the one thing that ought to be in our power, that, whatever may happen, we may bear it with courage and philosophy, remembering that we are but mortal, and console ourselves a good deal with literature and not least with the Ides of March. Now come to my aid in settling a point which is causing me anxiety. So much to be said on both sides occurs to me. If I set off, as I intended, on a free embassy to Greece, it seems as though I might avoid to some extent the danger of a massacre which is threatening, but I shall not escape some blame for deserting the state in such a crisis. On the other hand, if I stay, I see I shall be in danger, but I suspect there is a possibility that I may help the State. There are also private considerations, that I think it would be of great use in settling my son down if I went to Athens; and that was the only reason for my going, when I had the idea of getting the offer of an embassy from Caesar. So consider every side of the case, as you usually do in my affairs.

I return now to your letter. You say there are rumours that I am thinking of selling my house on the Lucrine lake and of handing over to Quintus my tiny villa at quite a fancy price, that he may bring the heiress Aquilia to it, as young Quintus says. I have had no thought of selling it, unless I

[1] In the original the last word is γάμοιο not λόγοιο.

nisi quid, quod magis me delectet, invenero. Quintus autem de emendo nihil curat hoc tempore. Satis enim torquetur debitione dotis, in qua mirificas Q. Egnatio gratias agit; a ducenda autem uxore sic abhorret, ut libero lectulo neget esse quicquam iucundius. Sed haec quoque hactenus.

Redeo enim ad miseram seu nullam potius rem publicam. M. Antonius ad me scripsit de restitutione Sex. Clodi; quam honorifice, quod ad me attinet, ex ipsius litteris cognosces (misi enim tibi exemplum), quam dissolute, quam turpiter quamque ita perniciose, ut non numquam Caesar desiderandus esse videatur, facile existimabis. Quae enim Caesar numquam neque fecisset neque passus esset, ea nunc ex falsis eius commentariis proferuntur. Ego autem Antonio facillimum me praebui. Etenim ille, quoniam semel induxit animum sibi licere, quod vellet, fecisset nihilo minus me invito. Itaque mearum quoque litterarum misi tibi exemplum.

XIIIa

ANTONIUS COS. S. D. M. CICERONI.

Scr. Romae inter a. d. X et VII K. Mai. a. 710

Occupationibus est factum meis et subita tua profectione, ne tecum coram de hac re agerem. Quam ob causam vereor, ne absentia mea levior sit apud te. Quodsi bonitas tua responderit iudicio meo, quod semper habui de te, gaudebo.

find something that suits me better, while Quintus is not thinking of buying it at the present time. He is quite bothered enough with repaying the dowry,[1] and is expressing the deepest gratitude to Egnatius for his assistance. To marrying again he is so averse that he declares a bachelor's couch is the most comfortable in the world. But enough of this also.

For now I return to the crushed or rather nonexistent republic. M. Antonius has written to me about the recall of Sex. Clodius. You will see from the copy I include that the tone of his letter, so far as concerns myself, is complimentary enough. But you can easily imagine the proposal is so unprincipled, so disgraceful, and so mischievous, that at times one almost wishes for Caesar back again. For things that Caesar never would have done, nor allowed to be done, are now being brought forward from forged notes of his. However, I have shown myself quite complaisant to Antonius. For when he has once got it into his head that he may do what he chooses, he would have done it just as readily against my will. So I have sent you a copy of my letter too.

XIIIa

ANTONY THE CONSUL SENDS HIS GREETINGS TO M. CICERO.

It was only because I was so busy and you departed so suddenly, that I did not see you personally about the following request. So I fear I may have less weight with you in my absence. But if your goodness of heart answers to the opinion I have always had of you, I shall be very glad.

Rome, April 22 to 25, B.C. 44

[1] To Pomponia, sister of Atticus, whom he had recently divorced.

MARCUS TULLIUS CICERO

A Caesare petii, ut Sex. Clodium restitueret; impetravi. Erat mihi in animo etiam tum sic uti beneficio eius, si tu concessisses. Quo magis laboro, ut tua voluntate id per me facere nunc liceat. Quodsi duriorem te eius miserae et adflictae fortunae praebes, non contendam ego adversus te, quamquam videor debere tueri commentarium Caesaris. Sed mehercule, si humaniter et sapienter et amabiliter in me cogitare vis, facilem profecto te praebebis, et voles P. Clodium, in optima spe puerum repositum, existimare non te insectatum esse, cum potueris, amicos paternos. Patere, obsecro, te pro re publica videri gessisse simultatem cum patre eius, non quod contempseris hanc familiam. Honestius enim et libentius deponimus inimicitias rei publicae nomine susceptas quam contumaciae. Me deinde sine ad hanc opinionem iam nunc dirigere puerum et tenero animo eius persuadere non esse tradendas posteris inimicitias. Quamquam tuam fortunam, Cicero, ab omni periculo abesse certum habeo, tamen arbitror malle te quietam senectutem et honorificam potius agere quam sollicitam. Postremo meo iure te hoc beneficium rogo. Nihil enim non tua causa feci. Quodsi non impetro, per me Clodio daturus non sum, ut intellegas, quanti apud me auctoritas tua sit, atque eo te placabiliorem praebeas.

LETTERS TO ATTICUS XIV. 13a

I petitioned Caesar for the return of Sex. Clodius, and obtained my request. It was my intention even then only to use his favour if you allowed. So I am now the more anxious that you may let me do it with your permission. But, if you show yourself hard-hearted to his affliction and misery, I will not contend with you, though I think I ought to observe Caesar's memoranda. But upon my word, if you are ready to take a generous, philosophical, and amiable view of my actions, you will, I am sure, show your indulgence, and will wish that most promising youth, P. Clodius, to think that you did not act spitefully to his father's friends when you had the chance. I beseech you to let it seem that your feud with his father was on public grounds, not because you despised the family. For we can lay aside quarrels we took up on public grounds with more honour and more readiness than those that come from a personal insult. So give me a chance of inculcating this lesson, and while the boy's mind is still receptive, let us convince him that quarrels should not be handed down from generation to generation. Though I know your fortune, Cicero, is above any danger, yet I think you would rather enjoy old age with peace and honour than with anxiety. Finally I feel a right to ask you this favour, for I have done all I could for your sake. If I do not gain it, I shall not grant Clodius his restoration, so that you may understand how much your authority weighs in my eyes, and that may make you all the more placable.

MARCUS TULLIUS CICERO

XIIIb

CICERO ANTONIO COS. S. D.

Scr. Puteolis VI K. Mai. 710

Quod mecum per litteras agis, unam ob causam mallem coram egisses. Non enim solum ex oratione, sed etiam ex vultu et oculis et fronte, ut aiunt, meum erga te amorem perspicere potuisses. Nam, cum te semper amavi, primum tuo studio, post etiam beneficio provocatus, tum his temporibus res publica te mihi ita commendavit, ut cariorem habeam neminem. Litterae vero tuae cum amantissime tum honorificentissime scriptae sic me adfecerunt, ut non dare tibi beneficium viderer, sed accipere a te ita petente, ut inimicum meum, necessarium tuum me invito servare nolles, cum id nullo negotio facere posses. Ego vero tibi istuc, mi Antoni, remitto, atque ita, ut me a te, cum iis verbis scripseris, liberalissime atque honorificentissime tractatum existimem, idque cum totum, quoquo modo se res haberet, tibi dandum putarem, tum do etiam humanitati et naturae meae. Nihil enim umquam non modo acerbum in me fuit, sed ne paulo quidem tristius aut severius, quam necessitas rei publicae postulavit. Accedit, ut ne in ipsum quidem Clodium meum insigne odium fuerit umquam, semperque ita statui, non esse insectandos inimicorum amicos, praesertim humiliores, nec his praesidiis nosmet ipsos esse spoliandos. Nam de puero Clodio tuas partes esse arbitror, ut eius animum tenerum, quem ad modum scribis, iis opinionibus imbuas, ut ne quas

XIIIb

CICERO TO ANTONY THE CONSUL, GREETING.

There is one reason why I wish you had made *Puteoli,* personally the request you are making by letter. *April 26,* Then you could have seen my affection for you not B.C. *44* only from what I said, but from my "expression, eyes and brow," as the phrase goes. For I have always had an affection for you, urged thereto at first by your attention to me and afterwards by benefits received, and in these days public affairs have so recommended you to me that there is no one for whom I have more regard. The letter you have written to me in such a friendly and flattering tone makes me feel as though I were receiving a favour from you, not granting one to you, since you refuse to recall your friend, who was my enemy, against my will, though you could quite easily do so. Of course I grant your request, my dear Antony, and I think myself, too, most liberally and honourably treated, when you address me in such a strain. I should have thought it right to grant it you freely, whatever the facts had been, and besides, I am gratifying my own natural kindliness. For I never had any bitterness or even the slightest sternness or severity in me, except what was demanded by public necessity. Besides, I never had any special grudge against Clodius himself, and I always laid down the rule that one should not attack one's enemies' friends, especially their humbler friends, nor should we ourselves be deprived of such supporters. As regards the boy Clodius I think it is your duty to imbue his "receptive mind," as you say, with the idea that

inimicitias residere in familiis nostris arbitretur. Contendi cum P. Clodio, cum ego publicam causam, ille suam defenderet. Nostras concertationes res publica diiudicavit. Si viveret, mihi cum illo nulla contentio iam maneret. Quare, quoniam hoc a me sic petis, ut, quae tua potestas est, ea neges te me invito usurum, puero quoque hoc a me dabis, si tibi videbitur, non quo aut aetas nostra ab illius aetate quicquam debeat periculi suspicari, aut dignitas mea ullam contentionem extimescat, sed ut nosmet ipsi inter nos coniunctiores simus, quam adhuc fuimus. Interpellantibus enim his inimicitiis animus tuus mihi magis patuit quam domus. Sed haec hactenus.

Illud extremum. Ego, quae te velle quaeque ad te pertinere arbitrabor, semper sine ulla dubitatione summo studio faciam. Hoc velim tibi penitus persuadeas.

XIV

CICERO ATTICO S. D.

Scr. in Puteolano a. d. V K. Mai. a. 710

"Iteradum eadem ista mihi." Coronatus Quintus noster Parilibus! Solusne? Etsi addis Lamiam. Quod demiror equidem: sed scire cupio, qui fuerint alii; quamquam satis scio nisi improbum neminem. Explanabis igitur hoc diligentius. Ego autem casu, cum dedissem ad te litteras VI Kal. satis multis verbis, tribus fere horis post accepi tuas et magni quidem ponderis. Itaque ioca tua plena facetiarum de haeresi

there is no enmity between our families. I fought P. Clodius because I was fighting for the State, he for his own hand; and the State decided the merits of our controversy. If he were alive now I should have no further quarrel with him. So, since in making your request you say you will not use the power you have against my will, you may make this concession to the boy too in my name, if you will; not that a man of my age has anything to fear from a youth of his, or that a person of my position needs shrink from any quarrel, but that we may be more intimate than we have been as yet. For these feuds have come between us, and so your heart has been more open to me than your house. But enough of this.

I have one thing to add, that, whatever I think you wish, and whatever is to your interest, I shall never have any hesitation in carrying out with all my heart and soul. Of that I hope you will feel fully persuaded.

XIV

CICERO TO ATTICUS, GREETING.

"Repeat your tale again to me."[1] Our nephew *Puteoli,* wearing a crown at the Parilia! Was he alone? *April 27,* Though you add Lamia, which astonishes me. But B.C. *44* I should like to know what others there were, though I am perfectly sure there were none but knaves. So please explain more in detail. As it happened, when I had sent you a pretty long letter on the 26th, about three hours afterwards I received yours, and a bulky one too. So there is no necessity for me to tell you that I had a good laugh at your witty

[1] From the *Iliona* of Pacuvius.

MARCUS TULLIUS CICERO

Vestoriana et de Pherionum more Puteolano risisse me satis nihil est necesse rescribere. Πολιτικώτερα illa videamus.

Ita Brutos Cassiumque defendis, quasi eos ego reprehendam; quos satis laudare non possum. Rerum ego vitia collegi, non hominum. Sublato enim tyranno tyrannida manere video. Nam, quae ille facturus non fuit, ea fiunt, ut de Clodio, de quo mihi exploratum est illum non modo non facturum, sed etiam ne passurum quidem fuisse. Sequetur Rufio Vestorianus, Victor numquam scriptus, ceteri, quis non? Cui servire ipsi non potuimus, eius libellis paremus. Nam Liberalibus quis potuit in senatum non venire? Fac id potuisse aliquo modo; num etiam, cum venissemus, libere potuimus sententiam dicere? nonne omni ratione veterani, qui armati aderant, cum praesidii nos nihil haberemus, defendendi fuerunt? Illam sessionem Capitolinam mihi non placuisse tu testis es. Quid ergo? ista culpa Brutorum? Minime illorum quidem, sed aliorum brutorum, qui se cautos ac sapientes putant; quibus satis fuit laetari, non nullis etiam gratulari, nullis permanere. Sed praeterita omittamus; istos omni cura praesidioque tueamur et, quem ad modum tu praecipis, contenti Idibus Martiis simus; quae quidem nostris amicis divinis viris aditum ad caelum dederunt, libertatem populo Romano non

---

[1] Vestorius was a banker (cf. XIV. 12), and Atticus had probably played on the two senses of αἵρεσις, "sect" and "grasping." The allusion to the Pheriones is inexplicable.

remarks about the sect of Vestorius[1] and the Puteolian custom of the Pheriones. Let us consider the more political part.

You defend Brutus and Cassius as though you thought I blamed them, though I cannot find praise enough for them. It is the weak points of the situation, not of the persons that I put together. For though the tyrant is dead, I see the tyranny persists. For things that he would not have done are being done now, as, for example, the recalling of Clodius —a thing I am sure he had no intention of doing and would not even have allowed to be done. Vestorius' enemy Rufio will follow, and Victor, whose name was never in Caesar's notes, and the rest; every one in fact. We could not be Caesar's slaves, but we bow down to his note-books. For who dared absent himself from the Senate on the Liberalia?[2] Suppose it had been possible somehow: even when we did come, could we speak our mind freely? Had we not to take precious good care of the veterans, who were there under arms, since we had no support ourselves. You can bear witness that that sitting still on the Capitol was not approved by me. Well, was that the fault of Brutus and the rest? Not a bit of it: it was the fault of the other brute beasts, who think themselves cautious and canny. They thought it enough to rejoice, some of them to go so far as to congratulate, none to stand their ground. But let us put the past behind us: let us guard our heroes with all our care and protection: and, as you say, let us be content with the Ides of March. That day gave our friends, who are more than men, an entrance to heaven, but it did not give freedom to

[2] March 17. Cf. *Att.* xiv. 10.

dederunt. Recordare tua. Nonne meministi clamare te omnia perisse, si ille funere elatus esset? Sapienter id quidem. Itaque, ex eo quae manarint, vides.

Quae scribis K. Iuniis Antonium de provinciis relaturum, ut et ipse Gallias habeat, et utrisque dies prorogetur, licebitne decerni libere? Si licuerit, libertatem esse recuperatam laetabor; si non licuerit, quid mihi attulerit ista domini mutatio praeter laetitiam, quam oculis cepi iusto interitu tyranni? Rapinas scribis ad Opis fieri; quas nos quoque tum videbamus. Ne nos et liberati ab egregiis viris nec liberi sumus. Ita laus illorum est, culpa nostra. Et hortaris me, ut historias scribam, ut colligam tanta eorum scelera, a quibus etiam nunc obsidemur! Poterone eos ipsos non laudare, qui te obsignatorem adhibuerunt? Nec mehercule me raudusculum movet, sed homines benevolos, qualescumque sunt, grave est insequi contumelia. Sed de omnibus meis consiliis, ut scribis, existimo exploratius nos ad K. Iunias statuere posse. Ad quas adero, et omni ope atque opera enitar adiuvante me scilicet auctoritate tua et gratia et summa aequitate causae, ut de Buthrotiis senatus consultum, quale scribis, fiat. Quod me cogitare iubes, cogitabo equidem, etsi tibi dederam superiore epistula cogitandum. Tu autem quasi iam recuperata re publica vicinis tuis Massiliensibus sua reddis. Haec armis,

---

[1] To wills in which legacies were left to Cicero. Cf. *Att.* XIV. 3.

the Roman people. Recall your words. Don't you remember how you exclaimed that all was lost if Caesar had a public funeral? And very wise it was. So you see what has come of the funeral.

You say Antony is going to bring a proposal before the Senate on the 1st of June about the allotment of provinces, that he should have Gaul and that both his own and his colleague's tenure should be prolonged. Will the House be allowed to vote freely? If so, I shall rejoice that liberty has been regained; if not, what has this change of masters brought me except the joy of feasting my eyes on the just death of a tyrant? You say there is plundering at the Temple of Ops: I saw it then with my own eyes. Yea, we have been set free by heroes and are not free after all. So theirs is the praise and ours the blame. And you advise me to write history, to collect all the crimes of those who even now have us under their thumb. Shall I be able to resist praising men who have called you in as a witness?[1] I give you my word it is not the petty gain that influences me, but it goes against the grain to heap contumely on the heads of benevolent persons whatever their character. But, as you say, I think we can make up our minds with more certainty about all my plans by the 1st of June. I shall be present then, and of course with the assistance of your authority and popularity, and the absolute justice of your case, I shall strive with all my might to obtain the senatorial decree that you mention about the people of Buthrotum. What you bid me consider, I will consider, though it is what I asked you to consider in a former letter. But here you are wanting to get back their rights for your neighbours the Massilians, as though we had recovered the republic. Perhaps they might be restored by arms—but how strong our

quae quam firma habeamus, ignoro, restitui fortasse possunt, auctoritate non possunt.

Epistula brevis, quae postea a te scripta est, sane mihi fuit iucunda de Bruti ad Antonium et de eiusdem ad te litteris. Posse videntur esse meliora, quam adhuc fuerunt. Sed nobis, ubi simus et quo iam nunc nos conferamus, providendum est.

XV

CICERO ATTICO SAL.

Scr. in Puteolano K. Mai. a. 710

O mirificum Dolabellam meum! iam enim dico meum; antea, crede mihi, subdubitabam. Magnam ἀναθεώρησιν res habet, de saxo, in crucem, columnam tollere, locum illum sternendum locare! Quid quaeris? heroica. Sustulisse mihi videtur simulationem desiderii, adhuc quae serpebat in dies et inveterata verebar ne periculosa nostris tyrannoctonis esset. Nunc prorsus adsentior tuis litteris speroque meliora. Quamquam istos ferre non possum, qui, dum se pacem velle simulant, acta nefaria defendunt. Sed non possunt omnia simul. Incipit res melius ire, quam putaram. Nec vero discedam, nisi cum tu me id honeste putabis facere posse. Bruto certe meo nullo loco deero, idque, etiamsi mihi cum illo nihil fuisset, facerem propter eius singularem incredibilemque virtutem.

---

[1] A column erected in honour of Caesar by the Pseudo-Marius. Riotous mass-meetings were held round it, and it

arms are I do not know—by influence they certainly cannot.

The short letter you wrote after the other, about Brutus' letter to Antony and also his to you, delighted me much. It looks as though things might be better than they have been at present. But we must look carefully into our present position and our immediate movements.

XV

CICERO TO ATTICUS, GREETING.

Well done my Dolabella! For now I call him mine: up to now, you know, I had some doubts. This will make people open their eyes—hurling from the rock, crucifixion, pulling down the column [1] and ordering the place to be paved. Why, these are heroic deeds. I take it he has put an end to this feigning of regret, which up to now was creeping on day by day, and, if it became a habit, I was afraid it might be dangerous to our tyrannicides. Now I quite agree with your letter and hope for better things. However I cannot put up with the people who under a pretence of wishing for peace defend criminal actions. But still we can't have everything at once. Things are beginning to get better than I had expected, and of course I will not go away, unless you think I can do so honourably. My friend Brutus certainly I will never desert; and I should act in the same way, even if there were no ties between us, on account of his extraordinary and incredible strength of character.

Puteoli, May 1, B.C. 44

was the people who took part in these who were summarily executed by Dolabella without any trial.

MARCUS TULLIUS CICERO

Piliae nostrae villam totam, quaeque in villa sunt,
trado, in Pompeianum ipse proficiscens K. Maiis.
Quam velim Bruto persuadeas, ut Asturae sit!

XVI

CICERO ATTICO SAL.

*Scr. Puteolis
in hortis
Cluvianis V
Non. Mai.
a. 710*

v Nonas conscendens ab hortis Cluvianis in phaselum epicopum has dedi litteras, cum Piliae nostrae villam ad Lucrinum, vilicos, procuratores tradidissem. Ipse autem eo die in Paeti nostri tyrotarichum imminebam; perpaucis diebus in Pompeianum, post in haec Puteolana et Cumana regna renavigare. O loca ceteroqui valde expetenda, interpellantium autem multitudine paene fugienda!

Sed ad rem ut veniam, o Dolabellae nostri magnam ἀριστείαν! Quanta est ἀναθεώρησις! Equidem laudare eum et hortari non desisto. Recte tu omnibus epistulis significas, quid de re, quid de viro sentias. Mihi quidem videtur Brutus noster iam vel coronam auream per forum ferre posse. Quis enim audeat laedere proposita cruce aut saxo, praesertim tantis plausibus, tanta approbatione infimorum?

Nunc, mi Attice, me fac ut expedias. Cupio, cum Bruto nostro adfatim satis fecerim, excurrere in Graeciam. Magni interest Ciceronis, vel mea potius vel mehercule utriusque, me intervenire discenti

LETTERS TO ATTICUS XIV. 15–16

I hand over the villa and all there is in it to our dear Pilia, as I am setting out for Pompeii on the 1st of May. How I wish you could persuade Brutus to come to Astura!

XVI

CICERO TO ATTICUS, GREETING.

I despatch this letter on the 3rd, as I embark in a rowing boat from Cluvius' gardens, after handing over the villa at the Lucrine lake to Pilia with its servants and care-takers. Myself I am threatening our friend Paetus' cheese and herrings for that day; in a few days I am going to Pompeii and after that sailing back to my domains here at Puteoli and Cumae. What very attractive places they are, if it were not that one almost has to shun them on account of the crowd of visitors.

Puteoli, May 3, B.C. 44

But to return to the point, what a magnificent stroke of our friend Dolabella! How it will make people open their eyes. For my part I keep on praising and encouraging him. You are right in what you say in every letter about the deed and about the man. To me it seems that our friend Brutus could walk through the forum with a golden crown on his head now. For who would dare to hurt him with the cross and rock before his eyes, especially when the rabble have shown such applause and approbation?

Now, my dear Atticus, do put things straight for me. I want to run over to Greece, as soon as I have quite satisfied Brutus. It is a matter of great concern to my son, or rather to me, or upon my word to both of us, that I should drop in upon

MARCUS TULLIUS CICERO

Nam epistula Leonidae, quam ad me misisti, quid habet, quaeso, in quo magno opere laetemur? Numquam ille mihi satis laudari videbitur, cum ita laudabitur: "Quo modo nunc est." Non est fidentis hoc testimonium, sed potius timentis. Herodi autem mandaram, ut mihi κατὰ μίτον scriberet. A quo adhuc nulla littera est. Vereor, ne nihil habuerit, quod mihi, cum cognossem, iucundum putaret fore.

Quod ad Xenonem scripsisti, valde mihi gratum est; nihil enim deesse Ciceroni cum ad officium tum ad existimationem meam pertinet. Flammam Flaminium audio Romae esse. Ad eum scripsi me tibi mandasse per litteras, ut de Montani negotio cum eo loquerere, et velim cures epistulam, quam ad eum misi, reddendam, et ipse, quod commodo tuo fiat, cum eo conloquare. Puto, si quid in homine pudoris est, praestaturum eum, ne sero cum damno dependatur. De Attica pergratum mihi fecisti quod curasti, ante scirem recte esse quam non belle fuisse.

XVII

CICERO ATTICO SAL.

Scr. in Pompeiano IV Non. Mai. a. 710

In Pompeianum veni v Nonas Maias, cum pridie, ut antea ad te scripsi, Piliam in Cumano conlocavissem. Ibi mihi cenanti litterae tuae sunt redditae, quas dederas Demetrio liberto pr. Kal.; in quibus multa sapienter, sed tamen talia, quem ad modum tute scribebas, ut omne consilium in fortuna positum

him at his studies. For what is there to give us any particular satisfaction in the letter of Leonidas, which you have sent to me? I shall never be content with his praise, when it is phrased, "as things go at present." There is no evidence of confidence, rather of anxiety in that. Again I had commissioned Herodes to write to me in detail: but as yet I have not had a single syllable from him. I am afraid he has had no news that he thought would gratify me, if I heard it.

I am very grateful to you for writing to Xeno; for that my son should not be short of money concerns both my duty and my reputation. I hear that Flaminius Flamma is in Rome. I have written to tell him that I have instructed you by letter to speak to him about Montanus' business: and, I should be glad if you would see that the letter I have sent for him is delivered, and would speak with him at your leisure. I think, if the man has any sense of shame, he will see that the payment is not deferred to my loss. You were very kind in informing me of Attica's recovery before I knew of her illness.

XVII

CICERO TO ATTICUS, GREETING.

I reached Pompeii on the 3rd of May, having *Pompeii,* established Pilia in my place at Cumae the day *May 4,* B.C. before, as I told you in a former letter. While I *44* was at dinner there, the letter you had given to the freedman Demetrius on the last of April was delivered. There was a lot of wise advice in it, but, as you admit yourself, with the reservation

videretur. Itaque his de rebus ex tempore et coram. De Buthrotio negotio utinam quidem Antonium conveniam! Multum profecto proficiam. Sed non arbitrantur eum a Capua declinaturum; quo quidem metuo ne magno rei publicae malo venerit. Quod idem L. Caesari videbatur, quem pridie Neapoli adfectum graviter videram. Quam ob rem ista nobis ad Kal. Iunias tractanda et perficienda sunt. Sed hactenus.

Quintus filius ad patrem acerbissimas litteras misit; quae sunt ei redditae, cum venissemus in Pompeianum. Quarum tamen erat caput Aquiliam novercam non esse laturum. Sed hoc tolerabile fortasse, illud vero, se a Caesare habuisse omnia, nihil a patre, reliqua sperare ab Antonio—o perditum hominem! Sed μελήσει.

Ad Brutum nostrum, ad Cassium, ad Dolabellam epistulas scripsi. Earum exempla tibi misi, non ut deliberarem, reddundaene essent. Plane enim iudico esse reddendas, quod non dubito quin tu idem existimaturus sis.

Ciceroni meo, mi Attice, suppeditabis, quantum videbitur, meque hoc tibi onus imponere patiere. Quae adhuc fecisti, mihi sunt gratissima. Librum meum illum ἀνέκδοτον nondum, ut volui, perpolivi; ista vero, quae tu contexi vis, aliud quoddam separatum volumen exspectant. Ego autem, credas mihi velim, minore periculo existimo contra illas nefarias

that everything seems to depend on chance. So these points we will discuss on the spot when we meet. As regards the business about Buthrotum I only wish I could meet Antony. I am sure I could make good headway with him. But people think he won't stir from Capua, and I fear his going there will do a great deal of harm to the state. L. Caesar, whom I saw yesterday very ill at Naples, thought the same too. So I shall have to handle this subject and get it settled on the 1st of June. But enough of this.

Young Quintus has sent his father a most unpleasant letter, which was delivered when we reached Pompeii. The chief point of it was that he would not put up with Aquilia as a step-mother: but that perhaps is excusable. But to say he owed everything to Caesar, nothing to his father, and for the future he looked to Antonius—what a scoundrel! However that shall be attended to.

I have written to Brutus, to Cassius and to Dolabella. I send you copies; not that I am in doubt whether to send the letters or not; for I feel sure that they ought to be sent, and I have no doubt you will agree with me.

Please, dear Atticus, supply my boy with as much money as you think fit, and forgive me for troubling you. For what you have done already I am most grateful. That unpublished book of mine[1] I have not yet polished up as I should wish: the points you want me to introduce must wait for a second volume. But I think—and I hope you will believe me—that one could have spoken against that disreputable party with less danger in the tyrant's

[1] Possibly his poem *De temporibus suis*; but it is not certain.

partes vivo tyranno dici potuisse quam mortuo. Ille
enim nescio quo pacto ferebat me quidem mirabiliter;
nunc, quacumque nos commovimus, ad Caesaris non
modo acta, verum etiam cogitata revocamur. De
Montano, quoniam Flamma venit, videbis. Puto rem
meliore loco esse debere.

XVIIa (= *Fam.* IX. 14)

CICERO DOLABELLAE COS. SUO SAL.

Scr. in Pompeiano V Non. Mai. a. 710

Etsi contentus eram, mi Dolabella, tua gloria, satisque ex ea magnam laetitiam voluptatemque capiebam, tamen non possum non confiteri cumulari me maximo gaudio, quod vulgo hominum opinio socium me ascribat tuis laudibus. Neminem conveni (convenio autem cotidie plurimos. Sunt enim permulti optimi viri, qui valetudinis causa in haec loca veniant; praeterea ex municipiis frequentes necessarii mei), quin omnes, cum te summis laudibus ad caelum extulerunt, mihi continuo maximas gratias agant. Negant enim se dubitare, quin tu meis praeceptis et consiliis obtemperans praestantissimum te civem et singularem consulem praebeas. Quibus ego quamquam verissime possum respondere te, quae facias, tuo iudicio et tua sponte facere, nec cuiusquam egere consilio, tamen neque plane adsentior, ne imminuam tuam laudem, si omnis a meis consiliis profecta videatur, neque valde nego. Sum enim avidior etiam, quam satis est, gloriae. Et tamen non alienum est dignitate tua,

LETTERS TO ATTICUS XIV. 17–17a

life than after his death. For he, somehow, was most patient with me; now, whichever way we turn, we are reminded not only of Caesar's enactments, but also of his intentions. Please see about Montanus, since Flamma has arrived. I think the matter ought to be put on a better footing.

XVIIa

CICERO TO HIS FRIEND DOLABELLA THE CONSUL, GREETING.

Though I feel content with the glory you have won, my dear Dolabella, and it affords me the greatest joy and pleasure, still I cannot help confessing that the crowning point of my joy is, that in the popular opinion my name is associated with yours in people's praise. I am daily meeting many people; for quite a number of persons of consideration come here for their health, besides many acquaintances of mine from the country towns; and I have not met anyone who does not extol you to the skies, and in the same breath offer me the sincerest congratulations. For they say they have no doubt that it is by following my precepts and advice that you are showing yourself a most distinguished citizen and an excellent consul. Though I can answer them with the fullest truth that what you do, you do acting on your own judgment and on your own initiative and that you need no advice, still I do not entirely assent, lest I should diminish your glory, if it all appears to have sprung from my advice, nor do I quite deny it; for I have more than my proper share of desire for glory. And yet it would not detract from your

Pompeii, May 3, B.C. *44*

quod ipsi Agamemnoni, regum regi, fuit honestum, habere aliquem in consiliis capiendis Nestorem, mihi vero gloriosum te iuvenem consulem florere laudibus quasi alumnum disciplinae meae. L. quidem Caesar, cum ad eum aegrotum Neapolim venissem, quamquam erat oppressus totius corporis doloribus, tamen, antequam me plane salutavit, "O mi Cicero," inquit, "gratulor tibi, cum tantum vales apud Dolabellam, quantum si ego apud sororis filium valerem, iam salvi esse possemus. Dolabellae vero tuo et gratulor et gratias ago; quem quidem post te consulem solum possumus vere consulem dicere." Dein multa de facto ac de re gesta tua; nihil magnificentius, nihil praeclarius actum umquam, nihil rei publicae salutarius. Atque haec una vox omnium est. A te autem peto, ut me hanc quasi falsam hereditatem alienae gloriae sinas cernere meque aliqua ex parte in societatem tuarum laudum venire patiare. Quamquam, mi Dolabella, (haec enim iocatus sum) libentius omnes meas, si modo sunt aliquae meae laudes, ad te transfuderim quam aliquam partem exhauserim ex tuis. Nam, cum te semper tantum dilexerim, quantum tu intellegere potuisti, tum his tuis factis sic incensus sum, ut nihil umquam in amore fuerit ardentius. Nihil est enim, mihi crede, virtute formosius, nihil pulchrius, nihil amabilius. Semper amavi, ut scis, M. Brutum propter eius summum ingenium, suavissimos mores, singularem probitatem atque constantiam.

dignity any more than it disgraced Agamemnon, the king of kings, to have some Nestor to assist in your plans; while it would redound to my glory that you with your brilliant reputation as a consul while still so young should be thought a pupil of my training. Indeed L. Caesar, when I paid him a visit on his sick bed at Naples, though he was racked with pains all over his body, had hardly finished his first greeting before he said: "My dear Cicero, I congratulate you on the influence you have with Dolabella. If I had had as much with my sister's son,[1] we might have been safe now. Dolabella himself I both congratulate and thank: indeed he is the first consul since yourself who can really be called a consul." Then he had much to say about the incident and your achievement. No more splendid and magnificent deed was ever done, nor any more salutary to the state: and that is what the whole world is saying with one voice. I beg you to let me enter into this false heritage of another's glory, and suffer me to share your praises in some slight degree. However, my dear Dolabella, so far I have only been joking, and, if I have any reputation myself, I would rather turn its full stream upon you, than divert any part of yours upon myself. For, though I have always been as fond of you as you must have realized, now by your actions my fondness has been fanned into the most ardent love that is possible. For, believe me, there is nothing fairer than virtue, nothing more beautiful, nothing more loveable. I have always loved M. Brutus, as you know, for his great ability, his most agreeable manners, his extraordinary upright-

[1] Julia, sister of L. Caesar, was mother of Antony by her first husband, Antonius Creticus.

MARCUS TULLIUS CICERO

Tamen Idibus Martiis tantum accessit ad amorem, ut mirarer locum fuisse augendi in eo, quod mihi iam pridem cumulatum etiam videbatur. Quis erat, qui putaret ad eum amorem, quem erga te habebam, posse aliquid accedere? Tantum accessit, ut mihi nunc denique amare videar, ante dilexisse. Quare quid est, quod ego te horter, ut dignitati et gloriae servias? Proponam tibi claros viros, quod facere solent, qui hortantur? Neminem habeo clariorem quam te ipsum. Te imitere oportet, tecum ipse certes. Ne licet quidem tibi iam tantis rebus gestis non tui similem esse. Quod cum ita sit, hortatio non est necessaria, gratulatione magis utendum est. Contigit enim tibi, quod haud scio an nemini, ut summa severitas animadversionis non modo non invidiosa, sed etiam popularis esset et cum bonis omnibus tum infimo cuique gratissima. Hoc si tibi fortuna quadam contigisset, gratularer felicitati tuae, sed contigit magnitudine cum animi tum etiam ingenii atque consilii. Legi enim contionem tuam. Nihil illa sapientius. Ita pedetemptim et gradatim tum accessus a te ad causam facti, tum recessus, ut res ipsa maturitatem tibi animadvertendi omnium concessu daret. Liberasti igitur et urbem periculo et civitatem metu, neque solum ad tempus maximam utilitatem attulisti, sed etiam ad exemplum. Quo facto intellegere debes in te positam esse rem publicam, tibique

ness and constancy. However on the Ides of March my affection was so enhanced that I wondered there was any room for increase in what I had long thought had reached its culminating point. Who would have thought that there could be any increase in the affection I have for you? But there has been such an increase that I seem to myself now to love, while before I only liked. So what need is there that I should exhort you to have a regard for your dignity and glory? Shall I do what people generally do when exhorting others, set before your eyes distinguished examples? There is none more distinguished than your own. You must imitate yourself and vie with yourself. Indeed, after such an achievement, you dare not fail to be like yourself. As that is so, exhortation is unnecessary and congratulation is more in place. For you have had the fortune, which I doubt if anyone else ever had, that great severity in punishment should not only bring no ill will, but should be popular and most pleasing to all, both of the upper and of the lower class. If this had happened to you by a stroke of fortune, I should congratulate you on your luck: but it has happened through your greatness of heart, yes, and of ability and of prudence. For I have read your harangue. Nothing could have been more skilful. You led up to the case so gradually and gently, and then left it again, that by universal consent the facts themselves showed it was high time to resort to punitive measures. So you freed the city from danger and the state from fear, and you performed a sound service not only to meet the emergency but to serve as a precedent. After that you ought to understand that the republic is in your hand, and

non modo tuendos, sed etiam ornandos illos viros, a quibus initium libertatis profectum est. Sed his de rebus coram plura prope diem, ut spero. Tu, quoniam rem publicam nosque conservas, fac, ut diligentissime te ipsum, mi Dolabella, custodias.

XVIII

CICERO ATTICO.

Scr. in Pompeiano VII Id. Mai. a. 710

Saepius me iam agitas, quod rem gestam Dolabellae nimis in caelum videar efferre. Ego autem, quamquam sane probo factum, tamen, ut tanto opere laudarem, adductus sum tuis et unis et alteris litteris. Sed totum se a te abalienavit Dolabella ea de causa, qua me quoque sibi inimicissimum reddidit. O hominem impudentem! Kal. Ian. debuit, adhuc non solvit, praesertim cum se maximo aere alieno Faberi manu liberarit et opem ab Ope petierit. Licet enim iocari, ne me valde conturbatum putes. Atque ego ad eum VIII Idus litteras dederam bene mane, eodem autem die tuas litteras vesperi acceperam in Pompeiano sane celeriter tertio abs te die. Sed, ut ad te eo ipso die scripseram, satis aculeatas ad Dolabellam litteras dedi; quae si nihil profecerint, puto fore ut me praesentem non sustineat.

Albianum te confecisse arbitror. De Patulciano

---

[1] Faberius was Caesar's secretary and was used by Antony to insert extra details in Caesar's memoranda. Here Dolabella is included in the accusation repeatedly brought by

that you should not only protect but honour the men who paved the way for freedom. But I hope we shall soon meet to discuss these things. Do you, my dear Dolabella, take the greatest care of yourself, since you preserve the state and all of us.

XVIII

CICERO TO ATTICUS.

You are continually reproaching me now with lauding Dolabella to the skies more than I ought. But, though I strongly approve of his action, still it was one and then another letter of yours which induced me to belaud it so highly. But Dolabella has entirely lost your good graces for the same reason that he has made me too a bitter enemy. What a shameless fellow! He has not paid yet, though he ought to have done so on the first of January, especially as he has freed himself from enormous debts by the handwriting of Faberius and has sought help from the goddess of help.[1] For I must have my joke, that you may not think I am seriously concerned. I had written too to him very early on the 8th, and on the same day in the evening I got a letter from you at Pompeii, delivered very quickly on the third day after you wrote it. But, as I told you then, my letter to Dolabella was sufficiently stinging. If it takes no effect, I don't think he will be able to face me.

I think you have settled Albius' business. With

Pompeii, May 9, B.C. 44

Cicero against Antony, that he used for his own purposes the large sum left by Caesar in the public treasury in the temple of Ops.

nomine, quod mihi suppetiatus es,[1] gratissimum est et simile tuorum omnium. Sed ego Erotem ad ista expedienda factum mihi videbar reliquisse; cuius non sine magna culpa vacillarunt. Sed cum ipso videro.

De Montano, ut saepe ad te scripsi, erit tibi tota res curae. Servius proficiscens quod desperanter tecum locutus est, minime miror neque ei quicquam in desperatione concedo. Brutus noster, singularis vir, si in senatum non est Kal. Iuniis venturus, quid facturus sit in foro, nescio. Sed hoc ipse melius. Ego ex iis, quae parari video, non multum Idibus Martiis profectum iudico. Itaque de Graecia cotidie magis et magis cogito. Nec enim Bruto meo, exsilium ut scribit ipse meditanti, video quid prodesse possim. Leonidae me litterae non satis delectarunt. De Herode tibi adsentior. Saufei legisse vellem. Ego ex Pompeiano vi Idus Mai. cogitabam.

XIX

CICERO ATTICO.

Scr. in Pompeiano VIII Id. Mai. a. 710

Nonis Maiis cum essem in Pompeiano, accepi binas a te litteras, alteras sexto die, alteras quarto. Ad superiores igitur prius. Quam mihi iucundum opportune tibi Barnaeum litteras reddidisse!

Tu vero cum Cassio ut cetera. Quam commode autem, quod id ipsum, quod me mones, quadriduo

[1] suppetiatus es *Montagnanus*: suspendiatus est *MSS.*

regard to Patulcius' debt, it was most kind of you and just like yourself to come to my aid. But I seem to have deserted Eros, who is just the man to clear the matter up: it was through a grave fault of his that it went wrong. But I will see to that with him.

Montanus' business, as I have often said, you must take charge of entirely. I am not surprised that Servius spoke to you in a tone of despair, when he was leaving; and my despair quite equals his. What our friend Brutus is going to do in the Forum, incomparable hero though he is, if he is not going to attend the Senate on the first of June, I do not know. But he should know this better himself. From the things I see in course of preparation I don't think the Ides of March are going to help much. So I am daily thinking more and more of Greece. For I fail to see what use I can be to Brutus, who, as he writes to me, is meditating exile. Leonidas' letter did not give me much pleasure. I agree about Herodes. I should like to have read Saufeius' note. I am thinking of leaving Pompeii on the tenth of May.

XIX

CICERO TO ATTICUS.

Here at Pompeii on the seventh of May I received *Pompeii,* two letters, one five days old, the other three. So *May 8,* B.C. I will answer the earlier first. How glad I am *44* Barnaeus delivered the letter so opportunely!

Take Cassius in hand like everything else. It is however very lucky that I had written to him

ante ad eum scripseram, exemplumque mearum litterarum ad te miseram! Sed, cum ex Dolabellae aritia (sic enim tu ad me scripseras) magna desperatione adfectus essem, ecce tibi et Bruti et tuae litterae! Ille exsilium meditari. Nos autem alium portum propiorem huic aetati videbamus; in quem mallem equidem pervehi florente Bruto nostro constitutaque re publica. Sed nunc quidem, ut scribis, non utrumvis. Adsentiris enim mihi nostram aetatem a castris, praesertim civilibus, abhorrere.

Antonius ad me tantum de Clodio rescripsit, meam lenitatem et clementiam et sibi esse gratam et mihi voluptati magnae fore. Sed Pansa furere videtur de Clodio itemque de Deiotaro, et loquitur severe, si velis credere. Illud tamen non belle, ut mihi quidem videtur, quod factum Dolabellae vehementer improbat. De coronatis, cum sororis tuae filius a patre accusatus esset, rescripsit se coronam habuisse honoris Caesaris causa, posuisse luctus gratia; postremo se libenter vituperationem subire, quod amaret etiam mortuum Caesarem.

Ad Dolabellam, quem ad modum tibi dicis placere, scripsi diligenter. Ego etiam ad Siccam; tibi hoc oneris non impono. Nolo te illum iratum habere. Servi orationem cognosco; in qua plus timoris video quam consilii. Sed, quoniam perterriti omnes sumus, adsentior Servio. Publilius tecum tricatus est. Huc enim Caerellia missa ab istis est legata ad me; cui

---

[1] Apparently a slip of the pen on the part of Atticus for *avaritia*, unless the text is corrupt.

LETTERS TO ATTICUS XIV. 19

four days ago, as you advise, and had sent a copy of my letter to you. But when I was in the depths of despair owing to Dolabella's arice[1] (for that is what you wrote), lo and behold your letter and Brutus'. Brutus is meditating exile. I however see another haven and a readier one for my age: but I would rather sail into it with Brutus in prosperity and the republic set in order. But now, as you say, I have not the choice. For you agree that age unfits me for a soldier's camp, especially in civil war.

Antony only answered about Clodius, that my leniency and clemency had been very gratifying to him and would be a source of pleasure to myself. But Pansa appears to be in a fury about Clodius and about Deiotarus too; and, if you are willing to believe him, he expresses himself very forcibly. But there is one thing that to my mind is shady, that he strongly disapproves of Dolabella's action. As for the people who wore garlands, when your nephew was reproved for it by his father, he answered that he wore a garland for Caesar's honour, and laid it aside for grief; and finally that he would willingly submit to reproaches for loving Caesar even after his death.

To Dolabella I have written carefully, as you advise: and I have written myself to Sicca too. I do not want to lay this burden on you, for I don't want him to be angry with you. I recognise Servius' way of talking; and there seems to me to be more fright than wisdom in it. But, since we are all frightened, I agree with Servius. Publilius has been hoaxing. For Caerellia was sent here by them as their ambassadress to me;[2] but

---

[2] To persuade Cicero to remarry his divorced wife Publilia.

facile persuasi mihi id, quod rogaret, ne licere quidem, non modo non lubere. Antonium si videro, accurate agam de Buthroto.

Venio ad recentiores litteras; quamquam de Servio iam rescripsi. "Me facere magnam πρᾶξιν Dolabellae." Mihi mehercule ita videtur, non potuisse maior tali re talique tempore. Sed tamen, quicquid ei tribuo, tribuo ex tuis litteris. Tibi vero adsentior maiorem πρᾶξιν eius fore, si mihi, quod debuit, dissolverit. Brutus velim sit Asturae. Quod autem laudas me, quod nihil ante de profectione constituam, quam, ista quo evasura sint, videro, muto sententiam. Neque quicquam tamen ante, quam te videro. Atticam meam gratias mihi agere de matre gaudeo; cui quidem ego totam villam cellamque tradidi eamque cogitabam v Idus videre. Tu Atticae salutem dices. Nos Piliam diligenter tuebimur.

XX

CICERO ATTICO.

Scr. in Puteolano V Id. Mai. a. 710

E Pompeiano navi advectus sum in Luculli nostri hospitium vi Idus hora fere tertia. Egressus autem e navi accepi tuas litteras, quas tuus tabellarius in Cumanum attulisse dicebatur Nonis Maiis datas. A Lucullo postridie eadem fere hora veni in Puteolanum. Ibi accepi duas epistulas, alteram Nonis,

---

[1] There is a play on the double sense of πρᾶξις, (1) exploit, (2) exaction of money. The money in question was an

278

LETTERS TO ATTICUS XIV. 19-20

I persuaded her easily that what she asked was not even lawful, besides being repugnant to me. If I see Antony, I will do my best for Buthrotum.

I come to your more recent letter, though I have answered already about Servius. You say I make much of Dolabella's score. Well, I don't see that he could have made a greater one considering the times and circumstances. However, all the credit I give him I give in accordance with your letter. But I agree with you that it would be still better, if he would pay off my score.[1] I hope Brutus will be at Astura. You praise me for not making up my mind about leaving the country before I see how things are going to turn out. I have changed my mind: however I won't do anything until I see you. I am gratified at Attica's thanking me for her mother. I have put the whole house and the store-rooms at her service and I am thinking of seeing her on the 11th. Give Attica my love. I will take great care of Pilia.

XX

CICERO TO ATTICUS.

Puteoli, May 11, B.C. 44

From Pompeii I came by sea to enjoy our friend Lucullus' hospitality on the 10th about nine o'clock in the morning. As I disembarked, I received your letter, which your messenger says was taken to Cumae, dated the 7th. I left Lucullus and reached Puteoli about the same hour the next day. There I received two letters, dated one the 7th the other

---

instalment of Tullia's dowry, which Dolabella had to repay.

alteram VII Idus Lanuvio datas. Audi igitur ad omnes. Primum, quae de re mea gesta et in solutione et in Albiano negotio, grata. De tuo autem Buthroto, cum in Pompeiano essem, Misenum venit Antonius. Inde ante discessit, quam illum venisse audissem in Samnium. A quo[1] vide quid speres. Romae igitur de Buthroto. L. Antoni horribilis contio, Dolabellae praeclara. Iam vel sibi habeat nummos, modo numeret Idibus. Tertullae nollem abortum. Tam enim Cassii sunt iam quam Bruti serendi. De regina velim atque etiam de Caesare filio. Persolvi primae epistulae, venio ad secundam.

De Quintis, Buthroto, cum venero, ut scribis. Quod Ciceroni suppeditas, gratum. Quod errare me putas, qui rem publicam putem pendere e Bruto, sic se res habet. Aut nulla erit aut ab isto istisve servabitur. Quod me hortaris, ut scriptam contionem mittam, accipe a me, mi Attice, καθολικὸν θεώρημα earum rerum, in quibus satis exercitati sumus. Nemo umquam neque poëta neque orator fuit, qui quemquam meliorem quam se arbitraretur. Hoc etiam malis contingit, quid tu Bruto putas et ingenioso et erudito? De quo etiam experti sumus nuper in edicto. Scripseram rogatu tuo. Meum mihi placebat, illi suum. Quin etiam, cum ipsius precibus paene adduc-

[1] a quo in Samnium *MSS*. *The words were transposed by Wesenberg.*

---

[1] An affectionate diminutive of the name of Tertia, half-sister of Brutus, and wife of Cassius.

the 9th, from Lanuvium. So listen to my answer to them all. First, my thanks for what you have done in my affairs both in payment and in the business with Albius. Then with regard to your Buthrotum, when I was at Pompeii, Antony came to Misenum: but he was gone again to Samnium, before I heard he had come. See that you do not build much hope on him. So I must see to Buthrotum at Rome. L. Antonius' speech is horrible, Dolabella's splendid. Let him keep his money now, provided he pays on the Ides. I am sorry about Tertulla's[1] miscarriage. For we want a crop of Cassii as much as one of Bruti. I hope it is true about Cleopatra and about Caesar's son[2] too. I have finished your first letter, now I come to your second.

The Quinti and Buthrotum we will leave till I come to Rome, as you say. Thanks for supplying my son's needs. You think I am wrong in thinking the republic hangs on Brutus: but it is a fact. There will be none, or he and his party will save it. You exhort me to send a written speech. You may take it from me, my dear Atticus, as a general axiom with regard to those matters, in which I have sufficient experience, that no one, whether poet or orator, ever thought anyone better than himself. This is so even in the case of bad ones: and what do you think it will be in the case of Brutus, who has talent and learning? Besides I have had experience of him lately over the edict. I had written one at your request. I liked mine, he liked his. Nay more, when I was induced by his entreaties to dedicate to him my book

[2] A child of Cleopatra, called Caesarion. Caesar denied the parentage.

tus scripsissem ad eum "de optimo genere dicendi," non modo mihi, sed etiam tibi scripsit sibi illud, quod mihi placeret, non probari. Quare sine, quaeso, sibi quemque scribere.

"Suam quoíque sponsam, míhi meam; suum
quoíque amorem, míhi meum."

Non scite. Hoc enim Atilius, poëta durissimus. Atque utinam liceat isti contionari! Cui si esse in urbe tuto licebit, vicimus. Ducem enim novi belli civilis aut nemo sequetur, aut ii sequentur, qui facile vincantur.

Venio ad tertiam. Gratas fuisse meas litteras Bruto et Cassio gaudeo. Itaque iis rescripsi. Quod Hirtium per me meliorem fieri volunt, do equidem operam, et ille optime loquitur, sed vivit habitatque cum Balbo, qui item bene loquitur. Quid credas, videris. Dolabellam valde placere tibi video; mihi quidem egregie. Cum Pansa vixi in Pompeiano. Is plane mihi probabat se bene sentire et cupere pacem. Causam armorum quaeri plane video. Edictum Bruti et Cassi probo. Quod vis, ut suscipiam cogitationem, quidnam istis agendum putem, consilia temporum sunt; quae in horas commutari vides. Dolabellae et prima illa actio et haec contra Antonium contio mihi profecisse permultum videtur. Prorsus ibat res; nunc autem videmur habituri ducem; quod unum municipia bonique desiderant. Epicuri mentionem facis et audes dicere μὴ πολιτεύεσθαι. Non te Bruti

"on the best oratorical style," he wrote not only to me but to you also, that what I found pleasing, he did not approve. So, pray, let every man do his writing for himself. "To each his own wife; mine for me. To each his own love; mine for me." It is not neatly put, for it is by Atilius, the most wooden of poets. I only hope Brutus may be able to deliver a speech. If he can enter the city in safety, we have won. For, as the leader in a new civil war, no one will follow him, or at least only those who can easily be conquered.

I come to your third letter. I am glad Brutus and Cassius were pleased with my letter: so I have sent an answer. They want me to turn Hirtius into a better citizen. I am doing my best, and his promises are fair enough, but he spends all his days and nights with Balbus, whose promises are fair enough too. How much of them you can believe, you must see for yourself. I observe you are very well satisfied with Dolabella, and I am more than satisfied. I saw a lot of Pansa at Pompeii: and he quite convinced me that he was well inclined and desirous of peace. I can see quite clearly that a pretext for war is being sought. The edict of Brutus and Cassius I approve. You want me to consider what I think they ought to do. One has to form one's plans according to circumstances; and, as you see, they are changing every hour. I think Dolabella's first move and this speech of his against Antony have both done a lot of good. Things are certainly advancing: and now we seem to be going to have a leader, which is the very thing the country towns and the well affected want. You mention Epicurus and dare to say "take no part in politics." Are you not

nostri vulticulus ab ista oratione deterret? Quintus filius, ut scribis, Antoni est dextella. Per eum igitur, quod volemus, facile auferemus. Exspecto, si, ut putas, L. Antonius produxit Octavium, qualis contio fuerit.

Haec scripsi raptim. Statim enim Cassi tabellarius. Eram continuo Piliam salutaturus, deinde ad epulas Vestori navicula. Atticae plurimam salutem.

XXI

CICERO ATTICO.

Scr. in Puteolano V Id. Mai. a. 710

Cum paulo ante dedissem ad te Cassi tabellario litteras, v Idus venit noster tabellarius, et quidem, portenti simile, sine tuis litteris. Sed cito conieci Lanuvi te fuisse. Eros autem festinavit, ut ad me litterae Dolabellae perferrentur non de re mea (nondum enim meas acceperat), sed rescripsit ad eas, quarum exemplum tibi miseram, sane luculente. Ad me autem, cum Cassi tabellarium dimisissem, statim Balbus. O dei boni, quam facile perspiceres timere otium! Et nosti virum, quam tectus. Sed tamen Antoni consilia narrabat; illum circumire veteranos, ut acta Caesaris sancirent idque se facturos esse iurarent, ut castra[1] omnes haberent, eaque duumviri omnibus mensibus inspicerent. Questus est etiam de sua invidia, eaque omnis eius oratio fuit, ut amare videretur Antonium. Quid quaeris? nihil sinceri.

[1] ut castra *Otto*: utram *M¹*.

frightened out of such talk by our friend Brutus' frown? Young Quintus, you tell me, is Antony's right hand man. So we shall easily get what we want through him. I am wondering what sort of speech Octavius made, if, as you thought, L. Antonius introduced him to a public meeting.

I am writing in haste: for Cassius' letter carrier is starting at once. I am going directly to call on Pilia, and then on to dinner with Vestorius by boat. My best love to Attica.

XXI

CICERO TO ATTICUS.

Just after I had given Cassius' messenger a letter for you on the 11th, came my messenger, and, to my extraordinary surprise, without a letter from you. But I quickly conjectured you had been at Lanuvium. Eros however had hastened to let me have a letter from Dolabella, not about my debt (for he had not got my letter yet), but an answer, and a very good answer too, to the one of which I sent you a copy. No sooner had I got rid of Cassius' messenger than Balbus came to see me. Good God, how easy it is to see that he is afraid of peace! And you know how secretive the man is. Yet he told me Antony's plans. He is canvassing all the veterans, asking them to support Caesar's acts and to take an oath to that effect, to keep in camps, which are to be inspected every month by two officials. He grumbled too about the prejudice against himself, and everything he said seemed to show affection for Antony. In fact there is no reliability in him. To me

Puteoli, May 11, B.C. 44

MARCUS TULLIUS CICERO

Mihi autem non est dubium, quin res spectet ad castra. Acta enim illa res est animo virili, consilio puerili. Quis enim hoc non vidit, regni heredem relictum? Quid autem absurdius?

"Hoc métuere, alterum ín metu non pónere!"

Quin etiam hoc ipso tempore multa ὑποσόλοικα. Ponti Neapolitanum a matre tyrannoctoni possideri! Legendus mihi saepius est "Cato maior" ad te missus. Amariorem enim me senectus facit. Stomachor omnia. Sed mihi quidem βεβίωται; viderint iuvenes. Tu mea curabis, ut curas.

Haec scripsi seu dictavi apposita secunda mensa apud Vestorium. Postridie apud Hirtium cogitabam et quidem πεντέλοιπον. Sic hominem traducere ad optumates paro. Λῆρος πολύς. Nemo est istorum, qui otium non timeat. Quare talaria videamus. Quidvis enim potius quam castra.

Atticae salutem plurimam velim dicas. Exspecto Octavi contionem et si quid aliud, maxime autem, ecquid Dolabella tinniat an in meo nomine tabulas novas fecerit.

---

[1] Servilia, mother of Brutus.
[2] If this reading is correct, which is very doubtful, it probably refers to Hirtius, Pansa, Octavius, Lentulus

there seems no doubt that things are tending towards war. For the deed was done with the courage of men, but with the blind policy of a child. For who did not see that the tyrant left an heir? And what could be more absurd than "to fear the one, and reck not of his friend"? Nay even now there are many absurdities. Think of the mother of the tyrannicide[1] occupying Pontius' house at Naples! I must keep on reading my *Cato Major* which is dedicated to you: for old age is beginning to make me ill-tempered. Everything puts me in a rage. However, my life is over: the young people must look out for themselves. Take care of my affairs for me, as you are doing.

This I have written or rather dictated when at dessert with Vestorius. To-morrow I am thinking of paying a visit to Hirtius, the only survivor of the band of five.[2] That is my way of trying to make him join the conservative party. It is all nonsense: there is none of Caesar's party who does not fear peace. So let us look for our seven-league boots. Anything is better than a camp.

Please pay my best respects to Attica. I am looking for Octavius' speech and any other news there may be, but especially whether we shall hear the ring of Dolabella's money or whether he repudiated his debts in my case.[3]

Spinther and Philippus, who had been together at Puteoli.
[3] Referring to Dolabella's action as a tribune.

MARCUS TULLIUS CICERO

XXII

CICERO ATTICO.

Scr. in Puteolano prid. Id. Mai. a. 710

Certior a Pilia factus mitti ad te Idibus tabellarios statim hoc nescio quid exaravi. Primum igitur scire te volui me hinc Arpinum xvi Kalend. Iun. Eo igitur mittes, si quid erit posthac; quamquam ipse iam iamque adero. Cupio enim, antequam Romam venio, odorari diligentius, quid futurum sit. Quamquam vereor, ne nihil coniectura aberrem. Minime enim obscurum est, quid isti moliantur; meus vero discipulus, qui hodie apud me cenat, valde amat illum, quem Brutus noster sauciavit. Et, si quaeris (perspexi enim plane), timent otium; ὑπόθεσιν autem hanc habent eamque prae se ferunt, clarissimum virum interfectum, totam rem publicam illius interitu perturbatam, inrita fore, quae ille egisset, simul ac desisteremus timere; clementiam illi malo fuisse; qua si usus non esset, nihil ei tale accidere potuisse. Mihi autem venit in mentem, si Pompeius cum exercitu firmo veniat, quod est εὔλογον, certe fore bellum. Haec me species cogitatioque perturbat. Neque enim iam, quod tibi tum licuit, nobis nunc licebit. Nam aperte laetati sumus. Deinde habent in ore nos ingratos. Nullo modo licebit, quod tum et tibi licuit et multis. Φαινοπροσωπητέον ergo et ἰτέον in castra?

XXII

CICERO TO ATTICUS.

As soon as I learned from Pilia that she was *Puteoli,* sending a messenger to you on the 15th, I scrawled *May 14,* B.C. this bit of a note. First then I want you to 44 know that I am leaving here for Arpinum on May 17th. So, if you have anything to send after that, you must send it there: though I shall be in Rome almost directly. For I want to scent out as clearly as possible what is going to happen before I come to town. However, I fear my suspicions are not far from the truth. For it is clear enough what they are doing. My pupil,[1] who dined with me to-day, is a warm admirer of the man who was wounded by our Brutus: and, if you want to know, I see quite clearly that they are afraid of peace. This is the theme on which they are always dwelling: that a most distinguished person has been killed, that by his death the whole state has been thrown into disorder; that his acts will be null and void as soon as we have ceased to fear; that his clemency was his destruction, and that, if he had not practised clemency, such a thing could not have happened to him. I cannot help thinking, then, that if Pompey comes with a strong force, which is quite possible, there will certainly be war. When I picture this and think of it, I am disturbed: for now we shall not have the choice you had before. For we have shown our joy openly. Again they speak of us as ingrates. What you and many others did then certainly will not be possible now. Must I put in an appearance,

[1] Hirtius.

MARCUS TULLIUS CICERO

Miliens mori melius, huic praesertim aetati. Itaque me Idus Martiae non tam consolantur quam antea. Magnum enim mendum continent. Etsi illi iuvenes

ἄλλοις ἐν ἐσθλοῖς τόνδ' ἀπωθοῦνται ψόγον.

Sed, si tu melius quidpiam speras, quod et plura audis et interes consiliis, scribas ad me velim simulque cogites, quid agendum nobis sit super legatione votiva. Equidem in his locis moneor a multis, ne in senatu Kalendis. Dicuntur enim occulte milites ad eam diem comparari et quidem in istos, qui mihi videntur ubivis tutius quam in senatu fore.

then, and join the army? A thousand times better to die, especially at my time of life. So now I am not so much consoled as I was with the thought of the Ides of March, for there was a grave mistake committed then. However, those youths "in other noble deeds wipe out their shame."[1] But, if you have any better hope, as you hear more news and are in the midst of affairs, please write, and at the same time consider what I ought to do about the votive legation. Here many people warn me against attending the Senate on the 1st. They say troops are being collected secretly for that occasion, and that too against your friends, who to my idea will be safer anywhere than in the Senate.

[1] Attributed to Sophocles.

M. TULLI CICERONIS
EPISTULARUM AD ATTICUM
LIBER QUINTUS DECIMUS

I

CICERO ATTICO SAL.

Scr. in Puteolano XVI Kal. Iun. a. 710

O factum male de Alexione! Incredibile est, quanta me molestia adfecerit, nec mehercule ex ea parte maxime, quod plerique mecum: "Ad quem igitur te medicum conferes?" Quid mihi iam medico? Aut, si opus est, tanta inopia est? Amorem erga me, humanitatem suavitatemque desidero. Etiam illud. Quid est, quod non pertimescendum sit, cum hominem temperantem, summum medicum tantus inproviso morbus oppresserit? Sed ad haec omnia una consolatio est, quod ea condicione nati sumus, ut nihil, quod homini accidere possit, recusare debeamus.

De Antonio iam antea tibi scripsi non esse eum a me conventum. Venit enim Misenum, cum ego essem in Pompeiano. Inde ante profectus est, quam ego eum venisse cognovi. Sed casu, cum legerem tuas litteras, Hirtius erat apud me in Puteolano. Ei legi et egi. Primum quod attinet, nihil mihi concedebat, deinde ad summam arbitrum me statuebat non modo huius rei, sed totius consulatus sui. Cum Antonio autem sic agemus, ut perspiciat, si in eo negotio

CICERO'S LETTERS TO ATTICUS
BOOK XV

I

CICERO TO ATTICUS, GREETING.

What a misfortune about Alexio! It has upset me *Puteoli,* more than you can believe, and not, I assure you, *May 17,* B.C. particularly on the score which most people seem to *44* think it has, asking to what doctor I shall turn now. What do I want with a doctor now? And, if I do want one, is there such a dearth of them? It is his love for me, his kindness and charming manner that I miss. There is another thing, too. What have we not to fear, when so temperate a person and so skilful a physician can be overcome suddenly by such a disease? But for all these things there is one consolation: we are born under this condition, that we may not refuse anything that fate has in store for mortals.

As for Antony, I have told you before that I have not met him. For he came to Misenum when I was at Pompeii, and he left before I knew he was there. But by chance, when I was reading your letter, Hirtius was with me at Puteoli. I read it to him and pleaded with him. At first he would not make any concession worth counting, but in the end he said I should direct not only this matter but all his consulship. With Antony I shall put the matter so that he may see that, if he obliges me in this par-

nobis satis fecerit, totum me futurum suum. Dolabellam spero domi esse.

Redeamus ad nostros. De quibus tu bonam spem te significas habere propter edictorum humanitatem. Ego autem perspexi, cum a me XVII Kal. de Puteolano Neapolim Pansae conveniendi causa proficisceretur Hirtius, omnem eius sensum. Seduxi enim et ad pacem sum cohortatus. Non poterat scilicet negare se velle pacem, sed non minus se nostrorum arma timere quam Antoni, et tamen utrosque non sine causa praesidium habere, se autem utraque arma metuere. Quid quaeris? οὐδὲν ὑγιές.

De Quinto filio tibi adsentior. Patri quidem certe gratissimae bellae tuae litterae fuerunt. Caerelliae vero facile satis feci; nec valde laborare mihi visa est, et, si illa, ego certe non laborarem. Istam vero, quam tibi molestam scribis esse, auditam a te esse omnino demiror. Nam, quod eam conlaudavi apud amicos audientibus tribus filiis eius et filia tua, τί ἐκ τούτου;[1]

"Quid est aútem, cur ego pérsonatus ámbulem?"

Parumne foeda persona est ipsius senectutis?

Quod Brutus rogat, ut ante Kalendas, ad me quoque scripsit, et fortasse faciam. Sed plane, quid velit, nescio. Quid enim illi adferre consilii possum, cum ipse egeam consilio, et cum ille suae inmortalitati melius quam nostro otio consuluerit? De regina rumor exstinguitur. De Flamma, obsecro te, si quid potes.

[1] τὸ ἐκ τούτου quid est hoc *MSS*. *The Latin words were excluded by Lambinus,* τί *suggested by Kayser.*

ticular matter, I shall be entirely his for the future. I hope Dolabella is at home.

Let us return to our heroes. You hint that you have good hopes for them in the moderate tone of the edicts. But, when Hirtius left me at Puteoli on the 16th of May to meet Pansa at Naples, his whole mind was revealed to me. For I took him aside and exhorted him to keep the peace. He could not, of course, say that he did not want peace, but he did say that he was as much afraid of armed action on our side as from Antony, and that after all both had reason for being on their guard, and for his part he was afraid of hostilities from both. In fact he is quite unreliable.

About young Quintus I agree with you. His father, at any rate, was most pleased with your nice letter. Caerellia I easily satisfied; she did not seem to me to bother herself much, and, if she had, I certainly should not have done so. As to the lady who you say is plaguing you, I wonder you listened to her at all. For, if I did compliment her before friends, when three of her own sons and your daughter were present, what is there in that?

"Why should I wear a mask before men's eyes?"

Is not old age itself a mask ugly enough?

You say Brutus asks me to come before the 1st. He has written to me too, and perhaps I shall do so. But I really don't know what he wants. What advice can I give him, when I want advice myself, and when he has thought of his immortality rather than our peace of mind? The rumour about Cleopatra is dying out. As to Flamma, pray do what you can.

MARCUS TULLIUS CICERO

Ia

CICERO ATTICO SAL.

Scr. in Sinuessano XV Kal. Iun. a. 710

Here dederam ad te litteras exiens e Puteolano deverteramque in Cumanum. Ibi bene valentem videram Piliam. Quin etiam paulo post Cumis eam vidi. Venerat enim in funus; cui funeri ego quoque operam dedi. Cn. Lucullus, familiaris noster, matrem efferebat. Mansi igitur eo die in Sinuessano atque inde mane postridie Arpinum proficiscens hanc epistulam exaravi. Erat autem nihil novi, quod aut scriberem aut ex te quaererem, nisi forte hoc ad rem putas pertinere. Brutus noster misit ad me orationem suam habitam in contione Capitolina, petivitque a me, ut eam ne ambitiose corrigerem, antequam ederet. Est autem oratio scripta elegantissime sententiis, verbis, ut nihil possit ultra. Ego tamen, si illam causam habuissem, scripsissem ardentius. Ὑπόθεσις vides quae sit et persona dicentis. Itaque eam corrigere non potui. Quo enim in genere Brutus noster esse vult et quod iudicium habet de optimo genere dicendi, id ita consecutus in ea oratione est, ut elegantius esse nihil possit; sed ego secutus aliud sum, sive hoc recte sive non recte. Tu tamen velim eam orationem legas, nisi forte iam legisti, certioremque me facias, quid iudices ipse. Quamquam vereor, ne cognomine tuo lapsus ὑπεραττικὸς sis in iudicando. Sed, si recordabere Δημοσθένους fulmina, tum intel-

Ia

CICERO TO ATTICUS, GREETING.

Yesterday I sent off a letter to you as I was leaving *Sinuessa,* Puteoli and stopped at my house at Cumae. There *May 18,* B.C. I found Pilia enjoying the best of health. Indeed, *44* I saw her again shortly afterwards at Cumae. For she had come for a funeral, which I also was attending. Our friend Cn. Lucullus was burying his mother. So I stayed that day at Sinuessa, and there I have scribbled this as I am starting early in the morning of the next day for Arpinum. However, I have no news either to write to you or to ask from you, unless you think this is to the point. Brutus has sent me the speech he delivered in the meeting on the Capitol, and has asked me to correct it without regarding his feelings, before he publishes it. Now the speech is most elegantly expressed as regards its sentiments, and its language could not be surpassed. But myself, if I had pleaded that cause, I should have written with more fire. You realize what the theme is and what the speaker is. So I could not alter it. For considering the style our friend Brutus affects and the opinion he holds of the best style of oratory, he has attained it in its highest elegance in this speech. But rightly or wrongly I have aimed at something different. However, I should like you to read the speech, if you have not done so already, and to let me know your opinion, though I am afraid that your name will lead you astray and you will be hyper-Attic in your criticism. However, if you will recall Demosthenes' thunder-bursts, you will be able to realize that one can use considerable force even in

leges posse vel ἀττικώτατα gravissime dici. Sed haec coram. Nunc nec sine epistula nec cum inani epistula volui ad te Metrodorum venire.

II

CICERO ATTICO SAL.

Scr. in Vesciano XV Kal. Iun. a. 710

xv Kal. e Sinuessano proficiscens cum dedissem ad te litteras devertissemque acutius,[1] in Vesciano accepi a tabellario tuas litteras; in quibus nimis multa de Buthroto. Non enim tibi ea res maiori curae aut est aut erit quam mihi. Sic enim decet te mea curare, tua me. Quam ob rem id quidem sic susceptum est mihi, ut nihil sim habiturus antiquius.

L. Antonium contionatum esse cognovi tuis litteris et aliis sordide; sed, id quale fuerit, nescio; nihil enim scripsisti. De Menedemo probe. Quintus certe ea dictitat, quae scribis. Consilium meum a te probari, quod ea non scribam, quae tu a me postularis, facile patior, multoque magis id probabis, si orationem eam, de qua hodie ad te scripsi, legeris. Quae de legionibus scribis, ea vera sunt. Sed non satis hoc mihi videris tibi persuasisse, qui de Buthrotiis nostris per senatum speres confici posse. Quod puto (tantum enim video) non videmur esse victuri, sed, ut iam nos hoc fallat, de Buthroto te non fallet. De Octavi contione idem sentio quod tu, ludorumque

---

[1] acutius *is probably a corruption of* ad *and a proper name.*

the purest Attic style. But of this when we meet.
At the present time all I wanted was that Metrodorus
should not come to you without a letter or with a
letter that had nothing in it.

II

CICERO TO ATTICUS, GREETING.

On the 18th I sent a letter to you as I was starting *Vescia,*
from Sinuessa, and stopped at . . . Then at Vescia *May 18,* B.C.
your messenger delivered your letter, which contained *44*
more than enough about Buthrotum. For you cannot
and will not have that business at heart more than I
have and shall have : that is the way that I ought to
care for your business, and you for mine. Accordingly, as I have undertaken it, I shall give it the
preference to everything else.

I hear from your letters and others that L. Antonius' speech was a poor thing; but what it was like
I do not know, as you have not told me. I am
glad to hear about Menedemus. Quintus certainly
keeps on reiterating what you mention. I am relieved to hear that you approve of my determination
not to write the sort of thing you asked me to write,
and you will approve of it much more, if you read
the speech about which I am writing to you to-day.
What you say about the legions is true. But you do
not seem to me to have taken the point sufficiently
to heart, if you hope we can settle the matter of
Buthrotum through the Senate. In my opinion (for
so much I can see) we have no chance of winning;
but supposing I am mistaken about that, you will
not be disappointed about Buthrotum. About Octavius' speech I think the same as you, and I don't like

eius apparatus et Matius ac Postumus mihi procuratores non placent; Saserna collega dignus. Sed isti omnes, quem ad modum sentis, non minus otium timent quam nos arma. Balbum levari invidia per nos velim, sed ne ipse quidem id fieri posse confidit. Itaque alia cogitat.

Quod prima disputatio Tusculana te confirmat, sane gaudeo; neque enim ullum est perfugium aut melius aut paratius. Flamma quod bene loquitur, non moleste fero. Tyndaritanorum causa, de qua causa laborat, quae sit, ignoro. Hos tamen . . .[1] Πεντέλοιπον movere ista videntur, in primis erogatio pecuniae. De Alexione doleo, sed, quoniam inciderat in tam gravem morbum, bene actum cum illo arbitror. Quos tamen secundos heredes, scire velim et diem testamenti.

III

CICERO ATTICO SAL.

Scr. in Arpinati XI Kal. Iun. a. 710

Undecimo Kal. accepi in Arpinati duas epistulas tuas, quibus duabus meis respondisti. Una erat xv Kal., altera xii data. Ad superiorem igitur prius. Accurres in Tusculanum, ut scribis; quo me vi Kal. venturum arbitrabar. Quod scribis parendum victoribus, non mihi quidem, cui sunt multa potiora

[1] hos tamen *MSS., which may be an aposiopesis, or some such word as* defendam *may be omitted*: noscum tamen *Reid*.

his preparations for the games or Matius and Postumus as his agents. Saserna is a colleague worthy of them. But all that party, as you realize, fear peace no less than we fear war. I should be glad if we could relieve Balbus of his unpopularity; but even he has no hope of that happening, so he is thinking of other things.

I am very glad if the first *Tusculan Disputation* gives you courage, for there is no other refuge either better or more available.[1] I am relieved that Flamma gives a good account of himself. What the case of the people of Tyndaris is, about which he is concerned, I do not know, but I am on their side. The "last of the five" seems to be upset by the things you wot of, especially the withdrawal of the money. I am grieved about Alexio, but, as he had contracted such a serious disease, I think he was fortunate. Whom he has appointed residuary heirs I should like to know, and the latest day for acceptance of the inheritance under his will.

III

CICERO TO ATTICUS, GREETING.

On the 22nd I received two letters from you at Arpinum, in which you answered two of mine. One was dated the 18th, the other the 21st. So I will answer the earlier first. Pray hasten to Tusculum, as you say: I think I shall get there on the 27th. You say we must obey the victors. I, for one, will not: there are many courses I should prefer to that. For

---

[1] *i.e.* than death, which is the subject of the book mentioned.

MARCUS TULLIUS CICERO

Nam illa quae recordaris Lentulo et Marcello consulibus acta in aede Apollinis, nec causa eadem est nec simile tempus, praesertim cum Marcellum scribas aliosque discedere. Erit igitur nobis coram odorandum, et constituendum, tutone Romae esse possimus. Novi conventus habitatores sane movent; in magnis enim versamur angustiis. Sed sunt ista parvi; quin vel maiora contemnimus. Calvae testamentum cognovi, hominis turpis ac sordidi. Tabula Demonici quod tibi curae est, gratum. De malo[1] scripsi iam pridem ad Dolabellam accuratissime, modo redditae litterae sint. Eius causa et cupio et debeo.

Venio ad propiorem. Cognovi de Alexione, quae desiderabam. Hirtius est tuus. Antonio, quoniam[2] est, volo peius esse. De Quinto filio, ut scribis, ἅλις.[3] De patre coram agemus. Brutum omni re, qua possum, cupio iuvare. Cuius de oratiuncula idem te quod me sentire video. Sed parum intellego, quid me velis scribere quasi a Bruto habita oratione, cum ille ediderit. Qui tandem convenit? an sic ut in tyrannum iure optimo caesum? Multa dicentur, multa scribentur a nobis, sed alio modo et tempore. De sella Caesaris bene tribuni; praeclaros etiam xiv ordines! Brutum apud me fuisse gaudeo, modo et libenter fuerit et sat diu.

---

[1] *For* malo *many suggestions have been made*: *e.g.* Mario *by* Manutius *and* Manlio *by Shuckburgh, who compares Att.* XIII. 9.
[2] quam iam *Orelli*: quoniam male *Alanus*. *Tyrrell suggests that* male *can be supplied in thought from the* peius *that follows.* [3] ἅλις *Turnebus*: A.M.C. *MSS.*

the case is not the same, nor is the occasion the same, as in the proceedings which you recall to my memory as taking place in the temple of Apollo in the consulship of Lentulus and Marcellus:[1] especially as you say Marcellus and others are leaving Rome. So when we meet we must scent out the facts and settle whether we can be safe at Rome. The inhabitants of the new community[2] trouble me a good deal, for I am in considerable difficulties. But these are small matters: I am treating even more important things than this with contempt. I know Calva's will. How disgracefully mean! I am grateful to you for attending to Demonicus' sale. About... I wrote to Dolabella long ago very fully, if only my letter was delivered. In his interests I am keen and devoted.

I come to your more recent letter. I have learned all I want about Alexio. Hirtius is devoted to you. With Antonius I wish things were going even worse than they are. About young Quintus, as you say, *assez*. About his father we will speak when we meet. I want to assist Brutus in every way that is possible. I see you have the same opinion of his harangue as I have. But I don't quite understand why you want me to write a speech attributing it to Brutus, when he has published his own. How could that be proper? Should I write as though against a tyrant justly executed? I shall have much to say and much to write, but in another way and at another time. Well done the tribunes about Caesar's chair, and well done the famous fourteen rows of equites! I am glad Brutus stayed at my house, and I only hope he enjoyed himself and stayed a long time.

[1] 49 B.C., when the Senate summoned all good citizens to Rome.
[2] A *colonia* of veterans planted by Antony at Casilinum.

MARCUS TULLIUS CICERO

IV

CICERO ATTICO SAL.

Scr. in Arpinati IX K Iun. a. 710

ix K. H. x fere a Q. Fufio venit tabellarius. Nescio quid ab eo litterularum, uti me sibi restituerem. Sane insulse, ut solet, nisi forte, quae non ames, omnia videntur insulse fieri. Scripsi ita, ut te probaturum existimo. Mihi duas a te epistulas reddidit, unam xi, alteram x. Ad recentiorem prius et pleniorem. Laudo; si vero etiam Carfulenus, "ἄνω ποταμῶν." Antoni consilia narras turbulenta. Atque utinam potius per populum agat quam per senatum! quod quidem ita credo. Sed mihi totum eius consilium ad bellum spectare videtur, si quidem D. Bruto provincia eripitur. Quoquo modo ego de illius nervis existimo, non videtur fieri posse sine bello. Sed non cupio, quoniam cavetur Buthrotiis. Rides? At ego doleo non mea potius adsiduitate, diligentia, gratia perfici. Quod scribis te nescire, quid nostris faciendum sit, iam pridem me illa ἀπορία sollicitat. Itaque stulta iam Iduum Martiarum est consolatio. Animis enim usi sumus virilibus, consiliis, mihi crede, puerilibus. Excisa enim est arbor, non evulsa. Itaque quam fruticetur, vides. Redeamus igitur, quoniam

---

[1] Presumably of the action of the Martian legion, which was reported to have deserted Antony and joined Octavius. Carfulenus, mentioned in the next sentence, was an officer in that legion.

IV

CICERO TO ATTICUS, GREETING.

On the 24th, about four o'clock, came a messenger from Q. Fufius bringing some sort of a note from him, begging me to make it up with him. A very silly letter as usual, unless one thinks that everything one does not like is very silly. I sent an answer of which I think you would approve. The messenger delivered two of your letters, one of the 22nd, the other of the 23rd. I answer the later and fuller one first. I approve.[1] Why, if even Carfulenus deserts him, it will be the end of the world[2] for him. Antony's plans, as you describe them, are revolutionary. And I only hope he will try to get his way through the people and not through the Senate, which I think is probable. But to me his whole policy seems to point to war, since D. Brutus is being robbed of his province. Whatever I may think of Brutus' resources, I don't think that can happen without war. But I don't want war, since the Buthrotians are all right as it is. You may smile: but I am sorry it was not rather accomplished by my persistence, diligence, and influence. You say you don't know what our friends are to do: that difficulty has been bothering me for a long time. So now I see it was folly to be consoled by the Ides of March: for though our courage was that of men, believe me we had no more sense than children. We have only cut down the tree, not rooted it up. So you see how it is shooting out.

Arpinum, May 24, B.C. *44*

---

[2] A quotation from Euripides, *Medea*, 409:—

ἄνω ποταμῶν ἱερῶν χωροῦσι παγαί,
καὶ δίκα καὶ πάντα πάλιν στρέφεται,

which had apparently passed into a proverb.

saepe usurpas, ad Tusculanas disputationes. Saufeium de te celemus; ego numquam indicabo. Quod te a Bruto scribis, ut certior fieret, quo die in Tusculanum essem venturus, ut ad te ante scripsi, VI Kal., et quidem ibi te quam primum per videre velim. Puto enim nobis Lanuvium eundum et quidem non sine multo sermone. Sed μελήσει.

Redeo ad superiorem. Ex qua praetereo illa prima de Buthrotiis; quae mihi sunt inclusa medullis, sit modo, ut scribis, locus agendi. De oratione Bruti prorsus contendis, cum iterum tam multis verbis agis. Egone ut eam causam, quam is scripsit? ego scribam non rogatus ab eo? Nulla παρεγχείρησις fieri potest contumeliosior. "At," inquis, "Ἡρακλείδειον aliquod." Non recuso id quidem, sed et componendum argumentum est et scribendi exspectandum tempus maturius. Licet enim de me, ut libet, existimes (velim quidem quam optime), si haec ita manant, ut videntur (feres, quod dicam), me Idus Martiae non delectant. Ille enim numquam revertisset, nos timor confirmare eius acta non coëgisset, aut, ut in Saufei eam relinquamque Tusculanas disputationes, ad quas tu etiam Vestorium hortaris, ita gratiosi eramus apud illum, quem di mortuum perduint! ut nostrae aetati,

---

[1] Atticus and Saufeius both professed the Epicurean philosophy, which was attacked in the first book of the *Tusculan Disputations*. The "secret" is Atticus' lapse from Epicureanism in approving of the views expressed in that book.

[2] To meet Brutus.

LETTERS TO ATTICUS XV. 4

Let us return, then, to the *Tusculan Disputations*, since you often refer to them. Let us keep your secret from Saufeius:[1] I will never betray it. You send a message from Brutus, asking me to let him know when I shall reach Tusculum. On the 27th, as I told you before; and I should very much like to see you there as soon as possible. For I think we shall have to go to Lanuvium,[2] and that not without a lot of talk. However, I will see to it.

I return to your earlier letter, and I pass over the first part about the Buthrotians. For that is engraved on my heart of hearts, if only, as you say, there is an opening for action. You are very insistent about Brutus' speech, since you say so much about it again. Am I really to plead the same case as that he has written about? Am I to write without being asked by him? One could not put one's oar in more rudely. "But," you say, "write something in the style of Heracleides."[3] That I don't refuse, but I should have to settle on a line of argument, and I should have to wait for more time to write it. For think what you will of me—though of course I should like you to think as well as possible, and not be offended at what I say—if affairs drift on as they seem to be doing, I can take no pleasure in the Ides of March. Caesar would never have come back,[4] and fear would not have compelled us to ratify his acts; or, if I join Saufeius' school and desert the *Tusculan Disputations*, which you would press even on Vestorius, I was so high in his favour (heaven confound him, though he is dead!) that to a person of my age he was not a

[3] Heracleides of Pontus, a pupil of Plato, who wrote on political subjects.

[4] From the Parthian war, in all probability; though some take it to refer to Antony, as a reincarnation of Caesar.

quoniam interfecto domino liberi non sumus, non fuerit dominus ille fugiendus. Rubeo, mihi crede, sed iam scripseram; delere nolui.

De Menedemo vellem verum fuisset, de regina velim verum sit. Cetera coram, et maxime quid nostris faciendum sit, quid etiam nobis, si Antonius militibus obsessurus est senatum. Hanc epistulam si illius tabellario dedissem, veritus sum, ne solveret. Itaque misi dedita. Erat enim rescribendum tuis.

IVa

CICERO ATTICO SAL.

Scr. in Tusculano VI K. Iun. a. 710

Quam vellem Bruto studium tuum navare potuisses! Ego igitur ad eum litteras. Ad Dolabellam Tironem misi cum mandatis et litteris. Eum ad te vocabis et, si quid habebis, quod placeat, scribes. Ecce autem de traverso L. Caesar ut veniam ad se rogat in Nemus aut scribam, quo se venire velim; Bruto enim placere se a me conveniri. O rem odiosam et inexplicabilem! Puto me ergo iturum et inde Romam, nisi quid mutaro. Summatim adhuc ad te; nihildum enim a Balbo. Tuas igitur exspecto nec actorum solum, sed etiam futurorum.

LETTERS TO ATTICUS XV. 4–4a

master to run away from, since the death of a master has not set us free. I blush, believe me; but I have written it, and I won't erase it.

I wish it had been true about Menedemus, and I hope it may be true about Cleopatra. The rest when we meet, and especially what our friends must do, and what even we must do, if Antony is going to surround the House with soldiers. I was afraid he might open this letter, if I gave it to his messengers, so I have sent it with special care, for I had to answer yours.

IVa

CICERO TO ATTICUS, GREETING.

How I wish you could have rendered your service *Tusculum,* to Brutus! So I am writing to him. I have sent *May 27,* B.C. Tiro to Dolabella with a message and a letter. *44* Summon him to you, and, if you have any pleasant news, write. But here is a letter from L. Caesar all of a sudden, asking me to come to him at the Grove[1] or write where I should like to meet him: Brutus wants me to see him. What a nuisance and what a surprise! I suppose then I must go, and from there on to Rome, unless I change my mind. At present I am only sending you a short note, for I have not heard yet from Balbus. So I am looking for a letter from you to tell me not only what has happened but what is going to happen.

[1] The Nemus Dianae at Aricia.

MARCUS TULLIUS CICERO

V

CICERO ATTICO SAL.

Scr. in Tus-   A Bruto tabellarius rediit; attulit et ab eo et a
culano V K.   Cassio. Consilium meum magno opere exquirunt,
Iun. a. 710   Brutus quidem, utrum de duobus. O rem miseram!
plane non habeo, quid scribam. Itaque silentio puto
me usurum, nisi quid aliud tibi videtur; sin tibi quid
venit in mentem, scribe, quaeso. Cassius vero vehementer
orat ac petit, ut Hirtium quam optimum
faciam. Sanum putas? ὁ θησαυρὸς ἄνθρακες.[1] Epistulam
tibi misi.

Ut tu de provincia Bruti et Cassi per senatus consultum,
ita scribit et Balbus et Oppius. Hirtius
quidem se afuturum (etenim iam in Tusculano est)
mihique, ut absim, vehementer auctor est, et ille
quidem periculi causa, quod sibi etiam fuisse dicit,
ego autem, etiam ut nullum periculum sit, tantum
abest, ut Antoni suspicionem fugere nunc curem, ne
videar eius secundis rebus non delectari, ut mihi
causa ea sit, cur Romam venire nolim, ne illum
videam. Varro autem noster ad me epistulam misit
sibi a nescio quo missam (nomen enim delerat); in
qua scriptum erat veteranos eos, qui reiciantur (nam
partem esse dimissam), improbissime loqui, ut magno
periculo Romae sint futuri, qui ab eorum partibus
dissentire videantur. Quis porro noster itus, reditus,
vultus, incessus inter istos? Quodsi, ut scribis,

[1] ὁ θησαυρὸς ἄνθρακες *Vict.*: OTENATCΔNΘPΔKEC *M.*

V

CICERO TO ATTICUS, GREETING.

My messenger has returned from Brutus, bringing a letter from him and from Cassius too. They want my advice badly, and Brutus asks which of two courses he ought to pursue. Alas! I have not the remotest idea what to say. So I think I shall keep silent, unless you think I must not. If anything occurs to you, please write. Cassius, indeed, begs and beseeches me to make Hirtius as sound as possible. Do you think he is in his senses? It's fairy gold![1] I am sending his letter.

Tusculum, May 28, B.C. 44

Balbus and Oppius tell me the same as you about the province to be assigned by the Senate to Brutus and Cassius, and Hirtius says he will not attend—he is here at Tusculum—and he strongly advises me to keep away. He does so on the strength of the danger which he says there has been even for him; but, even if there be no danger, I am so far from caring to avoid giving Antony a suspicion that I do not rejoice in his prosperity, that the very reason why I would rather not go to Rome is to avoid seeing him. But our friend Varro has sent me a letter from somebody or other—I don't know who, as he has erased the name—telling him that the veterans whose claims have been put off (for some of them have been disbanded) are using most criminal language, saying that those who seem not to favour their claims will be in great danger at Rome. What, I should like to know, can our goings and comings, our looks and our demeanour, be among them? If again, as you say,

[1] Lit. "the treasure is ashes," a proverbial expression for disappointment; cf. Lucian, *Zeuxis*, 2: *Timon*, 41.

MARCUS TULLIUS CICERO

L. Antonius in D. Brutum, reliqui in nostros, ego quid faciam aut quo me pacto geram? Mihi vero deliberatum est, ut nunc quidem est, abesse ex ea urbe, in qua non modo florui cum summa, verum etiam servivi cum aliqua dignitate; nec tam statui ex Italia exire, de quo tecum deliberabo, quam istuc non venire.

VI

CICERO ATTICO SAL.

Scr. in Tus-culano VI K. Iun. vesperi a. 710

Cum ad me Brutus noster scripsisset et Cassius, ut Hirtium, qui adhuc bonus fuisset (sciebam neque eum confidebam fore) mea auctoritate meliorem facerem (Antonio est enim fortasse iratior, causae vero amicissimus), tamen ad eum scripsi eique dignitatem Bruti et Cassi commendavi. Ille quid mihi rescripsisset, scire te volui, si forte idem tu quod ego existimares, istos etiam nunc vereri, ne forte ipsi nostri plus animi habeant quam habent.

" HIRTIUS CICERONI SUO SAL.

"Rurene iam redierim, quaeris. An ego, cum omnes caleant, ignaviter aliquid faciam? Etiam ex urbe sum profectus, utilius enim statui abesse. Has tibi litteras exiens in Tusculanum scripsi. Noli autem me tam strenuum putare, ut ad Nonas recurram. Nihil enim iam video opus esse nostra cura, quoniam

L. Antonius is attacking D. Brutus, and the others attacking our friends, what am I to do and how am I to bear myself? As things are now I have made up my mind to keep away from a city in which I have not only been distinguished in the highest position, but have even maintained some position in servitude. I have not quite made up my mind to leave Italy, a question which I will discuss with you, so much as not to go to Rome.

VI

CICERO TO ATTICUS, GREETING.

Our friend Brutus and Cassius had written to me *Tusculum,* to use my authority to improve Hirtius' patriotism, *May 27,* since he had at present shown some (I knew he had, B.C. *44* but I doubted if he would continue, for, although he is a little annoyed with Antony, he is very much devoted to the cause); in spite of my doubts I wrote to him and commended to his care the maintenance of Brutus' and Cassius' position. What his answer was I want you to know, to see whether you think the same as I do, that the Caesarians are even now afraid our friends have more courage than they really have.

"HIRTIUS TO HIS FRIEND CICERO, GREETING.

"You ask if I have returned from the country. Can I play the laggard, when all the world is so excited? In fact I have just left the city, for I thought my absence would be more useful than my presence. This letter I have written as I set out for Tusculum. Don't think I shall do anything so energetic as to hurry back for the 5th. I see no need for my protecting anyone, since proper precautions

praesidia sunt in tot annos provisa. Brutus et Cassius utinam, quam facile a te de me impetrare possunt, ita per te exorentur, ne quod calidius ineant consilium! Cedentes enim haec ais scripsisse—quo aut quare? Retine, obsecro te, Cicero, illos, et noli sinere haec omnia perire, quae funditus medius fidius rapinis, incendiis, caedibus pervertuntur. Tantum, si quid timent, caveant, nihil praeterea moliantur. Non medius fidius acerrimis consiliis plus quam etiam inertissimis, dum modo diligentibus, consequentur. Haec enim, quae fluunt, per se diuturna non sunt; in contentione praesentes ad nocendum habent vires. Quid speres de illis, in Tusculanum ad me scribe."

Habes Hirti epistulam. Cui rescripsi nil illos calidius cogitare idque confirmavi. Hoc, qualecumque esset, te scire volui.

Obsignata iam Balbus ad me Serviliam redisse, confirmare non discessuros. Nunc exspecto a te litteras.

VII

CICERO ATTICO SAL.

Scr. in Tusculano V aut IV K. Iun. a. 710

Gratum, quod mihi epistulas; quae quidem me delectarunt, in primis Sexti nostri. Dices: "quia te laudat." Puto mehercule id quoque esse causae, sed tamen, etiam antequam ad eum locum veni, valde mihi placebat cum sensus eius de re publica tum genus scribendi. Servius vero pacificator cum

have been taken for so many years. I wish you could obtain a promise from Brutus and Cassius, not to enter upon any hot-headed scheme, as easily as you can from me. For you say they wrote what you mention when on the point of leaving the country. Whither and why? Stop them, I beg you, Cicero, and do not let everything go to rack and ruin. For upon my honour things are already being upset by rapine, fire, and slaughter. If they have any fear, let them take some precaution merely, and not make any fresh move. Upon my honour they will not accomplish any more by violent measures than they will by quiet, provided they are careful. The present unsettled state of affairs cannot last long in the nature of things; if there is a struggle and they are here, they have power to do much harm. What your hopes for them are, write and tell me at Tusculum."

There is Hirtius' letter. I answered, affirming that they had no hot-headed scheme. I wanted you to know this for what it is worth.

Just as I had sealed this Balbus writes to me that Servilia has returned, and avers that they will not leave Italy. Now I look for a letter from you.

VII

CICERO TO ATTICUS, GREETING.

Thanks for sending the letters. They have given me much pleasure, especially that of our friend Sextus. You will say, "Because he praises you." Upon my word I think that is part of the reason: but even before I got to that passage I was very much pleased both by his sentiments on politics and by his style. Servius the peacemaker with a nobody,

Tusculum, May 28 or 29, B.C. 44

MARCUS TULLIUS CICERO

librariolo suo videtur obisse legationem et omnes captiunculas pertimescere. Debuerat autem non "ex iure manum consertum," sed quae sequuntur; tuque scribes.

VIII

CICERO ATTICO SAL.

Scr. in Tusculano prid. K. Iun. a. 710

Post tuum discessum binas a Balbo (nihil novi) itemque ab Hirtio, qui se scribit vehementer offensum esse veteranis. Exspectat animus, quidnam agam de K. Misi igitur Tironem et cum Tirone plures, quibus singulis, ut quicque accidisset, dares litteras, atque etiam scripsi ad Antonium de legatione, ne, si ad Dolabellam solum scripsissem, iracundus homo commoveretur. Quod autem aditus ad eum difficilior esse dicitur, scripsi ad Eutrapelum, ut is ei meas litteras redderet. Legatione mihi opus esse. Honestior est votiva, sed licet uti utraque.

De te, quaeso, etiam atque etiam vide. Velim possis coram; si minus, litteris idem consequemur. Graeceius ad me scripsit C. Cassium sibi scripsisse homines comparari, qui armati in Tusculanum mitterentur. Id quidem mihi non videbatur; sed cavendum tamen tutelaeque plures videndae. Sed aliquid crastinus dies ad cogitandum nobis dabit.

[1] The quotation from Ennius continues: *sed magi ferro Rem repetunt.* What Servius Sulpicius was undertaking is

his secretary, seems to have undertaken an embassy and to be on his guard against all the quips and quiddities of the law. But he ought to realize that it is not a case of "joining hands in legal claim," but of what follows."[1] Please write.

VIII

CICERO TO ATTICUS, GREETING.

After you had left came two letters from Balbus, with no news in them, and one from Hirtius, who says he is very annoyed with the veterans. My mind is still anxious about what I shall do about the 1st. So I have sent Tiro and some men with him—please give them letters one by one, as things happen—and I have written to Antony about the legation, for fear that, if I had written only to Dolabella, his quick temper might be aroused. But, as it is said to be rather difficult to get an audience with him, I have written to Eutrapelus, so that he may deliver my letter. I must have an embassy: a votive embassy is more honourable, but I could use either.

Your own position, I beg you, review most carefully. I wish we could do so together; if not, we must accomplish it by letters. Graeceius has written to me that he has heard from Cassius that armed men are being got ready to be sent to my house at Tusculum. I don't think that is the case; but still I must take care to have more safeguards ready. But to-morrow may give us some food for reflection.

Tusculum, May 31, B.C. 44

---

uncertain; possibly to patch up peace between Antony and Caesar's murderers.

MARCUS TULLIUS CICERO

IX

CICERO ATTICO SAL.

Scr. in Tusculano IV Non. Iun. a. 710

IIII Non. vesperi a Balbo redditae mihi litterae fore Nonis senatum, ut Brutus in Asia, Cassius in Sicilia frumentum emendum et ad urbem mittendum curarent. O rem miseram! primum ullam ab istis, dein, si aliquam, hanc legatoriam provinciam! Atque haud scio an melius sit quam ad Eurotam sedere. Sed haec casus gubernabit. Ait autem eodem tempore decretum iri, ut et iis et reliquis praetoriis provinciae decernantur. Hoc certe melius quam illa Περσικὴ porticus; nolo enim Lacedaemonem longinquiorem quam Lanuvium existimare. "Rides," inquies, "in talibus rebus?" Quid faciam? plorando fessus sum.

Di inmortales! quam me conturbatum tenuit epistulae tuae prior pagina! quid autem iste in domo tua casus armorum? Sed hunc quidem nimbum cito transisse laetor. Tu quid egeris tua cum tristi tum etiam difficili ad consiliandum legatione, vehementer exspecto; est enim inexplicabilis. Ita circumsedemur copiis omnibus. Me quidem Bruti litterae, quas ostendis a te lectas, ita perturbarunt, ut, quamquam ante egebam consilio, tamen animi dolore sim tardior. Sed plura, cum ista cognoro. Hoc autem tempore

---

[1] Lit. "which could be delegated to *legati*."

IX

CICERO TO ATTICUS, GREETING.

On the evening of the 2nd I received a letter from Balbus telling me there would be a meeting of the Senate on the 5th to send Brutus to Asia, and Cassius to Sicily, to buy corn and send it to Rome. What a shame! First that they should take any office from that party, and secondly, if any, that it should be this subordinate[1] position. Still, I don't know whether it is not better than for him to sit on the banks of his Eurotas.[2] But fate must have its way in this. He says that at the same time a decree will be passed assigning provinces to them and other ex-praetors. This is certainly better than his Persian porch. For I don't want you to think I am referring to a Sparta farther off than Lanuvium. "You can jest," you will say, "in such important matters?" What am I to do? I am tired of mourning.

Good God! how the first page of your note held me transfixed with horror! How did that violent brawl happen in your house? But I am glad this cloud passed away quickly. I am very eager to know how you have fared with your sad and very difficult conciliatory mission; for the knot cannot be unravelled. We are so surrounded by force of every kind. Brutus' letter, which you show that you have read, has so disturbed me, that, undecided as I was before, my sorrow makes me still slower at making up my mind. But I will write more when I have news from you. At present I have nothing to write,

Tusculum, June 2, B.C. 44

[2] Brutus apparently called a stream on his estate at Lanuvium "Eurotas," and a building there the "Persian porch," after the river Eurotas and the στοὰ Περσικὴ at Sparta.

quod scriberem, nihil erat eoque minus, quod dubitabam, tu has ipsas litteras essesne accepturus. Erat enim incertum, visurusne te esset tabellarius. Ego tuas litteras vehementer exspecto.

X

CICERO ATTICO SAL.

Scr. in Tusculano Non. Iun. aut postridie a. 710

O Bruti amanter scriptas litteras! o iniquum tuum tempus, qui ad eum ire non possis! Ego autem quid scribam? ut beneficio istorum utantur? Quid turpius? Ut moliantur aliquid? Nec audent nec iam possunt. Age, quiescant auctoribus nobis; quis incolumitatem praestat? Si vero aliquid de Decimo gravius, quae nostris vita, etiamsi nemo molestus sit? ludos vero non facere! quid foedius? frumentum imponere! quae est alia Dionis legatio aut quod munus in re publica sordidius? Prorsus quidem consilia tali in re ne iis quidem tuta sunt, qui dant; sed possim id neglegere proficiens; frustra vero qui ingrediar? Matris consilio cum utatur vel etiam precibus, quid me interponam? Sed tamen cogitabo, quo genere utar litterarum; nam silere non possum. Statim igitur mittam vel Antium vel Circeios.

---

[1] Brutus as *praetor urbanus* ought to have presided at the Ludi Apollinares, but fearing to go to Rome he left it to a colleague Gaius Antonius.

especially as I have doubts as to whether you may get this letter. For it is uncertain whether the messenger may see you. I am looking for a letter from you very eagerly.

X

CICERO TO ATTICUS, GREETING.

Tusculum, June 5 or 6, B.C. 44

What an affectionate letter from Brutus. And what hard luck that you cannot go to him! But what am I to say? That they should accept the other party's favours? That were the depth of shame. That they should try some new move? They dare not, and now they cannot. Well, suppose I advise them to keep quiet and they do, who can guarantee their safety? Indeed, if anything unpleasant happens to Decimus, what sort of life shall we lead, even if no one molests us? It is a sad disgrace not to preside at the games.[1] Fancy putting the burden of the corn-supply on them! What is this but promotion downwards,[2] and what state office is more contemptible? To give advice in such matters is certainly quite unsafe, even for those who give it. If I were doing good, I might overlook that; but why should I put my foot in it to no purpose? Since he is following his mother's advice, or rather her supplications, why should I interfere? However, I will consider what kind of letter I can write, for I must give some answer. So I will write at once either to Antium or to Circeii.

[2] The banishment of Dion from Syracuse by the younger Dionysius under the pretext of an embassy seems to have passed into a proverb in this sense.

MARCUS TULLIUS CICERO

XI

CICERO ATTICO SAL.

*Scr. in
Antiati a. d.
VI Id. Iun.
a. 710*

Antium veni a. d. vi Idus. Bruto iucundus noster adventus. Deinde multis audientibus, Servilia, Tertulla, Porcia, quaerere, quid placeret. Aderat etiam Favonius. Ego, quod eram meditatus in via, suadere, ut uteretur Asiatica curatione frumenti; nihil esse iam reliqui, quod ageremus, nisi ut salvus esset; in eo etiam ipsi rei publicae esse praesidium. Quam orationem cum ingressus essem, Cassius intervenit. Ego eadem illa repetivi. Hoc loco fortibus sane oculis Cassius (Martem spirare diceres) se in Siciliam non iturum. "Egone ut beneficium accepissem contumeliam?" "Quid ergo agis?" inquam. At ille in Achaiam se iturum. "Quid tu," inquam, "Brute?" "Romam," inquit, "si tibi videtur." "Mihi vero minime; tuto enim non eris." "Quid? si possem esse, placeretne?" "Atque ut omnino neque nunc neque ex praetura in provinciam ires; sed auctor non sum, ut te urbi committas." Dicebam ea, quae tibi profecto in mentem veniunt, cur non esset tuto futurus. Multo inde sermone querebantur, atque id quidem Cassius maxime, amissas occasiones Decimumque graviter accusabant. Ego negabam oportere praeterita, adsentiebar tamen. Cumque ingressus essem dicere, quid oportuisset, nec vero quicquam novi, sed ea,

XI

CICERO TO ATTICUS, GREETING.

I reached Antium on the 8th. Brutus was very glad *Antium,* to see me. Then before Servilia, Tertulla, Porcia,[1] *June 8,* B.C and a lot of others, he asked me for my opinion. *44* Favonius was present too. I had made up my mind on the journey, and advised him to accept the control of the corn supply from Asia: there was nothing else for us to do now except to keep him out of danger: by so doing we should have some safeguard for the republic too. When I was in the midst of my speech, in came Cassius. I said the same over again. Whereupon Cassius, with flashing eyes and fairly breathing war, declared he would not go to Sicily. "Am I to take an insult like a favour?" "What will you do then?" I asked; and he said he would go to Achaia. "What of you, Brutus?" I said. "To Rome," he answered, "if you think I ought." "I don't think so at all, for you won't be safe." "Well, if it were possible to be there in safety, would you approve?" "Yes, I would rather you did not go to a province either now or after your praetorship; but I don't advise you to trust yourself in Rome." I gave him the reasons that will occur to you, why it would not be safe. Then they kept on bewailing the chances that had been let slip, especially Cassius, and they complained bitterly of Decimus. I said they ought not to harp on the past, but I agreed with them. When I had gone on to explain what ought to have been done, saying nothing new, but what everybody is saying daily,

[1] Respectively mother, half-sister, and second wife of Brutus.

quae cotidie omnes, nec tamen illum locum attingerem, quemquam praeterea oportuisse tangi, sed senatum vocari, populum ardentem studio vehementius incitari, totam suscipi rem publicam, exclamat tua familiaris: "Hoc vero neminem umquam audivi!" Ego repressi. Sed et Cassius mihi videbatur iturus (etenim Servilia pollicebatur se curaturam, ut illa frumenti curatio de senatus consulto tolleretur), et noster cito deiectus est de illo inani sermone quo Romae[1] velle esse dixerat. Constituit igitur, ut ludi absente se fierent suo nomine. Proficisci autem mihi in Asiam videbatur ab Antio velle. Ne multa, nihil me in illo itinere praeter conscientiam meam delectavit. Non enim fuit committendum, ut ille ex Italia, priusquam a me conventus esset, discederet. Hoc dempto munere amoris atque officii sequebatur, ut mecum ipse:

"Ἡ δεῦρ' ὁδός σοι τί δύναται νῦν, θεοπρόπε;"

Prorsus dissolutum offendi navigium vel potius dissipatum. Nihil consilio, nihil ratione, nihil ordine. Itaque, etsi ne antea quidem dubitavi, tamen nunc eo minus evolare hinc idque quam primum,

"ubi nec Pélopidarum fácta neque famam aúdiam."

Et heus tu! ne forte sis nescius, Dolabella me sibi legavit a. d. IIII Nonas. Id mihi heri vesperi nun-

---

[1] quo Romae *added by Tyrrell*.

ize and not touching on the point as to whether anyone else ought to have been attacked, but saying that the Senate ought to have been called, the people in their violent excitement ought to have been roused to fury, and the whole conduct of affairs taken over by them, your friend Servilia exclaimed: "That I never heard anyone . . ." I interrupted her. But I think Cassius will go (for Servilia promises she will see that that appointment to the corn-supply shall be withdrawn from the senatorial decree): and our friend soon gave up his silly talk of wanting to go to Rome. So he has made up his mind that the games may be held in his absence under his name. I fancy, however, he wants to set out for Asia from Antium. To cut the matter short, I got nothing that satisfied me out of that journey except the satisfaction to my conscience. For I could not allow him to leave Italy before I had met him. Save for fulfilling the duty I owed to our affection, I could not help asking myself:

"What makest thou with thy journey hither, seer?"[1]

In fact I found a ship breaking up, or rather already in wreckage. No plan, no reason, no system. So, although I had no doubt even before, now I have still less that I must fly away from here as fast as possible,

"Where I may hear no bruit of Pelops' sons."[2]

And listen to this, if you have not heard it before: Dolabella has made me one of his legates on the 2nd of June. That I was told yesterday evening.

[1] The author of this line, which is quoted again in *Att.* XVI. 6, is unknown.   [2] From the *Pelops* of Accius.

tiatum est. Votiva ne tibi quidem placebat; etenim
erat absurdum, quae, si stetisset res publica, vovissem,
ea me eversa illa vota dissolvere. Et habent, opinor,
liberae legationes definitum tempus lege Iulia, nec
facile addi potest. Aveo genus legationis, ut, cum
velis, introire, exire liceat; quod nunc mihi additum
est. Bella est autem huius iuris quinquennii licentia.
Quamquam quid de quinquennio cogitem? Contrahi
mihi negotium videtur. Sed βλάσφημα mittamus.

XII

CICERO ATTICO SAL.

Scr. in
Antiati V
aut IV Id.
Iun. a. 710

Bene mehercule de Buthroto. At ego Tironem
ad Dolabellam cum litteris, quia iusseras, miseram.
Quid nocet? De nostris autem Antiatibus satis videbar plane scripsisse, ut non dubitares, quin essent
otiosi futuri, usurique beneficio Antoni contumelioso.
Cassius frumentariam rem aspernabatur; eam Servilia
sublaturam ex senatus consulto se esse dicebat. Noster vero καὶ μάλα σεμνῶς in Asiam, posteaquam mihi
est adsensus tuto se Romae esse non posse (ludos
enim absens facere malebat), statim ait se iturum,
simul ac ludorum apparatum iis, qui curaturi essent,
tradidisset. Navigia colligebat; erat animus in cursu.
Interea in isdem locis erant futuri. Brutus quidem
se aiebat Asturae. L. quidem Antonius liberaliter
litteris sine cura me esse iubet. Habeo unum beneficium, alterum fortasse, si in Tusculanum venerit.

LETTERS TO ATTICUS XV. 11-12

Even you did not like the idea of a votive legation; for indeed it was absurd for me to be fulfilling vows after the constitution was overthrown, which I had made in case it were maintained. I fancy, too, free legations have a limit of time set by one of Caesar's laws, and it is not easy to get it prolonged. I want the kind of legation that lets you come and go as you please, and that I have got now. It is a fine thing, too, to have the privilege for five years. Though why do I think of five years? Things seem to me to be drawing to a crisis: but *absit omen*.

XII

CICERO TO ATTICUS, GREETING.

Antium, June 9 or 10, B.C. 44

That's jolly good news about Buthrotum. But I had sent Tiro to Dolabella with a letter as you bade me. What harm is there in it? About our friends at Antium, I think I wrote plainly enough for you not to doubt that they are going to take things quietly and accept Antonius' insulting favour. Cassius rejects the corn-supply job, and Servilia says she will cut it out of the senatorial decree. Our friend is taking things very seriously, now he agrees with me that he cannot be safe in Rome (for he prefers the games to take place in his absence). He says he will go to Asia at once, as soon as he has handed over the management of the games to those who will attend to it. He is collecting vessels, and his heart is set on going. Meantime they will stay in the same places. Brutus says he will be at Astura. L. Antonius has sent a kind letter telling me to have no fear. That's one thing I have to thank him for; perhaps there will be another, if he comes to

MARCUS TULLIUS CICERO

O negotia non ferenda! quae feruntur tamen. Τῶνδε αἰτίαν τῶν Βρούτων τίς ἔχει; In Octaviano, ut perspexi, satis ingenii, satis animi, videbaturque erga nostros ἥρωας ita fore, ut nos vellemus, animatus. Sed quid aetati credendum sit, quid nomini, quid hereditati, quid κατηχήσει, magni consilii est. Vitricus quidem nihil censebat; quem Asturae vidimus. Sed tamen alendus est, et, ut nihil aliud, ab Antonio seiungendus. Marcellus praeclare, si praecipit nostro nostra. Cui quidem ille deditus mihi videbatur. Pansae autem et Hirtio non nimis credebat. Bona indoles, ἐὰν διαμείνῃ.

XIII

CICERO ATTICO SAL.

Scr. in Puteolano VIII K. Nov. a. 710

VIII Kal. duas a te accepi epistulas. Respondebo igitur priori prius. Adsentior tibi, ut nec duces simus nec agmen cogamus, faveamus tamen. Orationem tibi misi. Eius custodiendae et proferendae arbitrium tuum. Sed quando illum diem, cum tu edendam putes? Indutias quas scribis, non intellego fieri posse. Melior est ἀναντιφωνησία; qua me usurum arbitror. Quod scribis legiones duas Brundisium venisse, vos omnia prius. Scribes igitur, quicquid audieris. Varronis διάλογον exspecto. Iam probo Ἡρακλείδειον, praesertim cum tu tanto opere delec-

---

[1] The *Second Philippic*, an answer to Antony's speech of September 19, never actually delivered by Cicero.

Tusculum. What intolerable nuisances! Yet we put up with them. Which of the Bruti have we to thank for this? In Octavianus, as I have observed, there is plenty of wit and plenty of spirit, and he seems likely to be as well disposed to our heroes as we could wish. But it is a grave question how far we can trust one of his age, name, heritage, and bringing up. His father-in-law, whom I saw at Astura, thinks he is not to be trusted at all. However, we must look after him, and, if nothing else, dissociate him from Antonius. Marcellus will be doing well if he inculcates our views into Brutus, to whom Octavianus seems to be well affected. In Pansa and Hirtius, however, he has but little trust. His disposition is good, if it will last.

XIII

CICERO TO ATTICUS, GREETING.

On the 25th I received two letters from you. So *Puteoli,* I will answer the former first. I agree with you that *Oct. 25,* B.C. we need not be the first to move nor the last to *44* follow, but that we should incline to Brutus' side. I have sent you my speech,[1] and leave it to you to keep it or publish it. But when shall we see the day when you will think it right to publish it? I don't understand how the truce you mention can be possible. It is better to make no reply; and that, I think, is what I shall do. You say that two legions have arrived at Brundisium: you get all the news first. So you must write whatever you hear. I am expecting Varro's dialogue.[2] I agree now about writing something in Heracleides' style,[3] especially as you like it

---

[2] A promised dialogue in which Cicero was to take part, or which was to be dedicated to him.   [3] Cf. xv. 4.

tere; sed, quale velis, velim scire. Quod ad te antea atque adeo prius scripsi (sic enim mavis), ad scribendum (licet enim[1] tibi vere dicere) fecisti me acriorem. Ad tuum enim iudicium, quod mihi erat notum, addidisti Peducaei auctoritatem, magnam quidem apud me et in primis gravem. Enitar igitur, ne desideres aut industriam meam aut diligentiam. Vettienum, ut scribis, et Faberium foveo. Clodium nihil arbitror malitiose; quamquam — sed quod egerit. De libertate retinenda, qua certe nihil est dulcius, tibi adsentior. Itane Gallo Caninio? O hominem nequam! quid enim dicam aliud? Cautum Marcellum! me sic, sed non tamen cautissimum.

Longiori epistulae superiorique respondi. Nunc breviori propiorique quid respondeam, nisi eam fuisse dulcissimam? Res Hispanienses valde bonae, modo Balbilium incolumem videam, subsidium nostrae senectutis. De Anniano idem, quod me valde observat Visellia. Sed haec quidem humana. De Bruto te nihil scire dicis, sed Servilia venisse M. Scaptium, eumque non qua pompa adsuevisset, ad se tamen clam venturum sciturumque me omnia; quae ego statim. Interea narrat eadem Bassi servum venisse, qui nuntiaret legiones Alexandrinas in armis esse, Bassum arcessi, Cassium exspectari. Quid quaeris? videtur res publica ius suum recuperatura. Sed ne

[1] licet enim *added by Lehmann.*

---

[1] After *quod egerit* some such words as *id actum habebo* must be supplied. On this phrase, which occurs several times in Cicero's letters, cf. Lehmann, *De epp. ad Atticum recensendis*, 1892, p. 189.

so much; but I will write whatever you wish. As I told you before, or rather previously, as you prefer to say, I must confess you have made me more eager to write. For to your own opinion, which I knew, you have added Peducaeus' authority, which I count great and as weighty as any. So I will make an effort not to disappoint you in my industry or diligence. I am making much of Vettienus and Faberius, as you suggest. I don't think Clodius meant any harm, though—but it is nothing to me.[1] I agree with you about preserving our liberty, our most precious possession. So it is Gallus Caninius' turn now?[2] What a knave! For what else can one call him? How cautious Marcellus is. So am I, but not over-cautious.

I have answered your longer and earlier letter. Now what can I say to the shorter and more recent, except that it was most delightful? Affairs in Spain are going really well, if only I can see Balbilius in safety as a support for our old age. About Annianus[3] I agree, as Visellia is very polite to me. But that is the way of the world. You say you know nothing of Brutus, but Servilia says M. Scaptius has come, and that without any of his usual parade, and he will pay her a visit quietly, and I shall be told everything. I shall know soon. Meantime she says a slave of Bassus has come announcing that the legions in Alexandria are in arms, that Bassus has been summoned, and Cassius is expected with eagerness. In short it looks as though the republic was going to recover its rights. But don't let us anticipate. You

---

[2] From *Att.* XVI. 14 it appears that Gallus had just died. Probably Antony, to whom the next words apparently refer, threatened to confiscate his property.

[3] Or "the estate of Annius," as Shuckburgh.

quid ante. Nosti horum exercitationem in latrocinio et amentiam.

Dolabella, vir optimus, etsi, cum scribebam secunda mensa adposita, venisse eum ad Baias audiebam, tamen ad me ex Formiano scripsit, quas litteras, cum e balineo exissem, accepi, sese de attributione omnia summa fecisse. Vettienum accusat (tricatur scilicet ut monetalis), sed ait totum negotium Sestium nostrum suscepisse, optimum quidem illum virum nostrique amantissimum. Quaero autem, quid tandem Sestius in hac re facere possit, quod non quivis nostrum. Sed, si quid praeter spem erit, facies, ut sciam ; sin est, ut arbitror, negotium perditum, scribes tamen, neque ista res commovebit.

Nos hic φιλοσοφοῦμεν (quid enim aliud ?) et τὰ περὶ τοῦ καθήκοντος magnifice explicamus προσφωνοῦμενque Ciceroni ; qua de re enim potius pater filio ? Deinde alia. Quid quaeris ? exstabit opera peregrinationis huius. Varronem hodie aut cras venturum putabant; ego autem in Pompeianum properabam, non quo hoc loco quicquam pulchrius, sed interpellatores illic minus molesti. Sed perscribe, quaeso, quae causa sit Myrtilo (poenas quidem illum pependisse audivi), et satisne pateat, unde corruptus.

Haec cum scriberem, tantum quod existimabam ad te orationem esse perlatam. Hui, quam timeo, quid existimes ! Etsi quid ad me ? quae non sit foras proditura nisi re publica recuperata. De quo quid sperem, non audeo scribere.

---

[1] Cf. *Att.* XVI. 11. He was accused of attempting to murder Antony.

know what practice that lot have had in rascality, and how reckless they are.

That pretty fellow Dolabella has written to me from Formiae, though, when I was writing this letter at dessert, I heard he had arrived at Baiae, and I got his letter as I left my bath. He says he has done his level best about assigning debts to me. He blames Vettienus—of course he is up to some dodge like a true business man—but he says Sestius, who is a very honest fellow and a good friend of mine, has undertaken the whole affair. Still, I should like to know what on earth Sestius can do in this business that any of us could not have done. But if anything does happen contrary to my expectation, you must let me know; while, if it is, as I suspect, a hopeless business, write all the same: it will not disturb me.

I am philosophizing here (what else can I do?) and getting on splendidly with my *De Officiis*, which I am dedicating to my son. A father could not choose a more appropriate subject. Then I shall turn to other subjects. In fact this excursion will have some works to show for itself. Varro is expected either to-day or to-morrow; but I am hastening to Pompeii, not that anything could be prettier than this place, but I shall be less bothered by interruptions there. But please inform me what the charge was against Myrtilus,[1] for I hear he has been executed, and whether it has come out who suborned him.

As I am writing this, it just occurs to me that my speech is being delivered to you. How I fear your judgment on it! Though what does it matter to me, as it will not be published, unless the constitution is restored? And what hope I have of that I dare not say.

MARCUS TULLIUS CICERO

XIV

CICERO ATTICO SAL.

Scr. in Tus-
culano V K.
Quint. a. 710

vi Kalend. accepi a Dolabella litteras. Quarum exemplum tibi misi. In quibus erat omnia se fecisse, quae tu velles. Statim ei rescripsi et multis verbis gratias egi. Sed tamen, ne miraretur, cur idem iterum facerem, hoc causae sumpsi, quod ex te ipso coram antea nihil potuissem cognoscere. Sed quid multa? litteras hoc exemplo dedi:

"CICERO DOLABELLAE COS. SUO.

"Antea cum litteris Attici nostri de tua summa liberalitate summoque erga se beneficio certior factus essem, cumque tu ipse etiam ad me scripsisses te fecisse ea, quae nos voluissemus, egi tibi gratias per litteras iis verbis, ut intellegeres nihil te mihi gratius facere potuisse. Postea vero quam ipse Atticus ad me venit in Tusculanum huius unius rei causa, tibi ut apud me gratias ageret, cuius eximiam quandam et admirabilem in causa Buthrotia voluntatem et singularem erga se amorem perspexisset, teneri non potui, quin tibi apertius illud idem his litteris declararem. Ex omnibus enim, mi Dolabella, studiis in me et officiis, quae summa sunt, hoc scito mihi et amplissimum videri et gratissimum esse, quod perfeceris, ut Atticus intellegeret, quantum ego te, quantum tu me amares. Quod reliquum est, Buthrotiam et causam et civitatem, quamquam a te constituta est (beneficia autem nostra tueri solemus),

XIV

CICERO TO ATTICUS, GREETING.

On the 26th I received a letter from Dolabella, *Tusculum,* and I am sending you a copy of it. In it he says *June 27,* B.C. he has done everything you wanted. I answered at *44* once, thanking him profusely. However, to prevent his wondering why I should do so twice, I gave as a reason that I had not been able to get any information from you before when I met you. But, to cut it short, here is a copy of my letter :—

"CICERO TO HIS FRIEND DOLABELLA THE CONSUL.

"Once before, when our friend Atticus had informed me by letter of your great liberality and the great kindness you had shown him, and when you yourself had written that you had done all that we wished, I sent you my thanks couched in such terms that you might understand that you had done me the greatest favour. But afterwards, when Atticus came himself to me at Tusculum solely to declare his gratitude to you, as he had observed your remarkable and indeed wonderful kindness in the matter of the people of Buthrotum and your strong affection for himself, I could not help expressing my thanks again more clearly in this letter. For of all the favours and services you have done for me, and they are overwhelming, my dear Dolabella, let me assure you that the highest and the most gratifying is, that you have shown Atticus how great my affection is for you, and yours for me. For the rest, as one generally wishes to secure favours received, though the case of Buthrotum and its existence as a city have been set on a firm footing by you, I

tamen velim receptam in fidem tuam a meque etiam
atque etiam tibi commendatam auctoritate et auxilio
tuo tectam velis esse. Satis erit in perpetuum
Buthrotiis praesidii, magnaque cura et sollicitudine
Atticum et me liberaris, si hoc honoris mei causa
susceperis, ut eos semper a te defensos velis. Quod
ut facias, te vehementer etiam atque etiam rogo."

His litteris scriptis me ad συντάξεις dedi; quae
quidem vereor ne miniata cerula tua pluribus locis
notandae sint. Ita sum μετέωρος et magnis cogita-
tionibus impeditus.

XV

CICERO ATTICO SAL.

*Scr. in
Antiati Id.
Iun. a. 710*

L. Antonio male sit, si quidem Buthrotiis molestus
est! Ego testimonium composui, quod, cum voles,
obsignabitur. Nummos Arpinatium, si L. Fadius
aedilis petet, vel omnes reddito. Ego ad te alia
epistula scripsi de HS $\overline{\text{cx}}$, quae Statio curarentur.
Si ergo petet Fadius, ei volo reddi, praeter Fadium
nemini. Apud me idem puto depositum. Id scripsi
ad Erotem ut redderet.

Reginam odi. Id me iure facere scit sponsor pro-
missorum eius Ammonius, quae quidem erant φιλόλογα et dignitatis meae, ut vel in contione dicere
auderem. Saran autem, praeterquam quod nefarium
hominem, cognovi praeterea in me contumacem.
Semel eum omnino domi meae vidi. Cum φιλο-

should like you to use your authority and your power to protect it, as it was put in your care and repeatedly recommended to you by me. That will be sufficient to safeguard Buthrotum for ever, and, if in compliment to me you will undertake to see them always protected, you will relieve Atticus and me of a great care and anxiety: and this I beg and entreat you to do."

After finishing this letter I have devoted myself to my treatise. I fear you will run your red pencil under many passages in it. I have been so distracted and hindered by weighty thoughts.

XV

CICERO TO ATTICUS, GREETING.

Hang L. Antonius if he is obnoxious to the Buthrotians. I have drawn up a deposition, which shall be signed whenever you like. If the aedile L. Fadius asks for the money belonging to the people of Arpinum, pay it him back in full. In another letter I mentioned the 1,000 guineas to be paid to Statius. Well, if Fadius asks for them, I wish them to be paid to him, but to no one else. I think it was deposited with me. I have written to Eros to pay it.

Antium, June 13, B.C. 44

I detest Cleopatra; and the voucher for her promises, Ammonius, knows I have good reason to do so. Her promises were all things that had to do with learning and not derogatory to my dignity, so I could have mentioned them even in a public speech. Sara, besides being a knave, I have noticed is also impertinent to me. Once, and only once, have I

φρόνως ex eo quaererem, quid opus esset, Atticum se dixit quaerere. Superbiam autem ipsius reginae, cum esset trans Tiberim in hortis, commemorare sine magno dolore non possum. Nihil igitur cum istis; nec tam animum me quam vix stomachum habere arbitrantur.

Profectionem meam, ut video, Erotis dispensatio impedit. Nam, cum ex reliquis, quae Nonis Aprilibus fecit, abundare debeam, cogor mutuari, quodque ex istis fructuosis rebus receptum est, id ego ad illud fanum sepositum putabam. Sed haec Tironi mandavi, quem ob eam causam Romam misi; te nolui impeditum impedire. Cicero noster quo modestior est, eo me magis commovet. Ad me enim de hac re nihil scripsit, ad quem nimirum potissimum debuit; scripsit hoc autem ad Tironem, sibi post Kalend. Apriles (sic enim annuum tempus confici) nihil datum esse. Tibi pro tua natura semper placuisse teque existimasse scio, id etiam ad dignitatem meam pertinere eum non modo liberaliter a nobis, sed etiam ornate cumulateque tractari. Quare velim cures (nec tibi essem molestus, si per alium hoc agere possem), ut permutetur Athenas, quod sit in annuum sumptum ei. Scilicet Eros numerabit. Eius rei causa Tironem misi. Curabis igitur et ad me, si quid tibi de eo videbitur, scribes.

seen him in my house; and then, when I asked politely what he wanted, he said he wanted Atticus. But the insolence of the queen herself, when she was in her villa across the river, I cannot mention without great indignation. So no dealings with them. They don't credit me with any spirit or even any feelings at all.

My departure from Italy I see is hindered by Eros' management of my affairs. For, although from the balances he made on April 5 I ought to have plenty of cash, I have to borrow, and I think the receipts from those paying concerns are set aside for the shrine. But I have given Tiro orders about this, and am sending him to Rome on purpose. I did not want to add to your worries. The more moderate in his demands my son is, the more am I concerned about him. For he has not mentioned this point to me, the person of all others to whom of course he ought to have mentioned it; but in a letter to Tiro he said I had sent him nothing since April 1, which was the end of his financial year. Now I know that you, with your usual amiability, have always agreed and indeed thought that among other things my dignity demanded that he should be treated not only liberally, but even with excessive and extravagant liberality. So I should like you to see that he has a bill of exchange for his annual allowance payable at Athens. I would not trouble you, if I could manage it through anyone else. Eros, of course, will pay you. That is why I have sent Tiro. Please see about it and let me know if you have any views on the point.

MARCUS TULLIUS CICERO

XVI

CICERO ATTICO SAL.

Scr. in Antiati III aut prid. Id. Iun. a. 710

Tandem a Cicerone tabellarius, et mehercule litterae πεπινωμένως scriptae, quod ipsum προκοπὴν aliquam significat, itemque ceteri praeclara scribunt; Leonides tamen retinet suum illud "adhuc," summis vero laudibus Herodes. Quid quaeris? vel verba mihi dari facile patior in hoc, meque libenter praebeo credulum. Tu velim, si quid tibi est a Statio scriptum, quod pertineat ad me, certiorem me facias.

XVIa

CICERO ATTICO SAL.

Scr. in Arpinati XIV aut XIII K. Iun. a. 710

Narro tibi, haec loca venusta sunt, abdita certe, et, si quid scribere velis, ab arbitris libera. Sed nescio quo modo οἶκος φίλος. Itaque me referunt pedes in Tusculanum. Et tamen haec ῥωπογραφία ripulae videtur habitura celerem satietatem. Equidem etiam pluvias metuo, si Prognostica nostra vera sunt; ranae enim ῥητορεύουσιν. Tu, quaeso, fac sciam, ubi Brutum nostrum et quo die videre possim.

XVII

CICERO ATTICO SAL.

Scr. in Antiati postr. Id. Iun. a. 710

Duas accepi postridie Idus, alteram eo die datam, alteram Idibus. Prius igitur superiori. De D. Bruto, cum scies. De consulum ficto timore cognoveram.

---

[1] Cf. *Att.* xiv. 16.   [2] Apparently a proverb.
[3] Cicero translated the *Prognostica* of Aratus into Latin verse.

XVI

CICERO TO ATTICUS, GREETING.

At last a messenger from my son, and upon my word a letter written in first class style. That itself shows some advance, and other people send most favourable reports too. Leonides, however, still sticks to his "at present,"[1] while Herodes bestows the highest praise. Indeed, in this respect I gladly allow myself even to be hoodwinked, and am not sorry to be credulous. I should like you to let me know if Statius has written anything that concerns me.

Antium, June 11 or 12, B.C. 44

XVIa

CICERO TO ATTICUS, GREETING.

I tell you what, this place is lovely, and certainly it is retired and free from overlookers, if you want to write. But somehow or other there's no place like home.[2] So my feet are carrying me back to Tusculum. And after all the tameness of this bit of coast would probably soon cloy on one. Besides, I am afraid of rain, if our *Prognostics*[3] are right, for the frogs are holding forth. Please let me know where Brutus is and when I can see him.

Arpinum, May 19 or 20, B.C. 44

XVII

CICERO TO ATTICUS, GREETING.

I received two letters on the 14th, one dated the same day, one the day before. So I answer the earlier first. Tell me about D. Brutus, when you know. I had heard of the pretended terror of the consuls.[4]

Antias, June 14, B.C. 44

[4] They were afraid of violence on the part of Brutus and Cassius.

MARCUS TULLIUS CICERO

Sicca enim φιλοστόργως ille quidem, sed tumultuosius ad me etiam illam suspicionem pertulit. Quid tu autem? "τὰ μὲν διδόμενα —"? Nullum enim verbum a Siregio. Non placet. De Plaetorio, vicino tuo, permoleste tuli quemquam prius audisse quam me. De Syro prudenter. L. Antonium per Marcum fratrem, ut arbitror, facillime deterrebis. Antroni vetui; sed nondum acceperas litteras, ne cuiquam nisi L. Fadio aedili. Aliter enim nec caute nec iure fieri potest. Quod scribis tibi desse HS ϲ, quae Ciceroni curata sint, velim ab Erote quaeras, ubi sit merces insularum. Arabioni de Sittio nihil irascor. Ego de itinere nisi explicato Λ nihil cogito; quod idem tibi videri puto. Habes ad superiorem.

Nunc audi ad alteram. Tu vero facis ut omnia, quod Serviliae non dees, id est Bruto. De regina gaudeo te non laborare, testem etiam tibi probari. Erotis rationes et ex Tirone cognovi et vocavi ipsum. Gratissimum, quod polliceris Ciceroni nihil defuturum; de quo mirabilia Messalla, qui Lanuvio rediens ab illis venit ad me, et mehercule ipsius litterae sic et φιλοστόργως et πεπινωμένως scriptae, ut eas vel in acroasi audeam legere. Quo magis illi indulgendum puto. De Buciliano Sestium puto non moleste ferre. Ego, si Tiro ad me, cogito in Tusculanum. Tu vero, quicquid erit, quod me scire par sit, statim.

---

[1] A proverb presumably ending ἀνάγκη δέχεσθαι, "one must put up with," or something similar.

[2] If Λ stands for λοῖπῳ = *reliquiis* "balance," as was suggested by Gronovius.

For Sicca, in a very friendly but rather panic-stricken manner, has brought me word of that suspicion too. What do you say? "Take what the gods give"?[1] For I have not a word from Siregius. I don't like it. About your neighbour Plaetorius I was very annoyed that anyone heard before I did. About Syrus you did well. I fancy you will easily frighten L. Antonius through his brother Marcus. I told you not to pay Antro, but you had not yet received my letter forbidding you to pay anyone except L. Fadius the aedile. It is the only safe and proper thing. You say you are £1,000 out of pocket on the money sent to my son; please ask Eros what has become of the rents of the blocks of houses. I am not at all angry with Arabio about Sittius. I am not thinking of starting on my journey until my accounts[2] are all settled, and of that I think you approve. There is my answer to your first letter.

Now hear what I have to say to the second. You are acting as kindly as usual in standing by Servilia, that is to say, Brutus. As to Cleopatra, I am glad you are not anxious and that you accept the evidence. The state of Eros' accounts I have heard from Tiro, and I have sent for Eros himself. I am most grateful for your promise not to let my son lack in anything. Messalla, on his way back from our adversaries at Lanuvium, called on me with wonderfully good news about him, and upon my word his own letter is so affectionate and well-written that I should not be ashamed to read it before an audience. So I feel all the more indulgently disposed towards him. I don't think Sestius is annoyed about Bucilianus. As soon as Tiro returns home, I am thinking of going to Tusculum. Please let me know at once, if there is anything that I ought to know.

MARCUS TULLIUS CICERO

XVIII

CICERO ATTICO SAL.

Scr. in itinere ex Antiati in Tusculanum XVI K. Quint. a. 710

xvii Kal. etsi satis videbar scripsisse ad te, quid mihi opus esset, et quid te facere vellem, si tibi commodum esset, tamen, cum profectus essem et in lacu navigarem, Tironem statui ad te esse mittendum, ut iis negotiis, quae agerentur, interesset, atque etiam scripsi ad Dolabellam me, si ei videretur, velle proficisci, petiique ab eo de mulis vecturae. Ut in his (quoniam intellego te distentissimum esse qua de Buthrotiis, qua de Bruto, cuius etiam ludorum sumptuosorum[1] curam et administrationem suspicor ex magna parte ad te pertinere) ut ergo in eius modi re tribues nobis paulum operae; nec enim multum opus est.

Mihi res ad caedem et eam quidem propinquam spectare videtur. Vides homines, vides arma. Prorsus non mihi videor esse tutus. Sin tu aliter sentis, velim ad me scribas. Domi enim manere, si recte possum, multo malo.

XIX

CICERO ATTICO SAL.

Scr. in Tusculano inter a. d. XV et XI K. Quint. a. 710

Quidnam est, quod agendum amplius de Buthrotiis sit? Egisse[2] enim te frustra scribis. Quid autem se refert Brutus? Doleo mehercules te tam esse distentum, quod decem hominibus referendum est

---

[1] sumptuosorum *Lehmann*: suorum *MSS.*
[2] sit? egisse *Boot*: stetisse *MSS.*

XVIII

CICERO TO ATTICUS, GREETING.

Though I think I told you sufficiently what I wanted and what I wished you to do, if it was convenient to you, in my letter of the 15th, still, when I had started and was crossing the lake, I determined to send Tiro to you that he might attend to the necessary matters with you; and I wrote, too, to Dolabella, saying I wanted to start if he agreed, and asked him about baggage mules. So far as you can— I understand you are utterly distracted with business, what with the Buthrotians and what with Brutus, as I expect the care and arrangement of his sumptuous games fall largely to your share—still, so far as you can, give a little attention to my affairs. I shall not want much.

On the way to Tusculum, June 16, B.C. 44

To me things seem to foreshadow bloodshed, and that quite soon. You see the men, you see their warlike preparations. Indeed I do not count myself safe at all. If you think differently, I wish you would write. For, if I can with safety, I should much prefer to stay at home.

XIX

CICERO TO ATTICUS, GREETING.

What more can we possibly do about Buthrotum? For you say your labour has been in vain. Why too is Brutus returning to Rome? I am really very sorry you have been so overworked: you are indebted for

Tusculum, June 17 to 21, B.C. 44

acceptum. Est illud quidem ἐργῶδες, sed ἀνεκτὸν mihique gratissimum. De armis nihil vidi apertius. Fugiamus igitur, et ut ais. Coram Theophanes quid velit, nescio. Scripserat enim ad me. Cui rescripsi, ut potui. Mihi autem scribit venire ad me se velle, ut et de suis rebus et quaedam, quae ad me pertinerent. Tuas litteras exspecto. Vide, quaeso, ne quid temere fiat.

Statius scripsit ad me locutum secum esse Q. Ciceronem valde adseveranter se haec ferre non posse; certum sibi esse ad Brutum et Cassium transire. Hoc enim vero nunc discere aveo : hoc ego quid sit interpretari non possum. Potest aliquid iratus Antonio, potest gloriam iam novam quaerere, potest totum esse σχεδίασμα; et nimirum ita est. Sed tamen et ego vereor, et pater conturbatus est; scit enim, quae ille de hoc, mecum quidem ἄφατα olim. Plane, quid velit, nescio. A Dolabella mandata habebo, quae mihi videbuntur, id est nihil. Dic mihi, C. Antonius voluitne fieri septemvir? Fuit certe dignus. De Menedemo est, ut scribis. Facies omnia mihi nota.

---

[1] The commissioners for distributing land in Epirus.
[2] Seven commissioners were appointed to distribute land

LETTERS TO ATTICUS XV. 19

that to the ten commissioners.[1] That is certainly a tough piece of business, but one has to put up with it, and I am very thankful for it. As to the imminence of war I never saw anything more obvious. So let me flee, and in the way you suggest. I do not know why Theophanes wants to see me, for he wrote to me. I answered as best I could. But he writes saying he wants to come to me to discuss his own affairs and some that concern me. I am looking for a letter from you. Pray see that nothing is done rashly.

Statius has written to me saying my nephew Quintus has told him with emphasis that he cannot put up with things, and has resolved to go over to Brutus and Cassius. Here is something I am very eager to understand: here is a puzzle I can't interpret. Perhaps he is angry with Antony about something; perhaps he is looking for some new way of distinguishing himself; or perhaps it is all bunkum; and no doubt that is what it is. But for all that I am afraid, and his father is disturbed about him, for he knows what he used to say about Antony; indeed, he said things to me which I cannot repeat. What on earth he means I can't think. I shall only have such commissions as I choose from Dolabella, that is, none at all. Tell me if C. Antonius wanted to be put on the land commission.[2] He would certainly have been in his proper place. About Menedemus it is as you say. Please keep me posted up in all news.

in Italy among the soldiers. As the next sentence implies, several of them were nonentities.

MARCUS TULLIUS CICERO

XX

CICERO ATTICO SAL.

Scr. in Tusculano inter XV et XI K. Quint. a. 710

Egi gratias Vettieno; nihil enim potuit humanius. Dolabellae mandata sint quaelibet, mihi aliquid, vel quod Niciae nuntiem. Quis enim haec, ut scribis, ἀντερεῖ[1]? Nunc dubitare quemquam prudentem, quin meus discessus desperationis sit, non legationis? Quod ais extrema quaedam iam homines de re publica loqui et eos quidem viros bonos, ego, quo die audivi illum tyrannum in contione "clarissimum virum" appellari, subdiffidere coepi. Postea vero quam tecum Lanuvi vidi nostros tantum spei habere ad vivendum, quantum accepissent ab Antonio, desperavi. Itaque, mi Attice (fortiter hoc velim accipias, ut ego scribo), genus illud interitus, quo causae cursus[2] est, foedum ducens, et quasi denuntiatum nobis ab Antonio, ex hac nassa exire constitui non ad fugam, sed ad spem mortis melioris. Haec omnis culpa Bruti.

Pompeium Carteiae receptum scribis. Iam igitur contra hunc exercitum. Utra ergo castra? Media enim tollit Antonius. Illa infirma, haec nefaria. Properemus igitur. Sed iuva me consilio, Brundisione an Puteolis. Brutus quidem subito, sed sapienter. Πάσχω τι. Quando enim illum? Sed humana ferenda. Tu ipse eum videre non potes. Di illi mortuo, qui umquam Buthrotum! Sed acta missa; videamus, quae agenda sint.

---

[1] ἀντερεῖ *Tyrrell* : anteno *MSS.*: λεπτύνει *Gronovius and most editors.*

[2] causae cursus *Popma* : causa cursus *Z* : causurus *M¹* : casurus *M²* : Catulus usus est *Madvig, which gives a better sense but is not very near the reading of the MSS.*

XX

CICERO TO ATTICUS, GREETING.

Tusculum, June 17 to 21, B.C. 44

I have thanked Vettienus; for nothing could have been kinder. Let Dolabella give me what commissions he will, provided I have something, even a message to Nicias. For, as you say, who will deny it? Can any sane man doubt now that I am going away in despair, not on a mission? You say that people, aye, even good citizens, are talking of desperate political measures. I began to have my doubts on the day that I heard that tyrant called "a most distinguished man." Afterwards, when I was with you at Lanuvium and saw that our friends had precisely so much hope of life as Antony gave them, I lost all hope. So, my dear Atticus, I hope you will take what I am going to say with the same courage as I write it. As I think the kind of death towards which the current of affairs is setting is disgraceful and hold that we are practically condemned to it by Antony, I have decided to escape from the toils, not so much to escape as in hope of a better death. All this is Brutus' fault.

You say Pompeius has been received at Carteia. So there will soon be an army sent against him. Then which camp am I to choose? For Antony makes neutrality impossible. That is weak, this is criminal. So let me hasten away. But give me your counsel whether to sail from Brundisium or Puteoli. Brutus does wisely to go, but it is sudden. I am rather upset about it, for when shall I see him again? But such is life. You yourself cannot see him. Heaven confound that dead man for ever touching Buthrotum. But away with the past; let us see what has to be done.

Rationes **Erotis**, etsi ipsum nondum vidi, tamen et ex litteris eius et ex eo, quod Tiro cognovit, prope modum cognitas habeo. Versuram scribis esse faciendam mensum quinque, id est ad Kal. Nov., HS $\overline{cc}$; in eam diem cadere nummos, qui a Quinto debentur. Velim igitur, quoniam Tiro negat tibi placere me eius rei causa Romam venire, si ea te res nihil offendet, videas, unde nummi sint, mihi feras expensum. Hoc video in praesentia opus esse. Reliqua diligentius ex hoc ipso exquiram, in his de mercedibus dotalium praediorum. Quae si fideliter Ciceroni curabuntur, quamquam volo laxius, tamen ei prope modum nihil derit. Equidem video mihi quoque opus esse viaticum; sed ei ex praediis, ut cadet, ita solvetur, mihi autem opus est universo. Equidem, etsi mihi videtur iste, qui umbras timet, ad caedem spectare, tamen nisi explicata solutione non sum discessurus. Sitne autem explicata necne, tecum cognoscam. Hanc putavi mea manu scribendam itaque feci. De Fadio, ut scribis, utique alii nemini. Rescribas velim hodie.

XXI

CICERO ATTICO SAL.

Scr. in Tusculano X K. Quint. a. 710

Narro tibi, Quintus pater exsultat laetitia. Scripsit enim filius se idcirco profugere ad Brutum voluisse, quod, cum sibi negotium daret Antonius, ut eum dictatorem efficeret, praesidium occuparet, id recusasset;

Though I have not yet seen Eros, from his letters and from what Tiro found out I know pretty well how his accounts stand. You say I must raise a fresh loan for some £2,000 for five months, that is, till the 1st of November, when Quintus' debt falls due. So, since Tiro says you do not want me to come to Rome on purpose for that, if you do not mind, I should be glad if you would see where I can get the money, and put it down on my account. I see it is necessary for the present. I will enquire more closely into the rest from Eros himself, among other things about the rents of Terentia's dower property. If they are properly looked after for my son he will be pretty well provided for, though I want him to be more liberally treated. I see I shall want some journey-money myself; but he can get the rents of the property as they fall due, whereas I shall require a lump sum. I certainly shall not leave until the money has been paid, though that trembler at shadows[1] seems to me to be meditating a massacre. However, whether it has been arranged or not, I shall learn when I see you. I thought I had better write this myself, and so I have done so. As you say about Fadius: the money must not go to anyone else in any case. Please answer by return.

XXI

CICERO TO ATTICUS, GREETING.

Tusculum, June 22, B.C. 44

I must tell you my brother Quintus is jumping for joy. For his son has written saying that his reason for wanting to take refuge with Brutus is that, when Antony asked him to secure him the dictatorship and seize some fort, he refused, and the reason for

[1] Antony, who professed to be afraid of assassination at the hands of Brutus and Cassius. Cf. xv. 17.

recusasse autem se, ne patris animum offenderet; ex
eo sibi illum hostem. "Tum me," inquit, "collegi
verens, ne quid mihi ille iratus tibi noceret. Itaque
eum placavi. Et quidem $\overline{\text{cccc}}$ certa, reliqua in spe."
Scribit autem Statius illum cum patre habitare velle
(hoc vero mirum) et id gaudet. Ecquem tu illo
certiorem nebulonem?

Ἐποχὴν vestram de re Cani probo. Nihil eram
suspicatus de tabulis, ἀκεραίως restitutam arbitrabar.
Quae differs, ut mecum coram, exspectabo. Tabellarios, quoad voles, tenebis; es enim occupatus. Quod
ad Xenonem, probe. Quod scribo, cum absolvero.
Quinto scripsisti te ad eum litteras. Nemo attulerat.
Tiro negat iam tibi placere Brundisium et quidem
dicere aliquid de militibus. At ego iam destinaram
Hydruntem quidem. Movebant me tuae quinque
horae. Hic autem quantus πλοῦς! Sed videbimus. Nullas a te XI Kal. Quippe, quid enim
iam novi? Cum primum igitur poteris, venies.
Ego propero, ne ante Sextus; quem adventare
aiunt.

his refusal was that he did not want to hurt his father's feelings; and from that time Antony has been his enemy. "Then," he says, "I pulled myself together for fear that he should do you some mischief in his wrath with me; and so I smoothed him down, and indeed got £4,000[1] in cash, and have hopes of some more." Statius, however, says he wants to live with his father—which is a wonder—and my brother is delighted about it. Did you ever see a more thorough rascal?

I approve of your hesitation in the arrangement with Canus.[2] I had no idea about the documents; I thought her dowry had been paid back in full. I shall look forward to the matters you refrain from mentioning till we meet. Keep the messengers as long as you like, as you are busy. As to Xeno, quite right. What I am writing I will send when it is finished. You told Quintus you had sent him a letter, but none has been brought as yet. Tiro says you disapprove of Brundisium now, and indeed says something about soldiers. But I have already fixed upon Hydrus. Your saying that it was only a five-hour passage decided me. Think of the endless voyage from here. But we shall see. I had no letter from you on the 21st. Of course, for what news can there be now? Come, then, as soon as you can. I am in a hurry, for Sextus may get here before I leave. They say he is coming.

[1] 400 sestertia.
[2] Apparently there were negotiations for a marriage between young Quintus and Canus' daughter, who had lately been divorced. Cf. XIII. 41.

MARCUS TULLIUS CICERO

XXII

CICERO ATTICO SAL.

Scr. in Tus-
culano V K.
Quint. mane
a. 710

Gratulor nobis Quintum filium exisse. Molestus non erit. Pansam bene loqui credo. Semper enim coniunctum esse cum Hirtio scio; amicissimum Bruto et Cassio puto, si expediet (sed quando illos videbit?), inimicum Antonio, quando aut cur? Quousque ludemur? Ego autem scripsi Sextum adventare, non quo iam adesset, sed quia certe id ageret ab armisque nullus discederet. Certe, si pergit, bellum paratum est. Hic autem noster Cytherius nisi victorem neminem victurum. Quid ad haec Pansa? utrobi erit, si bellum erit? quod videtur fore. Sed et haec et alia coram hodie quidem, ut scribis, aut cras.

XXIII

CICERO ATTICO SAL.

Scr. in Tus-
culano VIII
aut VII K.
Quint. a. 710

Mirifice torqueor, sine dolore tamen; sed permulta mihi de nostro itinere in utramque partem occurrunt. "Quousque?" inquies. Quoad erit integrum; erit autem, usque dum ad navem. Pansa si tuae rescripserit, et meam tibi et illius epistulam mittam. Silium exspectabam; cui hypomnema compositum. Si quid novi. Ego litteras misi ad Brutum. Cuius de itinere etiam ex te velim, si quid scies, cognoscere.

---

[1] Antony. Cf. x. 10.

XXII
CICERO TO ATTICUS, GREETING.

Young Quintus' absence is a blessing; he won't be a nuisance to us. I believe Pansa is talking amiably. I know he is always hand in glove with Hirtius; I think he will be quite friendly with Brutus and Cassius, if it is expedient—but will he ever see them? —and that he will oppose Antony: but when and how? How long are we to be fooled? I said Sextus was coming, not because he was already near, but because he certainly has it in mind, and does not show the least sign of laying down his arms. Certainly, if he goes on, war must come. But our good lover of Cytheris[1] thinks no one sure of his life unless he gains a victory. What has Pansa to say to this? And which side will he take if there is war? So far as I can see, there will be. But more of this and other things when we meet—to-day, according to your letter, or to-morrow.

Tusculum, June 27, B.C. 44

XXIII
CICERO TO ATTICUS, GREETING.

I am absolutely on the rack, but not with pain. So many ideas for and against that journey of mine keep occurring to me. You will ask how long that is going to last. Until the matter is settled, and that won't be till I am on board ship. If Pansa sends an answer to your note, I will forward my letter and his. I am expecting Silius, and have a memorandum drawn up for him. If there is any news—— I have written to Brutus. If you know anything about his movements, I should be glad to hear that too.

Tusculum, June 24 or 25, B.C. 44

MARCUS TULLIUS CICERO

XXIV

CICERO ATTICO SAL.

Scr. in Tusculano VI K. Quint. mane a. 710

Tabellarius, quem ad Brutum miseram, ex itinere rediit VII Kal. Ei Servilia dixit eo die Brutum H.IS [1] profectum. Sane dolui meas litteras redditas non esse. Silius ad me non venerat. Causam composui; eum libellum tibi misi. Te quo die exspectem, velim scire.

XXV

CICERO ATTICO SAL.

Scr. in Tusculano III K. Quint. a. 710

De meo itinere variae sententiae; multi enim ad me. Sed tu incumbe, quaeso, in eam curam. Magna res est. An probas, si ad Kal. Ian. cogitamus? Meus animus est aequus, sic tamen, ut si nihil offensionis sit. Et tu etiam, scisne,[2] quo die olim piaculum?[3] Ut ut est res,[4] casus consilium nostri itineris iudicabit. Dubitemus igitur. Est enim hiberna navigatio odiosa, eoque ex te quaesieram mysteriorum diem. Brutum, ut scribis, visum iri a me puto. Ego hinc volo pr. Kal.

[1] H.IS (= hora prima semis) *Orelli*: his *most MSS.*
[2] scisne] scire *MSS.*
[3] olim piaculum *Bosius*: Olympiacum mysteria scilicet *MSS.* (*the last two words are rejected as a gloss by Boot*): Olympia *Shuckburgh*.
[4] ut ut est res *Moser*: ut tu scires *MSS.*: ut tu scribis *Lambinus*.

XXIV

CICERO TO ATTICUS, GREETING.

The messenger I sent to Brutus just got back yesterday. Servilia told him Brutus had started at half-past six in the morning. I was very sorry he did not get my letters. Silius has not come yet. I have drawn up a statement of his case, and am sending the pamphlet to you. I should like to know when to expect you.

Tusculum, June 26, B.C. 44

XXV

CICERO TO ATTICUS, GREETING.

Opinions differ about my journey, for I have had a lot of visitors. But please apply yourself to the question. It is a serious matter. Do you approve of my idea of returning by the 1st of January? I am open-minded on the subject, provided I don't give any offence. By the way, too, do you know the date of the sacrilege of yore?[1] However that may be, chance will decide the plan of my journey. So let us leave it in doubt. For a winter journey is most unpleasant, and that was why I asked you the date of the mysteries.[2] Brutus, as you say, I think I shall see. I want to leave here on the last of the month.

Tusculum, June 29, B.C. 44

---

[1] If the reading is right, which is very uncertain, this must refer to the violation of the rites of Bona Dea by Clodius in Cicero's consulship. It may, however, refer to the Olympic games as Shuckburgh suggests.

[2] *Vide* last note. Shuckburgh, however, thinks it refers to the Eleusinian mysteries.

MARCUS TULLIUS CICERO

XXVI

CICERO ATTICO SAL.

Scr. in Arpinati VI Non. Quint. a. 710

De Quinti negotio video a te omnia facta. Ille tamen dolet dubitans, utrum morem gerat Leptae an fidem infirmet filio. Inaudivi L. Pisonem velle exire legatum ψευδεγγράφῳ senatus consulto. Velim scire, quid sit. Tabellarius ille, quem tibi dixeram a me ad Brutum esse missum, in Anagninum ad me venit ea nocte, quae proxima ante Kal. fuit, litterasque ad me attulit; in quibus unum alienum summa sua prudentia, idem illud, ut spectem ludos suos. Rescripsi scilicet primum me iam profectum, ut non integrum sit; deinde ἀτοπώτατον esse me, qui Romam omnino post haec arma non accesserim neque id tam periculi mei causa fecerim quam dignitatis, subito ad ludos venire. Tali enim tempore ludos facere illi honestum est, cui necesse est, spectare mihi, ut non est necesse, sic ne honestum quidem est. Equidem illos celebrari et esse quam gratissimos mirabiliter cupio, idque ita futurum esse confido, et tecum ago, ut iam ab ipsa commissione ad me, quem ad modum accipiantur hi ludi, deinde omnia reliquorum ludorum in dies singulos persequare. Sed de ludis hactenus. Reliqua pars epistulae est illa quidem in utramque partem, sed tamen non nullos interdum iacit igniculos viriles. Quod quale tibi videretur, ut posses interpretari, misi ad te exemplum epistulae; quamquam mihi tabella-

XXVI

CICERO TO ATTICUS, GREETING.

I see you have done all you could in Quintus' *Arpinum*, business. He, however, is in distress and doubt as *July 2,* B.C. to whether he shall oblige Lepta or damage his son's *44* credit. I have heard a rumour that L. Piso wants to go on a mission with a forged decree of the senate. I should like to know what there is in it. The messenger I told you I had sent to Brutus came to me at Anagnia on the night of the 30th of June, and brought me a letter in which there was one request quite unlike his usual common-sense, the same old request that I should be present at his games. I answered, of course, firstly that I had already set out now, so that it was not in my power to do so, and secondly that it would be most out of place for me, who have not been near Rome at all since the outbreak of war—not so much to preserve my safety as to preserve my dignity—suddenly to go to the games. For at such a time it was honourable for him to give the games, since he had to do so, but, as there was no necessity for me to attend them, it would not be honourable for me to do so. Of course I keenly desire that they should be well attended and very popular, and I trust they will be; and I beg you to send me a description of how these games and all the other games are received day by day from the very beginning. But enough of the games. The rest of the letter is, one must confess, of rather a dubious kind, but still he does at times emit some sparks of manly courage. That you may judge for yourself what it is like, I have sent you a copy of the letter, although

MARCUS TULLIUS CICERO

rius noster dixerat tibi quoque se attulisse litteras a Bruto easque ad te e Tusculano esse delatas.

Ego itinera sic composueram, ut Nonis Quinctilibus Puteolis essem; valde enim festino, ita tamen, ut, quantum homo possit, quam cautissime navigem. M. Aelium cura liberabis; me paucos pedes[1] in extremo fundo et eos quidem subterraneos servitutis putasse aliquid[2] habituros. Id me iamiam nolle neque mihi aquam[3] esse tanti. Sed, ut mihi dicebas, quam lenissime, potius ut cura liberetur, quam ut me suscensere aliquid suspicetur. Item de illo Tulliano capite libere cum Cascellio loquere. Parva res est, sed tu bene attendisti. Nimis callide agebatur. Ego autem, si mihi imposuisset aliquid, quod paene fecit, nisi tua malitia affuisset, animo iniquo tulissem. Itaque, utut erit, rem impediri malo. Octavam partem Tullianarum aedium ad Streniae[4] memineris deberi Caerelliae.[5] Videris mancipio dare ad eam summam, quae sub praecone fuit maxima. Id opinor esse $\overline{\text{CCCLXXX}}$.

Novi si quid erit, atque etiam si quid prospicies, quod futurum putes, scribas ad me quam saepissime.

[1] pedes *Turnebus*: pe Z^t: spe M: specus Z^l, *Lambinus*.
[2] putasse aliquid *Madvig*: apud tale quid M.
[3] aquam *Turnebus*: quam MZ^t.
[4] Tullianarum aedium ad Streniae *Lambinus and Turnebus*: tuli luminarum medium ad strane *MSS*.
[5] deberi Caerelliae *Shuckburgh*: cui Caerellia *MSS*.

---

[1] The reading and meaning of this passage is uncertain. Apparently either Cicero had asserted some claim on some

my messenger tells me he brought you a letter, too, from Brutus, and that it was forwarded to you from Tusculum.

I have arranged my journeys so that I shall be at Puteoli on the 7th of July; for, though I am in a great hurry, I mean to take every care humanly possible in my voyage. Please relieve M. Aelius of his anxiety. Tell him I thought that on a few feet at the far end of the land there might be some sort of claims, and those only underground. Also that I have not the slightest desire for it, and that I don't value water at that price.[1] But, as you suggested, do it as mildly as possible, rather to relieve him of anxiety than to suggest that I am in the least annoyed. Again, about that debt of Tullius: speak to Cascellius frankly. It is a small matter, but I am glad you attended to it. There was too much trickery about it: and, if he had cheated me at all, which he very nearly did if you had not been too sharp for him, I should have been very much annoyed. So, whatever happens, I would rather the matter were broken off. Remember that an eighth share of the houses of Tullius near the temple of Strenia is due to Caerellia, and see that it is conveyed to her at the highest price bid at the auction. I think that was some 3,000 guineas.[2]

If there is any news, and, even if you foresee anything you think likely to happen, I should like you to write to me as often as possible. To Varro

underground water-pipes on property of Aelius adjoining his own, and was now disclaiming it; or Aelius had been commissioned to buy property for Cicero on which there was a disputed claim to such pipes, and Cicero refuses to purchase on that ground. *Servitus* above is used in the technical legal sense of an "easement" or liability on property.

[2] 380 sestertia. On this debt cf. *Att.* XII. 51.

Velim Varroni, quem ad modum tibi mandavi, memineris excusare tarditatem litterarum mearum. Mundus iste cum M. Ennio quid egerit de testamento (curiosus enim), facias me velim certiorem. Ex Arpinati vi Non.

XXVII
CICERO ATTICO SAL.

Scr. in Arpinati V Non. Quint. a. 710

Gaudeo id te mihi suadere, quod ego mea sponte pridie feceram. Nam, cum ad te vi Nonas darem, eidem tabellario dedi etiam ad Sestium scriptas πάνυ φιλοστόργως. Ille autem, quod Puteolos persequitur, humane, quod queritur, iniuste. Non enim ego tam illum expectare, dum de Cosano rediret, debui, quam ille aut non ire, antequam me vidisset, aut citius reverti. Sciebat enim me celeriter velle proficisci seseque ad me in Tusculanum scripserat esse venturum. Te, ut a me discesseris, lacrimasse moleste ferebam. Quod si me praesente fecisses, consilium totius itineris fortasse mutassem. Sed illud praeclare, quod te consolata est spes brevi tempore congrediendi; quae quidem exspectatio me maxime sustentat. Meae tibi litterae non derunt. De Bruto scribam ad te omnia. Librum tibi celeriter mittam " de gloria." Excudam aliquid Ἡρακλείδειον, quod lateat in thesauris tuis. De Planco memini. Attica iure queritur. Quod me de Bacchi,[1] de statuarum coronis certiorem fecisti, valde gratum; nec quicquam posthac non modo tantum, sed ne tantulum quidem praeterieris.

[1] Bacchide *Graevius, assuming it to be the name of an actress appearing at the games given by Brutus.*

remember to plead my excuses for my slowness in writing, as I told you. What your friend Mundus has done with M. Ennius about the will, please let me know, for I am inquisitive. *Arpinum, July 2.*

XXVII

CICERO TO ATTICUS, GREETING.

Arpinum, July 3, B.C. 44

I am glad you recommend me to do what I did of my own accord yesterday. For to the same messenger, to whom I gave the letter I sent you on the 2nd, I also gave another for Sestius, written in very friendly terms. It is very good of him to follow me to Puteoli, but he has no grounds for his complaint. For it was not my business to wait for his return from Cosa, so much as it was his not to go until he had seen me, or to return more quickly. He knew I wanted to start in a hurry, and he told me he would come to me at Tusculum. I am grieved that you wept when you left me. If you had done so in my presence, I might have changed all my plans about going. But there is one good thing, that you were consoled by the thought of meeting me again soon; and that, indeed, is the hope that buoys me up. I will not stint you of letters, and will give you full news about Brutus. I will send you my book *On Glory* soon. I will hammer out something in the style of Heracleides to be stored up in your treasure-house. I remember about Plancus. Attica has good reason for grumbling. I am much obliged to you for telling me about the garlands for Bacchus and the statues. Please don't omit any detail of the same importance, or even of the smallest importance in the future. I

MARCUS TULLIUS CICERO

Et de Herode et Mettio meminero et de omnibus,
quae te velle suspicabor modo. O turpem sororis
tuae filium! Cum haec scriberem, adventabat αὐτῇ
βουλύσει cenantibus nobis.

XXVIII

CICERO ATTICO SAL.

*Scr. in
Arpinati V
Non. Quint.
a. 710*

Ego, ut ad te pridie scripseram, Nonis constitueram venire in Puteolanum. Ibi igitur cotidie tuas
litteras exspectabo et maxime de ludis; de quibus
etiam ad Brutum tibi scribendum est. Cuius epistulae, quam interpretari ipse vix poteram, exemplum
pridie tibi miseram. Atticae meae velim me ita
excuses, ut omnem culpam in te transferas, et ei
tamen confirmes me immutatum amorem meum mecum abstulisse.

XXIX

CICERO ATTICO SAL.

*Scr. in Formiano III
Non. Quint.
a. 710*

Bruti ad te epistulam misi. Di boni, quanta ἀμηχανία! Cognosces, cum legeris. De celebratione
ludorum Bruti tibi adsentior. Ad M. Aelium nullus
tu quidem domum, sed sicubi inciderit. De Tulliano
semisse M. Axianum adhibebis, ut scribis. Quod cum
Cosiano egisti, optime. Quod non solum mea, verum
etiam tua eadem expedis, gratum. Legationem probari meam gaudeo. Quod promittis, di faxint! Quid
enim mihi meis iucundius? Sed istam, quam tu

won't forget about Herodes or Mettius, or anything that I have the least suspicion you would like. What disgraceful conduct of your sister's son! Here he is coming as the shades of night are falling, just as I am writing this at the dinner-table.

XXVIII

CICERO TO ATTICUS, GREETING.

As I told you in my letter yesterday, I have arranged to be at Puteoli on the 7th. So I shall look for a daily letter from you there, especially about the games. You ought to write to Brutus too about them. I sent you a copy yesterday of a letter of his, of which I can hardly make anything myself. Please make my excuses to Attica by taking the blame on yourself and assuring her that I depart with undiminished affection for her.

Arpinum, July 3, B.C. 44

XXIX

CICERO TO ATTICUS, GREETING.

I am sending you Brutus' letter. Heavens, what a helpless condition he is in! You will understand when you have read it. I agree about the celebration of his games. Don't go to Aelius' house on any account, but speak to him if you happen to meet him. Take M. Axianus' advice about the half of Tullius' debt, as you suggest. What you have done with Cosianus is excellent. Thanks for clearing up my affairs and your own at the same time. I am glad you approve of my appointment. God grant your promises come true. For what could suit me and mine better? But I am afraid of the proviso

Formiae, July 5, B.C. 44

excipis, metuo. Brutum cum convenero, perscribam omnia. De Planco et Decimo sane velim. Sextum scutum abicere nolebam. De Mundo, si quid scies. Rescripsi ad omnia tua; nunc nostra accipe.

Quintus filius usque Puteolos (mirus civis, ut tu Favonium Asinium dicas) et quidem duas ob causas, et ut mecum et σπείσασθαι vult cum Bruto et Cassio. Sed tu quid ais? Scio enim te familiarem esse Othonum. Ait hic sibi Iuliam ferre; constitutum enim esse discidium. Quaesivit ex me pater, qualis esset fama. Dixi nihil sane me audisse (nesciebam enim, cur quaereret) nisi de ore et patre. "Sed quorsus?" inquam. At ille filium velle. Tum ego, etsi ἐβδελυττόμην, tamen negavi putare me illa esse vera. Σκοπὸς est enim huic nostro nihil praebere, illa autem οὐ παρὰ τοῦτον. Ego tamen suspicor hunc, ut solet, alucinari. Sed velim quaeras (facile autem potes) et me certiorem.

Obsecro te, quid est hoc? Signata iam epistula Formiani, qui apud me cenabant, Plancum se aiebant hunc Buthrotium pridie, quam hoc scribebam, id est III Non., vidisse demissum, sine phaleris; servulos autem dicere eum et agripetas eiectos a Buthrotiis. Macte! Sed, amabo te, perscribe mihi totum negotium.

---

[1] Favonius was a follower of Cato; Asinius Pollio a Caesarian. Possibly Cicero may mean that Quintus sided with both parties; but the exact meaning is doubtful.

[2] Head of the land-commissioners in Epirus.

LETTERS TO ATTICUS XV. 29

you make about Attica's ill-health. When I meet Brutus I will tell you all about him. I hope you are right about Plancus and Decimus. I am sorry if Sextus is throwing down his shield. Give me news of Mundus if you have any. I have answered all your points: now for my own news.

Young Quintus is coming with me as far as Puteoli —what a noble citizen! you might call him a Favonius Asinius.[1] He has two reasons: he wants to be with me and to make peace with Brutus and Cassius. But what have you to say to this? For I know you are intimate with the Othones. He says that Julia proposed it herself, for a divorce has been arranged. His father has asked me what sort of reputation she has. Not knowing why he asked, I said I had never heard anything about her except about her looks and her father. "But why?" I asked: and he said his son wanted her. Then, though I was disgusted, I said I did not believe in those reports. My brother's point is not to offer him any allowance, but she says it is no business of his. I suspect, however, he is indulging in fairy tales as usual. Still I should like you to make enquiries, which will be no trouble to you, and let me know.

What, pray, is this? When I had already sealed this letter, some Formians, who were dining with me, said they had seen Plancus[2]—the one from Buthrotus—the day before I wrote this, that is, on the 5th, with downcast mien and unapparelled steed;[3] and that his boys said he and the land-grabbers had been ejected by the Buthrotians. Well done they! But pray let me know all the circumstances.

[3] As Tyrrell suggests, this is probably a quotation from a play.

M. TULLI CICERONIS
EPISTULARUM AD ATTICUM
LIBER SEXTUS DECIMUS

I

CICERO ATTICO SAL.

Scr. in Puteolano VIII Id. Quint. a. 710

Nonis Quinctilibus veni in Puteolanum. Postridie iens ad Brutum in Nesidem haec scripsi. Sed eo die, quo veneram, cenanti Eros tuas litteras. Itane? NONIS IULIIS? Di hercule istis! Sed stomachari totum diem licet. Quicquamne turpius quam Bruto IULIIS? Redeo ad meum igitur "ἔτ' ἐῶμεν;" Nihil vidi. Sed quid est, quaeso, quod agripetas Buthroti concisos audio? Quid autem Plancus tam cursim (ita enim inaudiebam) diem et noctem? Sane cupio scire, quid sit. Meam profectionem laudari gaudeo. Videndum est, ut mansio laudetur. Dymaeos agro pulsos mare infestum habere nil mirum. Ἐν ὁμοπλοίᾳ Bruti videtur aliquid praesidii esse, sed, opinor, minuta navigia. Sed iam sciam et ad te cras. De Ventidio πανικὸν puto. De Sexto pro certo habebatur abicere[1] arma. Quod si verum est, sine bello

---

[1] abicere *Klotz*: ad *MSS.*: haud *Orelli*.

---

[1] The month Quinctilis had recently been renamed Julius after Caesar, who was born in that month.

CICERO'S LETTERS
TO ATTICUS
BOOK XVI

I

CICERO TO ATTICUS, GREETING.

On the 7th of Quinctilis I arrived at Puteoli, and I am writing this on the following day as I am crossing to Brutus at Nesis. The day I arrived Eros brought me your letter as I was dining. Is it really so? The 9th of *July*?[1] Heaven confound them! But I could go on cursing all day. Could they have insulted Brutus worse than with their *July*? So I must fall back on my old cry, "How long, O Lord?" I have never seen anything like that. But what, pray, is this I hear about the land-grabbers being cut to pieces at Buthrotum? And why has Plancus been on the run, as they tell me he has, day and night? I am very eager to know what it means. I am glad my departure is approved; I must see whether my staying may be approved too. That the people of Dyme, now they have been expelled from their land, should take to piracy is no wonder. There may be some safeguard in having Brutus as a fellow-passenger, but I think his vessels are small. I shall know soon and will tell you to-morrow. I think the report about Ventidius is a false alarm. It is held pretty certain that Sextus is laying down his arms; and, if that is so, it looks as though we should be reduced to slavery without even a civil war. What

Puteoli, July 8, B.C. *44*

civili video serviendum.   Quid ergo? ad Kal. Ian.
in Pansa spes?   Λῆρος πολὺς in vino et in somno
istorum.

De C̄C̄X̄ optime.   Ciceronis rationes explicentur.
Ovius enim recens.   Is multa, quae vellem, in iis ne
hoc quidem malum[1] HS L̄X̄X̄ĪĪ satis esse, adfatim
prorsus, sed Xenonem perexigue et γλίσχρως praebere.   Quo plus permutasti quam ad fructum insularum, id ille annus habeat, in quem itineris sumptus
accessit.   Hinc ex Kal. Apr. ad HS L̄X̄X̄X̄ accommo-
detur.   Nunc enim insulae tantum.   Videndum enim
est quid, cum Romae erit.   Non enim puto socrum
illam ferendam.   Pindaro de Cumano negaram.   Nunc,
cuius rei causa tabellarium miserim, accipe.   Quintus
filius mihi pollicetur se Catonem.   Egit autem et
pater et filius, ut tibi sponderem, sed ita, ut tum
crederes, cum ipse cognosses.   Huic ego litteras ipsius
arbitratu dabo.   Eae te ne moverint.   Has scripsi in
eam partem, ne me motum putares.   Di faxint, ut
faciat ea, quae promittit!   Commune enim gaudium.
Sed ego — nihil dico amplius.   Is hinc VII Idus.   Ait
enim attributionem in Idus, se autem urgeri acriter.
Tu ex meis litteris, quo modo respondeas, modera-
bere.   Plura, cum et Brutum videro et Erotem re-
mittam.   Atticae meae excusationem accipio eamque
amo plurimum; cui et Piliae salutem.

[1] *After* malum *the MSS. have* in mandatis si abunde, *which was deleted by Lambinus as a gloss.*

LETTERS TO ATTICUS XVI. 1

hope have we, then? In Pansa, when he enters office? There is nothing but midsummer madness in their drunken dreams.

About the £2,000[1]: well done! Put my son's affairs straight. For Ovius has just come, and brings much satisfactory news; among other things, what is no bad hearing, that £700[2] is enough, quite enough, but that Xeno treats him very sparingly and niggardly. The excess over the rental of the town houses that your bill of exchange cost, may be reckoned to the year, in which there was the additional expense of the journey. From the 1st of April on let him have up to £800,[3] for that is the rent of the city property now. Some sort of provision must be made for him when he comes to Rome. For I don't think he could endure that woman as a mother-in-law. I refused Pindarus' offer for the villa at Cumae. Now let me tell you why I have sent a messenger. Young Quintus is promising to be as prim as a puritan: and both he and his father have begged me to go bail to you for him, but on the condition that you only believe it when you see it. I shall give him a letter according to his fancy, but don't take any notice of it. I am writing now to prevent you from thinking that I do. God grant he keeps his promise. It would be a satisfaction to everybody. But I—I won't say any more. He is leaving on the 9th. For he says there is some money to be paid over on the 15th, but that he is very hard pressed. You will judge from my letter how to answer. More when I have seen Brutus and am sending Eros back. I accept dear Attica's apology and send her my best love. Give my regards to her and Pilia.

[1] 210 sestertia.   [2] 72 sestertia.   [3] 80 sestertia.

MARCUS TULLIUS CICERO

II

CICERO ATTICO SAL.

Scr. in Puteolano V Idus Quint. a. 710

vi Idus duas epistulas accepi, unam a meo tabellario, alteram a Bruti. De Buthrotiis longe alia fama in his locis fuerat, sed cum aliis multis hoc ferendum. Erotem remisi citius, quam constitueram, ut esset, qui Hortensio et Ouiae[1] quibus quidem ait se Idibus constituisse. Hortensius vero impudenter. Nihil enim debetur ei nisi ex tertia pensione, quae est Kal. Sext.; ex qua pensione ipsa maior pars est ei soluta aliquanto ante diem. Sed haec Eros videbit Idibus.

De Publilio autem, quod perscribi oportet, moram non puto esse faciendam. Sed, cum videas, quantum de iure nostro decesserimus, qui de residuis $\overline{\mathrm{CCCC}}$ HS $\overline{\mathrm{CC}}$ praesentia solverimus, reliqua rescribamus, loqui cum eo, si tibi videbitur, poteris eum commodum nostrum exspectare debere, cum tanta sit a nobis iactura facta iuris. Sed, amabo te, mi Attice (videsne, quam blande?), omnia nostra, quoad eris Romae, ita gerito, regito, gubernato, ut nihil a me exspectes. Quamquam enim reliqua satis apta sunt ad solvendum, tamen fit saepe, ut ii, qui debent, non respondeant ad tempus. Si quid eius modi acciderit, ne quid tibi sit fama mea potius. Non modo versura, verum etiam venditione, si ita res coget, nos vindicabis.

[1] Ouiae *Gurlitt*: quia e *MSS.*: coheredibus *Junius*.

II

CICERO TO ATTICUS, GREETING.

On the 10th I received two letters, one by my own messenger, another from Brutus'. Here the story about the Buthrotians was very different; but that, like many other things, we must put up with. I have sent Eros back sooner than I intended, that there may be someone to pay Hortensius and Ovia, with whom, indeed, he says he had made an appointment for the 15th. It is really shameless of Hortensius, for there is nothing owing to him except on the third instalment, which is due on the 1st of August; and the greater part of that instalment has been paid some time before the proper date. But Eros will see to that on the 15th.

Puteoli, July 11, B.C. 44

In Publilius' case I don't think there ought to be any delay in letting him have a draft for what is owing. But, when you see how much I have yielded my rights in paying up half of a balance of £4,000[1] in ready money, and now giving a bill for the rest, you may, if you think fit, tell him that he ought to await my convenience, when I have waived so much of my rights. But please, my dear Atticus—see how coaxingly I put it—do transact, regulate, and manage all my affairs while you are in Rome, without waiting for a hint from me. For though I have sufficient outstanding debts to meet my creditors, it often happens that the debtors don't pay at the proper time. If anything of that sort happens, consider nothing so much as my credit. Preserve it not only by raising a fresh loan, but by selling if necessary.

---

[1] 400 *sestertia*. The money was a repayment of the dowry Cicero had received with his second wife, whom he had since divorced.

MARCUS TULLIUS CICERO

Bruto tuae litterae gratae erant. Fui enim apud illum multas horas in Neside, cum paulo ante tuas litteras accepissem. Delectari mihi Tereo videbatur et habere maiorem Accio quam Antonio gratiam. Mihi autem quo laetiora sunt, eo plus stomachi et molestiae est populum Romanum manus suas non in defendenda re publica, sed in plaudendo consumere. Mihi quidem videntur istorum animi incendi etiam ad repraesentandam improbitatem suam. Sed tamen,

"dúm modo doleant áliquid, doleant quídlibet."

Consilium meum quod ais cotidie magis laudari, non moleste fero, exspectabamque, si quid de eo ad me scriberes. Ego enim in varios sermones incidebam. Quin etiam idcirco trahebam, ut quam diutissime integrum esset. Sed, quoniam furcilla extrudimur, Brundisium cogito. Facilior enim et exploratior devitatio legionum fore videtur quam piratarum, qui apparere dicuntur.

Sestius VI Idus exspectabatur, sed non venerat, quod sciam. Cassius cum classicula sua venerat. Ego, cum eum vidissem, V Id. in Pompeianum cogitabam, inde Aeculanum. Nosti reliqua. De Tutia ita putaram. De Aebutio non credo nec tamen curo plus quam tu. Planco et Oppio scripsi equidem, quoniam rogaras, sed, si tibi videbitur, ne necesse habueris reddere. Cum enim tua causa fecerint omnia, vereor, ne meas litteras supervacaneas arbi-

---

[1] There had been some exhibition of public feeling at the performance of Accius' *Tereus* at the games given by Brutus. Here, and in the *Philippics* II. 31, it is implied that it was

Brutus was pleased with your letter. I spent several hours with him at Nesis, just after I received your letter. He seemed to be delighted at the Tereus incident, and to feel more grateful to Accius than to Antonius.[1] For my part the better the news is, the more it annoys and pains me, that the Roman people use their hands not for defending the constitution but for clapping. It seems to me that the Caesarian party is possessed of a positive mania for parading its disloyalty. However, "so they but suffer, be it what it will."[2]

You say my plan is daily more commended. I am not sorry, and I am looking forward to anything you may say about it. For I have met with various opinions; and, indeed, for that reason I am hesitating as long as possible before committing myself. But since I am being turned out with a pitchfork, I am thinking of Brundisium. For it seems to me to be more certain and easier to avoid the soldiers than the pirates, who are said to be in evidence.

I expected Sestius on the 10th, but he has not come, so far as I know. Cassius has arrived with his little fleet. When I have seen him, I am thinking of going on the 11th to Pompeii, and thence to Aeculanum. You know the rest. About Tutia, that is what I thought. As for Aebutius, I don't believe it; nor do I care any more than you do. I have written of course to Plancus and Oppius, as you asked me: but, if you think better of it, don't hold yourself bound to deliver the letters. For, since they have done it all for your sake, I fear my letters

---

favourable to Brutus, but Appius, *B.C.* III. 24, states that outbursts against Caesar's assassins drove them to decide on leaving Italy.

[2] From Afranius. Cf. Cicero, *Tusc. Disp.* IV. 45 and 55.

trentur, Oppio quidem utique, quem tibi amicissimum cognovi. Verum, ut voles.

Tu, quoniam scribis hiematurum te in Epiro, feceris mihi gratum, si ante eo veneris, quam mihi in Italiam te auctore veniendum est. Litteras ad me quam saepissime; si de rebus minus necessariis, aliquem nanctus; sin autem erit quid maius, domo mittito.

Ἡρακλείδειον, si Brundisium salvi, adoriemur. "De gloria" misi tibi. Custodies igitur, ut soles, sed notentur eclogarii, quos Salvius bonos auditores nactus in convivio dumtaxat legat. Mihi valde placent, mallem tibi. Etiam atque etiam vale.

III

CICERO ATTICO SAL.

Scr. in Pompeiano XVI K. Sext. a. 710

Tu vero sapienter (nunc demum enim rescribo iis litteris, quas mihi misisti convento Antonio Tiburi) sapienter igitur, quod manus dedisti, quodque etiam ultro gratias egisti. Certe enim, ut scribis, deseremur ocius a re publica quam a re familiari. Quod vero scribis te cotidie magis delectare "O Tite, si quid," auges mihi scribendi alacritatem. Quod Erotem non sine munusculo exspectare te dicis, gaudeo non fefellisse eam rem opinionem tuam; sed tamen idem σύνταγμα misi ad te retractatius, et quidem ἀρχέτυπον ipsum crebris locis inculcatum et refectum. Hunc tu tralatum in macrocollum lege arcano convivis tuis,

may appear superfluous to them—to Oppius at any rate, as I know he is a great admirer of yours. But just as you please.

As you say you are going to spend the winter in Epirus, I shall take it kindly if you will come before the time at which you advise me to return to Italy. Send me letters as often as possible; if on matters of little importance, by any messenger you can find; but if on important affairs, send some one of your own.

I will attempt a work in Heracleides' style, if I get safe to Brundisium. I am sending you my *De Gloria*. Please keep it as usual, but have select passages marked for Salvius to read when he has an appropriate party to dinner. I am very pleased with them, and I hope you will be too. Farewell, and yet again farewell.

III

CICERO TO ATTICUS, GREETING.

At last I am answering the letter you sent me *Pompeii,* after meeting Antony at Tibur. Well, then, you *July 17,* B.C. were wise in giving in and even going so far as to *44* thank him. For certainly, as you say, we shall be robbed of our constitution before we are robbed of our private property. So you take more and more delight in my treatise on Old Age daily. That increases my energy in writing. You say you expect Eros not to come to you empty-handed. I am glad you have not been disappointed in the event; but at the same time I am sending you the same composition more carefully revised, indeed the original copy, with plenty of additions between the lines and corrections. Have it copied on large paper and read it privately to your guests; but, if you love me, do it

sed, si me amas, hilaris et bene acceptis, ne in me stomachum erumpant, cum sint tibi irati.

De Cicerone velim ita sit, ut audimus. De Xenone coram cognoscam; quamquam nihil ab eo arbitror neque indiligenter neque inliberaliter. De Herode faciam, ut mandas, et ea, quae scribis, ex Saufeio et e Xenone cognoscam. De Quinto filio gaudeo tibi meas litteras prius a tabellario meo quam ab ipso redditas; quamquam te nihil fefellisset. Verum tamen —. Sed exspecto, quid ille tecum, quid tu vicissim, nec dubito, quin suo more uterque. Sed eas litteras Curium mihi spero redditurum. Qui quidem etsi per se est amabilis a meque diligitur, tamen accedet magnus cumulus commendationis tuae.

Litteris tuis satis responsum est; nunc audi, quod, etsi intellego scribi necesse non esse, scribo tamen. Multa me movent in discessu, in primis mehercule quod diiungor a te. Movet etiam navigationis labor alienus non ab aetate solum nostra, verum etiam a dignitate tempusque discessus subabsurdum. Relinquimus enim pacem, ut ad bellum revertamur, quodque temporis in praediolis nostris et belle aedificatis et satis amoenis consumi potuit, in peregrinatione consumimus. Consolantur haec: aut proderimus aliquid Ciceroni, aut quantum profici possit, iudicabimus. Deinde tu iam, ut spero, et ut promittis, aderis. Quod quidem si acciderit, omnia nobis erunt meliora. Maxime autem me angit ratio reliquorum meorum. Quae quamquam ex-

when they are in a good temper and have had a good dinner, for I don't want them to vent on me the anger they feel towards you.

In my son's case I hope things may be as we hear. About Xeno I shall know when I see him, though I don't suppose he is neglecting his duty or acting meanly. I will do as you say about Herodes, and will find out what you mention from Saufeius and Xeno. As for young Quintus, I am glad my letter was delivered by my messenger sooner than the one he took himself, though you would not have been taken in anyhow. However—but I am anxious to hear what he said to you and what you answered, though I have no doubt you both behaved characteristically. I hope Curius will deliver that letter to me. Though he is pleasant enough and I like him myself, still your recommendation will add the crowning grace.

I have answered your letter sufficiently; now hear what I am going to say, though I know there is no necessity for me to say it. In regard to my journey I am distressed about many things, the chief being that I am separated from you. Then again there is the fatigue of the voyage, a thing unsuitable not only to my age but to my rank too, and the time of my departure is rather ridiculous. For I am leaving peace to return to war, and wasting in travelling time that might be spent in my country houses, which are comfortably built and pleasantly situated. My consolations are these. I shall either benefit my son or see how much he can be benefited. Then again, as I hope and as you promise, you will soon be coming too; and if that happens it will make me far happier. But the thing that worries me most is the arranging of my balances; for, though things

plicata sunt, tamen, quod et Dolabellae nomen in iis est et in attributione mihi nomina ignota, conturbor, nec me ulla res magis angit ex omnibus. Itaque non mihi videor errasse, quod ad Balbum scripsi apertius, ut, si quid tale accidisset, ut non concurrerent nomina, subveniret, meque tibi etiam mandasse, ut, si quid eius modi accidisset, cum eo communicares. Quod facies, si tibi videbitur, eoque magis, si proficisceris in Epirum.

Haec ego conscendens e Pompeiano tribus actuariolis decemscalmis. Brutus erat in Neside etiam nunc, Neapoli Cassius. Ecquid amas Deiotarum et non amas Hieram? Qui, ut Blesamius venit ad me, cum ei praescriptum esset, ne quid sine Sexti nostri sententia ageret, neque ad illum neque ad quemquam nostrum rettulit. Atticam nostram cupio absentem suaviari. Ita mi dulcis salus visa est per te missa ab illa. Referes igitur ei plurimam itemque Piliae dicas velim.

IV

CICERO ATTICO SAL.

Scr. in Puteolano VI Id. Quint. a. 710

Ita ut heri tibi narravi vel fortasse hodie (Quintus enim altero die se aiebat), in Nesida VIII Idus. Ibi Brutus. Quam ille doluit de NONIS IULIIS! mirifice est conturbatus. Itaque sese scripturum aiebat, ut venationem eam, quae postridie ludos Apollinares futura est, proscriberent in III IDUS QUINCTILES. Libo

---

[1] Hieras and Blesamius were agents of Deiotarus in bribing Antony to restore Armenia to him, and apparently were

LETTERS TO ATTICUS XVI. 3–4

have been put straight, I am anxious when I see Dolabella's name among them, and drafts on people that I do not know among my assets: and that makes me more uneasy than anything else. So I don't think I was wrong in applying to Balbus quite openly to assist me, if such a thing should happen as my debts not coming in properly, and telling him that I had commissioned you to communicate with him in any such event. Do so, if you think fit, especially if you are starting for Epirus.

This I have written just as I was embarking from Pompeii with three ten-oared pinnaces. Brutus is still in Nesis, Cassius at Naples. Can you like Deiotarus and not like Hieras?[1] When Blesamius came to me Hieras was commissioned not to do anything without Sextus Peducaeus' advice, but he never consulted him or any of our friends. I should like to kiss Attica, far off as she is: I was so pleased with the good wishes she sent me through you. So please give her my best thanks, and the same to Pilia.

IV

CICERO TO ATTICUS, GREETING.

As I told you yesterday or perhaps to-day—for Quintus said he would take two days going—I went to Nesis on the 8th: and there was Brutus. How annoyed he was about the "7th of *July*."[2] It quite upset him. So he said he would send orders for them to advertise the beast-hunt, which is to take place on the day after the games to Apollo, as on the "13th of *Quinctilis*." Libo came in, and he

Puteoli, July 10, B.C. 44

now disowned by him after he had succeeded in getting it back.   [2] Cf. *Att.* XVI. 1.

intervenit. Is Philonem, Pompei libertum, et Hilarum, suum libertum, venisse a Sexto cum litteris ad consules, "sive quo alio nomine sunt." Earum exemplum nobis legit, si quid videretur. Pauca παρὰ λέξιν, ceteroqui et satis graviter et non contumaciter. Tantum addi placuit, quod erat "coss." solum, ut esset "PRAETT., TRIBB. PL., SENATVI," ne illi non proferrent eas, quae ad ipsos missae essent. Sextum autem nuntiant cum una solum legione fuisse Karthagine, eique eo ipso die, quo oppidum Baream cepisset, nuntiatum esse de Caesare, capto oppido miram laetitiam commutationemque animorum concursumque undique; sed illum ad sex legiones, quas in ulteriore reliquisset, revertisse. Ad ipsum autem Libonem scripsit nihil esse, nisi ad larem suum liceret. Summa postulatorum, ut omnes exercitus dimittantur, qui ubique sint. Haec fere de Sexto.

De Buthrotiis undique quaerens nihil reperiebam. Alii concisos agripetas, alii Plancum acceptis nummis relictis illis aufugisse. Itaque non video sciturum me, quid eius sit, ni statim aliquid litterarum.

Iter illud Brundisium, de quo dubitabam, sublatum videtur. Legiones enim adventare dicuntur. Haec autem navigatio habet quasdam suspiciones periculi. Itaque constituebam uti ὁμοπλοίᾳ. Paratiorem enim offendi Brutum, quam audiebam. Nam et ipse et Domitius bona plane habet dicrota, suntque navigia praeterea luculenta Sesti, Buciliani, cete-

told us that Philo, a freedman of Pompey, and Hilarus, one of his own, had come from Sextus with a letter for the consuls, "or whatever they call them." He read us a copy to see what we thought of it. There were a few odd expressions, but in other respects it was sufficiently dignified and not aggressive. We only thought it better to make an addition of "Praetors, Tribunes of the People, and Senate" to the simple address to the "Consuls," for fear they should not publish a letter sent to them. They say that Sextus has been at Carthage with only one legion, and that he received the news about Caesar on the very day that he took the town of Barea. After the capture there were great rejoicings and a change of sentiment, and people flocked to him from every side, but he returned to the six legions he had left in lower Spain. He has written to Libo himself saying it is all nothing to him if he cannot get home. The upshot of his demands is, that all the armies everywhere should be disbanded. That is all about Sextus.

I have been making enquiries in every direction about the Buthrotians, and discover nothing. Some say the land-grabbers were cut to pieces, others that Plancus pocketed the money and fled, leaving them in the lurch. So I don't see how I can find out what there is in it, unless I get a letter at once.

The route to Brundisium, about which I was hesitating, seems to be out of the question. They say the troops are arriving there. But the voyage from here has some suspicion of danger, so I have made up my mind to sail in company with Brutus. I found him better prepared than I had heard he was. For both he and Domitius have quite good two-banked galleys, and there are also some good ships belonging

rorum. Nam Cassi classem, quae plane bella est, non numero ultra fretum. Illud est mihi submolestum, quod parum Brutus properare videtur. Primum confectorum ludorum nuntios exspectat; deinde, quantum intellego, tarde est navigaturus consistens in locis pluribus. Tamen arbitror esse commodius tarde navigare quam omnino non navigare; et, si, cum processerimus, exploratiora videbuntur, etesiis utemur.

V

CICERO ATTICO SAL.

Scr. in Puteolano VII Id. Quint. a. 710

Tuas iam litteras Brutus exspectabat. Cui quidem ego non novum attuleram de Tereo Acci. Ille Brutum putabat. Sed tamen rumoris nescio quid adflaverat commissione Graecorum frequentiam non fuisse; quod quidem me minime fefellit; scis enim, quid ego de Graecis ludis existimem.

Nunc audi, quod pluris est quam omnia. Quintus fuit mecum dies complures, et, si ego cuperem, ille vel plures fuisset; sed, quam diu fuit, incredibile est, quam me in omni genere delectarit, in eoque maxime, in quo minime satis faciebat. Sic enim commutatus est totus et scriptis meis quibusdam, quae in manibus habebam, et adsiduitate orationis et praeceptis, ut tali animo in rem publicam, quali nos volumus, futurus sit. Hoc cum mihi non modo confirmasset, sed

LETTERS TO ATTICUS XVI. 4–5

to Sestus, Bucilianus, and others. For I don't count on Cassius' fleet, which is quite a fine one, beyond the straits of Sicily. There is one point that annoys me a little, Brutus seems in no hurry. First he is waiting for news of the completion of his games; then, so far as I can understand, he is going to sail slowly, stopping at several places. Still I think it will be better to sail slowly than not to sail at all; and if, when we have got some distance, things seem clearer, we shall take advantage of the Etesian winds.

V

CICERO TO ATTICUS, GREETING.

Brutus is expecting a letter from you. The news I brought him about Accius' *Tereus* was no news. He thought it was the *Brutus*.[1] There had, however, been some breath of rumour that at the opening of the Greek games the audience was small, at which, indeed, I was not at all surprised; for you know what I think of Greek games.

Now hear the most important point of all. Quintus has been with me several days, and, if I had desired, he would have stayed longer; but, so far as his visit went, you would not believe how pleased I was with him in every way, and especially in that in which I used most to disapprove of him. For he is so totally changed, partly by some works of mine, which I have in hand, and partly by my continual advice and exhortation, that he will in the future be as loyal as we could wish to the constitution. After he had not only asseverated this, but convinced me

Puteoli, July 9, B.C. 44

---

[1] Cf. *Att.* xvi. 2. Not being present Brutus had supposed it was the play called *Brutus,* whereas it was the *Tereus.*

etiam persuasisset, egit mecum accurate multis verbis, tibi ut sponderem se dignum et te et nobis futurum; neque se postulare, ut statim crederes, sed, cum ipse perspexisses, tum ut se amares. Quodnisi fidem mihi fecisset, iudicassemque hoc, quod dico, firmum fore, non fecissem id, quod dicturus sum. Duxi enim mecum adulescentem ad Brutum. Sic ei probatum est, quod ad te scribo, ut ipse crediderit, me sponsorem accipere noluerit, eumque laudans amicissime mentionem tui fecerit, complexus osculatusque dimiserit. Quam ob rem, etsi magis est, quod gratuler tibi, quam quod te rogem, tamen etiam rogo, ut, si quae minus antea propter infirmitatem aetatis constanter ab eo fieri videbantur, ea iudices illum abiecisse, mihique credas multum allaturam, vel plurimum potius, ad illius iudicium confirmandum auctoritatem tuam.

Bruto cum saepe iniecissem de ὁμοπλοίᾳ, non perinde, atque ego putaram, arripere visus est. Existimabam μετεωρότερον esse, et hercule erat et maxime de ludis. At mihi, cum ad villam redissem, Cn. Lucceius, qui multum utitur Bruto, narravit illum valde morari, non tergiversantem, sed exspectantem, si qui forte casus. Itaque dubito, an Venusiam tendam et ibi exspectem de legionibus. Si aberunt, ut quidam arbitrantur, Hydruntem, si neutrum erit ἀσφαλές, eodem revertar. Iocari me putas? Moriar, si quisquam me tenet praeter te. Etenim circumspice, sed antequam erubesco. O dies in auspiciis Lepidi

---

[1] Possibly there is some corruption in the text here, as the remark seems senseless.

of it, he was very pressing for me to go bail to you that he will come up to your and our expectations for the future; and he did not ask you to believe this at once, but that you should restore your affection to him, when you had seen it for yourself. If he had not convinced me of it, and I did not think that what I am saying is trustworthy, I should not have done what I am going to tell you. I took the young man with me to Brutus, and he was so convinced of what I mention that he believed it on his own account, refusing to hold me sponsor for Quintus. He praised him and mentioned you in the most friendly way, and dismissed him with an embrace and a kiss. So, although there is more reason for congratulating you than asking favours of you, still I do ask you, if you have regarded his actions up to now as showing some of the flightiness of youth, to believe that he has got rid of that, and to trust me that your influence will contribute much, or rather everything, towards making his decision permanent.

I have frequently thrown out a hint to Brutus about sailing with him, but he does not seem to jump at it as I thought he would. He seemed to me rather distrait, and indeed he was, especially about the games. But when I got back home, Lucceius, who is very intimate with him, said he was hesitating a good deal, not because he has changed his mind, but in the hope that something may turn up. So I am wondering whether to make for Venusia and there await news of the troops. If they are not there, as some think, I shall go to Hydrus; if neither road is safe, I will come back here. Do you think I am joking? Upon my life you are the only person who keeps me here. Just look round you, but do it before I blush.[1] Lepidus' choice of his day

lepide descriptos et apte ad consilium reditus nostri!
Magna ῥοπή ad proficiscendum in tuis litteris. Atque
utinam te illic! Sed ut conducere putabis.

Nepotis epistulam exspecto. Cupidus ille meorum?
qui ea, quibus maxime γαυριῶ, legenda non putet.
Et ais "μετ' ἀμύμονα"! Tu vero "ἀμύμων," ille quidem "ἄμβροτος." Mearum epistularum nulla est
συναγωγή; sed habet Tiro instar septuaginta; et quidem sunt a te quaedam sumendae. Eas ego oportet
perspiciam, corrigam. Tum denique edentur.

VI

CICERO ATTICO SAL.

Scr. Vibone
VIII K.
Sext. a. 710

Ego adhuc (perveni enim Vibonem ad Siccam)
magis commode quam strenue navigavi; remis enim
magnam partem, prodromi nulli. Illud satis opportune, duo sinus fuerunt, quos tramitti oporteret,
Paestanus et Vibonensis. Utrumque pedibus aequis
tramisimus. Veni igitur ad Siccam octavo die e Pompeiano, cum unum diem Veliae constitissem. Ubi
quidem fui sane libenter apud Talnam nostrum, nec
potui accipi, illo absente praesertim, liberalius. VIIII
Kal. igitur ad Siccam. Ibi tamquam domi meae scilicet. Itaque obduxi posterum diem. Sed putabam,

---

[1] Cf. *Odyssey* XI. 169, where Ajax is said to rank next after
"the blameless son of Peleus" (μετ' ἀμύμονα Πηλείωνα).

[2] North-north-east winds, called "fore-runners," because

LETTERS TO ATTICUS XVI. 5–6

of inauguration is as happy as his name, and fits excellently with my plan for returning. Your letter supplies a strong incentive for going. I only wish you were there: but that must be as you think best for yourself.

I am expecting a letter from Nepos. Does he really want my books, when he thinks the subjects I am keenest on not worth reading. You call him an Achilles to your Ajax.[1] No, you are the Achilles and he is one of the immortals. There is no collection of my letters, but Tiro has about seventy, and some can be got from you. Those I ought to see and correct, and then they may be published.

VI

CICERO TO ATTICUS, GREETING.

I have got as far as Sicca's house at Vibo, and at present I have taken it easy and not exerted myself. We have rowed most of the way, as there have been none of the usual north winds.[2] That was rather lucky, as there were two bays to cross, that of Paestum and that of Vibo. We crossed both with the wind behind us.[3] So I got to Sicca's place eight days after leaving Pompeii, having stopped one day at Velia. There I stayed at Talna's house very enjoyably, and I could not have been more liberally entertained, especially as he was away. So I got to Sicca on the 24th, and here I am quite at home. So I have stayed a day longer than I meant. But I think, when I get to

Vibo, July 25, B.C. 44

they usually prevailed for eight days before the rising of the Dog-star.

[3] The *pedes* were ropes attached to the sail to set it to the wind. Both would be let out to an equal length when sailing before the wind.

MARCUS TULLIUS CICERO

cum Regium venissem, fore ut illic "δολιχὸν πλόον ὁρμαίνοντες" cogitaremus, corbitane Patras an actuariolis ad Leucopetras Tarentinorum atque inde Corcyram, et, si oneraria, statimne freto an Syracusis. Hac super re scribam ad te Regio.

Mehercule, mi Attice, saepe mecum:

"Ἡ δεῦρ᾽ ὁδός σοι τί δύναται;"

Cur ego tecum non sum? cur ocellos Italiae, villulas meas, non video? Sed id satis superque, tecum me non esse, quid fugientem? periculumne? At id nunc quidem, nisi fallor, nullum est. Ad ipsum enim revocat me auctoritas tua; scribis enim in caelum ferri profectionem meam, sed ita, si ante K. Ianuar. redeam; quod quidem certe enitar. Malo enim vel cum timore domi esse quam sine timore Athenis tuis. Sed tamen perspice, quo ista vergant, mihique aut scribe, aut, quod multo malim, adfer ipse. Haec hactenus.

Illud velim in bonam partem accipias me agere tecum, quod tibi maiori curae sciam esse quam ipsi mihi. Nomina mea, per deos, expedi, exsolve. Bella reliqua reliqui; sed opus est diligentia, coheredibus pro Cluviano Kal. Sextil. persolutum ut sit. Cum Publilio quo modo agendum sit, videbis. Non debet urgere, quoniam iure non utimur. Sed tamen ei quoque satis fieri plane volo. Terentiae vero quid ego dicam? Etiam ante diem, si potes. Quin, si, ut spero, celeriter in Epirum, hoc, quod satisdato debeo,

---

[1] A verse from an unknown author, quoted in a fuller form in *Att.* xv. 11.

LETTERS TO ATTICUS XVI. 6

Regium, there, being "on a far voyage bent," I shall have to consider whether to proceed by a merchant vessel to Patrae or by packet-boats to Tarentine Leucopetra, and thence to Corcyra; and, if by a merchant ship, whether direct from the Sicilian strait or from Syracuse. On this point I will write to you from Regium.

Upon my word, Atticus, I often say to myself: "Why, what avails thee all thy journey here?"[1] Why am not I with you? Why may I not see my country houses, the jewels of Italy? But that alone is enough and more than enough, that I am not with you. And what am I fleeing from? Danger? Nay, unless I am mistaken, there is no danger now. For it is precisely at the hour of danger that you bid me come back. For you say my departure is praised to the skies, provided I return by the end of the year; and that I will certainly strive to do. For I had rather be at home in fear and trembling, than in your loved Athens without a fear. However, keep your eye on the trend of events, and write to me, or what I should much prefer, bring the news yourself. Enough of this.

Please take my next request in good part. I know you devote more care to it than I do myself. For mercy's sake keep my accounts clear and pay my debts. I have left a handsome balance; but it requires care to see to the payment of my fellow-heirs for the Cluvian property on the 1st of August. You will see how to manage about Publilius. He ought not to be pressing, as I am not insisting upon my legal rights. Still I should much like him also to be satisfied. As to Terentia, what am I to say? Pay her even before the proper date, if you can. But if, as I hope, you are coming soon to Epirus, pray make

peto a te ut ante provideas planeque expedias et
solutum relinquas. Sed de his satis, metuoque, ne tu
nimium putes.

Nunc neglegentiam meam cognosce. "De gloria"
librum ad te misi. At in eo prohoemium idem est
quod in Academico tertio. Id evenit ob eam rem,
quod habeo volumen prohoemiorum. Ex eo eligere
soleo, cum aliquod σύγγραμμα institui. Itaque iam
in Tusculano, qui non meminissem me abusum isto
prohoemio, conieci id in eum librum, quem tibi misi.
Cum autem in navi legerem Academicos, adgnovi
erratum meum. Itaque statim novum prohoemium
exaravi et tibi misi. Tu illud desecabis, hoc adgluti-
nabis. Piliae salutem dices et Atticae, deliciis atque
amoribus meis.

VII

CICERO ATTICO SAL.

Scripsit navi-
gans ad
Pompeianum
XIV K.
Sept. a. 710

VIII Idus Sextil. cum a Leucopetra profectus (inde
enim tramittebam) stadia circiter CCC processissem,
reiectus sum austro vehementi ad eandem Leucope-
tram. Ibi cum ventum exspectarem (erat enim villa
Valeri nostri, ut familiariter essem et libenter), Regini quidam illustres homines eo venerunt Roma
sane recentes, in iis Bruti nostri hospes, qui Brutum
Neapoli reliquisset. Haec adferebant, edictum Bruti
et Cassi, et fore frequentem senatum Kalendis, a
Bruto et Cassio litteras missas ad consulares et prae-

arrangements first for any bills I have put my name to, and put affairs straight and leave them paid. But of this enough, and I fear you may think too much.

Now I must confess my carelessness. I sent you the work *On Glory*. But the preface to it is the same as that to the third book of the *Academics*. That is due to my having a volume of prefaces, from which I select one when I have begun a composition. So, when I was at Tusculum, forgetting I had used that preface, I put it into the book I sent you. But when I was reading the *Academics* on the boat I noticed my mistake. So I dashed off a new preface at once, and have sent it to you. Please cut the other off and glue this on. Pay my respects to Pilia and to my pet and darling Attica.

VII

CICERO TO ATTICUS, GREETING.

On shipboard on the way to Pompeii, Aug. 19, B.C. 44

When I had started from Leucopetra—for that was where I began my crossing—on the sixth of August and gone some forty miles, I was driven back to Leucopetra again by a strong south wind. While I was waiting there for the wind—our friend Valerius has a house there, so I was at home and enjoying myself—there came some men of mark of Regium, fresh from Rome, among them a guest of our friend Brutus, who said he had left Brutus at Naples. They brought an edict of Brutus and Cassius and news that there would be a full meeting of the House on the first of the month and that a letter had been sent by Brutus and Cassius to the ex-

torios, ut adessent, rogare. Summam spem nuntiabant fore ut Antonius cederet, res conveniret, nostri Romam redirent. Addebant etiam me desiderari, subaccusari.

Quae cum audissem, sine ulla dubitatione abieci consilium profectionis, quo mehercule ne antea quidem delectabar. Lectis vero tuis litteris admiratus equidem sum te tam vehementer sententiam commutasse, sed non sine causa arbitrabar. Etsi, quamvis non fueris suasor et impulsor profectionis meae, adprobator certe fuisti, dum modo Kal. Ian. Romae essem. Ita fiebat, ut, dum minus periculi videretur, abessem, in flammam ipsam venirem. Sed haec, etiamsi non prudenter, tamen ἀνεμέσητα sunt, primum quod de mea sententia acta sunt, deinde, etiamsi te auctore, quid debet, qui consilium dat, praestare praeter fidem? Illud admirari satis non potui, quod scripsisti his verbis: "Bene igitur tu, qui εὐθανασίαν, bene! relinque patriam." An ego relinquebam aut tibi tum relinquere videbar? Tu id non modo non inhibebas, verum etiam adprobabas. Graviora, quae restant. "Velim σχόλιον aliquod elimes ad me oportuisse te istuc facere." Itane, mi Attice? defensione eget meum factum, praesertim apud te, qui id mirabiliter adprobasti? Ego vero istum ἀπολογισμὸν συντάξομαι, sed ad eorum aliquem, quibus invitis et

consuls and ex-praetors asking them to be present. They said there were great hopes that Antony might yield, some agreement be arrived at, and our friends allowed to return to Rome; and they added that I was missed and people were inclined to blame me.

When I heard that, I had no hesitation about giving up my idea of going away, which to be sure I had never fancied even before that: and when I read your letter, I was certainly surprised that you had so utterly changed your opinion; but there seemed to me to be good reason for it. However, though it was not you who persuaded and urged me to go, you certainly approved of my going, if I got back by the end of the year. That would have meant, that, when there was little danger, I should have been away, and should return when it was in full blaze. But that, although it was not a counsel of prudence, I have no right to resent, first because it happened by my own wish, and secondly, even if you had advised me, an adviser need not guarantee anything but his sincerity. What did astonish me beyond measure was that you should use the words: "A fine thing for you, who talk of a noble death, a fine thing, i' faith. Go, desert your country." Was I deserting it, or did you at the time think I was deserting it? You not only raised no finger against it, you even approved of it. The rest is even more severe: "I wish you would write me an explanatory note showing that it was your duty to do it?" So, my dear Atticus? Does my action need defending, especially to you, who expressed strong approval? Yes, I will write a defence, but for some of those who opposed my going and spoke against it.

MARCUS TULLIUS CICERO

dissuadentibus profectus sum. Etsi quid iam opus est σχολίῳ? si perseverassem, opus fuisset. "At hoc ipsum non constanter." Nemo doctus umquam (multa autem de hoc genere scripta sunt) mutationem consilii inconstantiam dixit esse. Deinceps igitur haec: "Nam, si a Phaedro nostro esses, expedita excusatio esset; nunc quid respondemus?" Ergo id erat meum factum, quod Catoni probare non possem? flagitii scilicet plenum et dedecoris. Utinam a primo ita tibi esset visum! tu mihi, sicut esse soles, fuisses Cato. Extremum illud vel molestissimum: "Nam Brutus noster silet," hoc est: non audet hominem id aetatis monere. Aliud nihil habeo, quod ex iis a te verbis significari putem, et hercule ita est. Nam, XVI Kal. Sept. cum venissem Veliam, Brutus audivit; erat enim cum suis navibus apud Heletem fluvium citra Veliam mil. pass. III. Pedibus ad me statim. Dei immortales, quam valde ille reditu vel potius reversione mea laetatus effudit illa omnia, quae tacuerat! ut recordarer illud tuum "Nam Brutus noster silet." Maxime autem dolebat me Kal. Sext. in senatu non fuisse. Pisonem ferebat in caelum; se autem laetari, quod effugissem duas maximas vituperationes, unam, quam itinere faciendo me intellegebam suscipere, desperationis ac relictionis rei publicae (flentes mecum vulgo querebantur, quibus de meo celeri reditu non probabam), alteram, de qua Brutus, et qui una erant (multi autem erant), laetabantur,

LETTERS TO ATTICUS XVI. 7

Though what need is there of an explanatory note? If I had gone on, there would have been. "But coming back is not consistent." No philosopher ever called a change of plan inconsistency, though there has been a good deal written on the point. So you add: "If you were a follower of our friend Phaedrus,[1] one would have a defence ready: but, as it is, what answer can one give?" So my deed was one Cato would not approve of, was it? Of course then it was criminal and disgraceful. Would to heaven you had thought so at first; you should have been my Cato, as you usually are. Your last cut is the most unkind of all: "For our friend Brutus holds his peace," that is to say, he does not dare remonstrate with a man of my age. I see no other meaning that I can attach to your words, and no doubt that is it. For on the 17th, when I reached Velia, Brutus heard of it—he was with his boats on the river Heles about three miles from Velia; and he came at once on foot to see me. Great heavens, how he let out all his pent-up silence in joy at my return or rather my turning back. I could not help thinking of your "Our friend Brutus holds his peace." But what he regretted most was that I was not in the House on the first of August. Piso he lauded to the skies: and he expressed his delight that I had escaped two grounds for reproach. One of these was that of despairing and abandoning the country—and that I knew I might incur in undertaking the voyage; for many had complained to me with tears in their eyes, and I could not convince them of my speedy return. The other point that rejoiced Brutus and those who were with him—and there

[1] An Epicurean philosopher at Athens; cf. *Ad Fam.* XIII. 1.

quod eam vituperationem effugissem, me existimari ad Olympia. Hoc vero nihil turpius quovis rei publicae tempore, sed hoc ἀναπολόγητον. Ego vero austro gratias miras, qui me a tanta infamia averterit.

Reversionis has speciosas causas habes, iustas illas quidem et magnas; sed nulla iustior, quam quod tu idem aliis litteris: "Provide, si cui quid debetur, ut sit, unde par pari respondeatur. Mirifica enim δυσχρηστία est propter metum armorum." In freto medio hanc epistulam legi, ut, quid possem providere, in mentem mihi non veniret, nisi ut praesens me ipse defenderem. Sed haec hactenus; reliqua coram.

Antoni edictum legi a Bruto et horum contra scriptum praeclare; sed, quid ista edicta valeant aut quo spectent, plane non video. Nec ego nunc, ut Brutus censebat, istuc ad rem publicam capessendam venio. Quid enim fieri potest? Num quis Pisoni est adsensus? num rediit ipse postridie? Sed abesse hanc aetatem longe a sepulcro negant oportere.

Sed, obsecro te, quid est, quod audivi de Bruto? Piliam πειράζεσθαι παραλύσει te scripsisse aiebat. Valde sum commotus. Etsi idem te scribere sperare melius. Ita plane velim, et ei dicas plurimam salutem et suavissimae Atticae. Haec scripsi navigans, cum prope Pompeianum accederem, XIIII Kal.

were a lot of them—was that I had escaped the reproach of being thought to be going to the Olympian games. Nothing could be more disgraceful than that in any political circumstances, but at the present time it would be inexcusable. I of course felt very grateful to the south wind, which had saved me from such infamy.

There you have the ostensible reasons for my return; and they are good and sufficient reasons too; but none of them is better than one you mention in your letter: "If you owe anything to anyone, take measures to provide yourself with the means to pay each his due. For the money market is wonderfully tight owing to fear of war." I was in the middle of the straits when I read this letter, and I could not think of any way of taking measures, unless I came to look after it myself. But enough of this; more when we meet.

I got a sight of Antony's edict from Brutus, and of our friends' magnificent answer; but I don't quite see the use or the object of these edicts. Nor have I come as Brutus thought, to take part in the management of affairs. For what can be done? Did anybody agree with Piso? Did he himself come back the next day? But, as the saying goes, a man of my time of life ought not to go far from his grave.

But for mercy's sake what is this that I hear from Brutus! He says you told him Pilia had had an attack of paralysis. I am very much disturbed about it, though he tells me you say you hope she is better. I sincerely hope she is; give her and darling Attica my best regards. This I have written on ship-board, as I was getting near to Pompeii, Aug. 19.

MARCUS TULLIUS CICERO

VIII

CICERO ATTICO SAL.

Scr. in Puteolano IV Non. Nov. a. 710

Cum sciam, quo die venturus sim, faciam, ut scias. Impedimenta exspectanda sunt, quae Anagnia veniunt, et familia aegra est. Kal. vesperi litterae mihi ab Octaviano. Magna molitur. Veteranos, qui sunt Casilini et Calatiae, perduxit ad suam sententiam. Nec mirum, quingenos denarios dat. Cogitat reliquas colonias obire. Plane hoc spectat, ut se duce bellum geratur cum Antonio. Itaque video paucis diebus nos in armis fore. Quem autem sequamur? Vide nomen, vide aetatem. Atque a me postulat, primum ut clam conloquatur mecum vel Capuae vel non longe a Capua. Puerile hoc quidem, si id putat clam fieri posse. Docui per litteras id nec opus esse nec fieri posse. Misit ad me Caecinam quendam Volaterranum familiarem suum; qui haec pertulit, Antonium cum legione Alaudarum ad urbem pergere, pecunias municipiis imperare, legionem sub signis ducere. Consultabat, utrum Romam cum ↀↀↀ veteranorum proficisceretur an Capuam teneret et Antonium venientem excluderet, an iret ad tres legiones Macedonicas, quae iter secundum mare Superum faciunt; quas sperat suas esse. Eae congiarium ab Antonio accipere noluerunt, ut hic quidem narrat, et ei convicium grave fecerunt contionantemque reliquerunt. Quid quaeris? ducem se profitetur

---

[1] 500 denarii.

VIII

CICERO TO ATTICUS, GREETING.

When I know what day I shall arrive, I will let you know. I must wait for my heavy baggage, which is coming from Anagnia, and there is illness in my household. On the evening of the 1st I got a letter from Octavian. He is setting about a heavy task. He has brought over the veterans, who are at Casilinum and Calatia, to his views; and no wonder, when he is giving them £20[1] apiece. He thinks of visiting the other colonies. Obviously his idea is a war with Antony under his leadership. So I see that before long we shall be in arms. But whom are we to follow? Look at his name, and at his age. And his first request of me is that I should meet him secretly at Capua or somewhere near Capua. That is quite childish, if he thinks it can be done secretly. I have told him by letter that there is no necessity for it and no possibility of it. He sent me one Caecina of Volaterra, an intimate friend of his, who brought this news, that Antony is making for Rome with the legion Alauda, raising a forced contribution from towns, and marching with his soldiers under colours. He asked my advice about setting out for Rome with 3,000 veterans or holding Capua and intercepting Antony's advance, or going to the three Macedonian legions, which are making for the northern Adriatic. Those he hopes are on his side; they refused to take Antony's bounty, or so he says, heaped insults on him and left him still haranguing. Of course, he offers himself as our leader, and thinks we ought not to fail

MARCUS TULLIUS CICERO

nec nos sibi putat deesse oportere. Equidem suasi, ut Romam pergeret. Videtur enim mihi et plebeculam urbanam, et, si fidem fecerit, etiam bonos viros secum habiturus. O Brute, ubi es? quantam εὐκαιρίαν amittis! Non equidem hoc divinavi, sed aliquid tale putavi fore. Nunc tuum consilium exquiro. Romamne venio an hic maneo an Arpinum (ἀσφάλειαν habet is locus) fugiam? Romam, ne desideremur, si quid actum videbitur. Hoc igitur explica. Numquam in maiore ἀπορίᾳ fui.

IX

CICERO ATTICO SAL.

Scr. in Puteolano prid. Non. Nov. a. 710

Binae uno die mihi litterae ab Octaviano, nunc quidem, ut Romam statim veniam; velle se rem agere per senatum. Cui ego non posse senatum ante K. Ianuar., quod quidem ita credo. Ille autem addit "consilio tuo." Quid multa? ille urget, ego autem σκήπτομαι. Non confido aetati, ignoro, quo animo. Nil sine Pansa tuo volo. Vereor, ne valeat Antonius, nec a mari discedere libet, et metuo, ne quae ἀριστεία me absente. Varroni quidem displicet consilium pueri, mihi non. Si firmas copias habet, Brutum habere potest, et rem gerit palam. Centuriat Capuae, dinumerat. Iam iamque video bellum. Ad haec rescribe. Tabellarium meum Kalend. Roma profectum sine tuis litteris miror.

him. I advised that he should make for Rome. For it seems to me he ought to have the city rabble, and, if he succeeds in inspiring them with confidence, even the loyalists on his side. O Brutus, where are you? What a golden opportunity you are missing! I never foresaw this, but I thought something of the kind would happen. Now, I want your advice. Shall I come to Rome, or stay here, or flee to Arpinum, which would be a harbour of refuge? Rome I think, for fear I be missed, if people think a blow has been struck. Read me this riddle. I never was in a greater quandary.

IX

CICERO TO ATTICUS, GREETING.

Two letters on one day from Octavian, now asking me to come to Rome at once, as he wishes to act through the Senate. I told him I did not think the Senate could meet before January, and I really believe that is so. But he adds "with your advice." In short he is pressing, while I am temporizing. I do not trust his age: I do not know his disposition. I do not want to do anything without your friend Pansa's advice. I am afraid Antony may succeed, and I don't like going away from the sea, and I fear some great deed may be done in my absence. Varro, for his part, dislikes the boy's plan; I do not. If he can trust his army, he can have Brutus, and he is playing his game openly. He is dividing his men into companies at Capua, and paying over their bounty money. I see war close upon us. Please answer this letter. I am surprised my messenger left Rome on the 1st without a letter from you.

Puteoli,
Nov. 4, B.C.
44

MARCUS TULLIUS CICERO

X

CICERO ATTICO SAL.

Scr. in Sinu-
essano VI
Id. Nov. a.
710

VII Id. veni ad me in Sinuessanum. Eodem die vulgo loquebantur Antonium mansurum esse Casilini. Itaque mutavi consilium; statueram enim recta Appia Romam. Facile me ille esset adsecutus. Aiunt enim eum Caesarina uti celeritate. Verti igitur me a Menturnis Arpinum versus. Constitueram, ut V Idus aut Aquini manerem aut in Arcano. Nunc, mi Attice, tota mente incumbe in hanc curam; magna enim res est. Tria sunt autem, maneamne Arpini an propius accedam an veniam Romam. Quod censueris, faciam. Sed quam primum. Avide exspecto tuas litteras. VI Idus mane in Sinuessano.

XI

CICERO ATTICO SAL.

Scr. in
Puteolano
Non. Nov. a.
710

Nonis accepi a te duas epistulas, quarum alteram Kal. dederas, alteram pridie. Igitur prius ad superiorem. Nostrum opus tibi probari laetor; ex quo ἄνθη ipsa posuisti. Quae mihi florentiora sunt visa tuo iudicio; cerulas enim tuas miniatulas illas extimescebam. De Sicca ita est, ut scribis: ab¹ asta ea aegre me tenui. Itaque perstringam sine ulla contumelia

Iliad, xx. 308   Siccae aut Septimiae, tantum ut sciant " παῖδες παί-

---

¹ ab *added by Reid*: asta (= hasta, sensu obscoeno; cf. *Priapea*, 43, 1).

LETTERS TO ATTICUS XVI. 10–11

X

CICERO TO ATTICUS, GREETING.

On the 7th I reached my house at Sinuessa, and on that day it was generally said that Antony was going to stay at Casilinum. So I changed my plan, for I had intended to go straight on by the Appian way to Rome. He would easily have caught me up, for they say he travels as fast as Caesar. So from Menturnae I am turning off towards Arpinum, and I have made up my mind to stay at Aquinum or in Arcanum on the 9th. Now, my dear Atticus, throw yourself heart and soul into this question, for it is an important matter. There are three things open to me: to stay at Arpinum, to come nearer to Rome, or to go to Rome. What you advise, I will do? But answer at once. I am eagerly expecting a letter from you. Sinuessa, Nov. 8 in the morning.

Sinuessa, Nov. 8, B.C. 44

XI

CICERO TO ATTICUS, GREETING.

On the 5th I received two letters from you, dated the first, the other a day earlier. So I am answering the earlier first. I am glad you like my book, from which you quoted the very gems; and they seemed to me all the more sparkling for your judgment on them. For I was afraid of those red pencils[1] of yours. As for Sicca, it is as you say: I could hardly hold myself in about Antony's lust. So I will touch on it lightly without any opprobrium for Sicca and Septimia, and only let our children's

Puteoli, Nov. 5, B.C. 44

---

[1] Cf. *Att.* xv. 14, 4.

δων" sine φαλλῷ Luciliano eum ex C. Fadi filia liberos habuisse. Atque utinam eum diem videam, cum ista oratio ita libere vagetur, ut etiam in Siccae domum introeat! Sed "illo tempore opus est, quod fuit illis III viris." Moriar, nisi facete! Tu vero leges Sexto eiusque iudicium mihi perscribes. "Εἷς ἐμοὶ μύριοι." Caleni interventum et Calvenae cavebis.

Quod vereris, ne ἀδόλεσχος mihi tu, quis minus? Cui, ut Aristophani Archilochi iambus, sic epistula tua longissima quaeque optima videtur. Quod me admones, tu vero etiamsi reprenderes, non modo facile paterer, sed etiam laetarer, quippe cum in reprensione sit prudentia cum εὐμενείᾳ. Ita libenter ea corrigam, quae a te animadversa sunt, "eodem iure quo Rubriana" potius quam "quo Scipionis," et de laudibus Dolabellae deruam cumulum. Ac tamen est isto loco bella, ut mihi videtur, εἰρωνεία, quod eum ter contra cives in acie. Illud etiam malo: "indignissimum est hunc vivere" quam "quid indignius?" Πεπλογραφίαν Varronis tibi probari non moleste fero;

---

[1] The point of this sentence is not obvious. The translation follows Watson, who suggests that the pleasantry lies in calling the days of the triumvirate free in comparison with the date at which Cicero was writing. Other suggestions are (a) that there is a play on the triumvirate and the fact that Caesar and Pompey each had three wives; (b) that Septimia had three husbands; or (c) that it refers to some earlier date, possibly Cicero's consulate, when Fadia had three lovers. (Cf. Gurlitt, in *Philologus*, LVII. (1898) pp. 403–8).

[2] The Alexandrine grammarian, not the comic poet.

[3] 2 *Phil.* 103, where Cicero accuses Antony of obtaining possession of property by underhand means.

children know, without taking Lucilian licence, that Antony had children by a daughter of Fadius. I only wish I could see the day when my second *Philippic* could be sufficiently freely circulated to enter even Sicca's door. "But we want back the days of freedom under the triumvirs."[1] Upon my life that was a neat touch of yours. Please read my book to Sextus and let me know his opinion. I would take his word against all the world. Keep your eyes open for the appearance of Calenus and Calvena.

You fear I may think you a gas-bag. Who is less of one? I am like Aristophanes[2] with Archilochus' iambics—the longest letter of yours ever seems the best to me. As for your giving me advice, why, if you found fault with me, I should not only put up with it cheerfully, but even be glad of it, since in your fault-finding there is both wisdom and kindly purpose. So I will willingly correct the point you mention, and write "by the same right as you did the property of Rubrius" instead of "the property of Scipio";[3] and I will take the pinnacle off my praises of Dolabella. And yet to my thinking there is fine irony in the passage where I say he had thrice stood up in arms against his fellow-citizens.[4] Again I prefer your "it is most unjust that such a man should live" to "what can be more unjust?"[5] I am not sorry to hear you praise the *Peplographia*[6]

[4] 2 *Phil.* 75, with Caesar in Thessaly, Africa, and Spain.

[5] 2 *Phil.* 86. But the original reading is still found in our MSS.

[6] A "book of worthies," so-called from the sacred robe, embroidered with mythological and historical figures, offered once a year to Athene at Athens. The book was possibly identical with that generally known as the *Hebdomades sive Imagines*, but that is doubtful.

MARCUS TULLIUS CICERO

a quo adhuc Ἡρακλείδειον illud non abstuli. Quod me hortaris ad scribendum, amice tu quidem, sed me scito agere nihil aliud. Gravedo tua mihi molesta est. Quaeso, adhibe, quam soles diligentiam. "O Tite" tibi prodesse laetor. "Anagnini" sunt Mustela ταξιάρχης et Laco, qui plurimum bibit. Librum, quem rogas, perpoliam et mittam.

Haec ad posteriorem. "Τὰ περὶ τοῦ καθήκοντος," quatenus Panaetius, absolvi duobus. Illius tres sunt; sed, cum initio divisisset ita, tria genera exquirendi officii esse, unum, cum deliberemus, honestum an turpe sit, alterum, utile an inutile, tertium, cum haec inter se pugnare videantur, quo modo iudicandum sit, qualis causa Reguli, redire honestum, manere utile, de duobus primis praeclare disseruit, de tertio pollicetur se deinceps, sed nihil scripsit. Eum locum Posidonius persecutus est. Ego autem et eius librum accersivi et ad Athenodorum Calvum scripsi, ut ad me τὰ κεφάλαια mitteret; quae exspecto. Quem velim cohortere et roges, ut quam primum. In eo est περὶ τοῦ κατὰ περίστασιν καθήκοντος. Quod de inscriptione quaeris, non dubito, quin καθῆκον "officium" sit, nisi quid tu aliud; sed inscriptio plenior "de officiis." Προσφωνῶ autem Ciceroni filio. Visum est non ἀνοίκειον.

---

[1] *O Tite* are the opening words of the *De Senectute*.

LETTERS TO ATTICUS XVI. 11

of Varro; I have not yet managed to get the book in the style of Heracleides from him. You exhort me to go on writing. That is friendly of you; but let me tell you I do nothing else. I am sorry to hear of your cold. Please take as much care as usual of it. I am glad my book *On Old Age*[1] does you good. The "men of Anagnia"[2] are Mustela, the swashbuckler, and Laco, the champion toper. The book you ask for I will polish up and send.

Now for the second letter. The *De Officiis*, so far as Panaetius is concerned, I have finished in two books. He has three: but, though at the beginning he makes a three-fold division of cases in which duty has to be determined, one when the question is between right or wrong, another when it is between expediency and inexpediency, and the third, how we are to decide when it is a conflict between duty and expediency—for example, in Regulus' case to return would be right, to stay expedient—he treated of the first two brilliantly; the third he promises to add, but never wrote it. Posidonius took up that topic: but I have ordered his book and written to Athenodorus Calvus to send me an analysis of it, and that I am expecting. I wish you would spur him on and beg him to let me have it as soon as possible. In it duties under given circumstances are handled. As to your query about the title, I have no doubt that καθῆκον (duty) corresponds with *officium*, unless you have any other suggestion to make. But the fuller title is *De Officiis*. I am dedicating it to my son. It seems to me not inappropriate.

---

[2] 2 *Phil.* 106. The names have been inserted, as they are given in our MSS.

MARCUS TULLIUS CICERO

De Myrtilo dilucide. O quales tu semper istos. Itane? in D. Brutum? Di istis! Ego me, ut scripseram, in Pompeianum non abdidi, primo tempestatibus, quibus nil taetrius; deinde ab Octaviano cotidie litterae, ut negotium susciperem, Capuam venirem, iterum rem publicam servarem, Romam utique statim.

Iliad, vii. 93

"Αἴδεσθεν μὲν ἀνήνασθαι, δεῖσαν δ' ὑποδέχθαι."

Is tamen egit sane strenue et agit. Romam veniet cum manu magna, sed est plane puer. Putat senatum statim. Quis veniet? Si venerit, quis incertis rebus offendet Antonium? Kal. Ianuar. erit fortasse praesidio, aut quidem ante depugnabitur. Puero municipia mire favent. Iter enim faciens in Samnium venit Cales, mansit Teani. Mirifica ἀπάντησις et cohortatio. Hoc tu putares? Ob hoc ego citius Romam, quam constitueram. Simul et constituero, scribam.

Etsi nondum stipulationes legeram (nec enim Eros venerat), tamen rem pridie Idus velim conficias. Epistulas Catinam, Tauromenium, Syracusas commodius mittere potero, si Valerius interpres ad me nomina gratiosorum scripserit. Alii enim sunt alias, nostrique familiares fere demortui. Publice tamen scripsi, si uti vellet eis Valerius; aut mihi nomina mitteret.

---

[1] Of attempting Antony's life.

LETTERS TO ATTICUS XVI. 11

You make it as plain as daylight about Myrtilus. How well you can always take that lot off! Is it so? Do they accuse D. Brutus?[1] A malison on them! I have not hidden myself in Pompeii, as I said I should; first because of the weather, which has been abominable, and secondly because I get a letter from Octavian every day, asking me to take a hand in affairs, to come to Capua, to save the Republic again, and anyhow to go to Rome at once. It is a case of "ashamed to shirk, but yet afraid to take." He, however, has been acting, and still is acting, with great vigour. He will come to Rome with a big army; but he is such a boy. He thinks he can call a Senate at once. Who will come? If anyone comes, who will offend Antony in this uncertainty? Perhaps he may act as a safeguard on the 1st of January, or the battle may be over before then. The country towns are wonderfully enthusiastic for the boy. For, as he was making his way to Samnium, he came to Cales and stopped at Teanum. There was a marvellous crowd to meet him and cheers for him. Should you have thought it? That will make me come to Rome sooner than I had intended. As soon as I have arranged, I will write.

Though I have not yet read the agreements—for Eros has not come yet—still I wish you would get the business settled on the 12th. It will make it easier for me to send letters to Catina, Tauromenium, and Syracuse, if Valerius the interpreter will let me know the names of the influential people. For such people vary with the times, and most of my particular friends are dead. However, I have written general letters, if Valerius will content himself with them; otherwise he must send me names.

MARCUS TULLIUS CICERO

De Lepidianis feriis Balbus ad me usque ad III Kal. Exspectabo tuas litteras meque de Torquati negotiolo sciturum puto. Quinti litteras ad te misi, ut scires, quam valde eum amaret, quem dolet a te minus amari. Atticae, quoniam, quod optimum in pueris est, hilarula est, meis verbis suavium des volo.

XII

CICERO ATTICO SAL.

Scr. in Puteolano VIII Id. Nov. a. 710

Oppi epistulae, quia perhumana erat, tibi misi exemplum. De Ocella, dum tu muginaris nec mihi quicquam rescribis, cepi consilium domesticum itaque me pr. Idus arbitror Romae futurum. Commodius est visum frustra me istic esse, cum id non necesse esset, quam, si opus esset, non adesse, et simul, ne intercluderer, metuebam. Ille enim iam adventare potest. Etsi varii rumores multique, quos cuperem veros; nihil tamen certi. Ego vero, quicquid est, tecum potius, quam animi pendeam, cum a te absim, et de te et de me. Sed quid tibi dicam? Bonum animum. De Ἡρακλειδείῳ Varronis negotia salsa. Me quidem nihil umquam sic delectavit. Sed haec et alia maiora coram.

LETTERS TO ATTICUS XVI. 11–12

About the holidays for Lepidus' inauguration,[1] Balbus tells me they will last till the 29th. I am looking for a letter from you, and hope I shall hear about that little affair of Torquatus. I am sending Quintus' letter to show you how strong his affection is for the youth for whom he regrets you have so little. Please give Attica a kiss in my name for being such a merry little thing. It is the best sign in children.

XII

CICERO TO ATTICUS, GREETING.

I am sending you a copy of Oppius' letter, because it is so very courteous. About Ocella, while you are messing about and not writing me a line, I have consulted my own wits, and so I think I shall be in Rome on the 12th. I think it better for me to come there to no purpose, even if it is not necessary, than not to be there if it is, and at the same time I am afraid of being shut in there. For Antony may always be getting near. However, there are plenty of different rumours, which I hope may be true; there is no definite news. For my part, whatever it may be, I would rather be with you, than be in suspense both about you and about myself, when I am away from you. But what am I to say to you? Keep up your heart. About Varro's work in Heracleides' vein, that's an amusing business. I was never so pleased with anything. But of this and more important things when we meet.

Puteoli, Nov. 6, B.C. 44

[1] As *Pontifex Maximus*.

MARCUS TULLIUS CICERO

XIIIa

CICERO ATTICO SAL.

Scr. Aquini
IV Id. Nov.
a. 710

O casum mirificum! **v** Idus cum ante lucem de Sinuessano surrexissem venissemque diluculo ad pontem Tirenum, qui est Menturnis, in quo flexus est ad iter Arpinas, obviam mihi fit tabellarius; qui me offendit " δολιχὸν πλόον ὁρμαίνοντα." At ego statim "Cedo," inquam, "si quid ab Attico." Nondum legere poteramus; nam et lumina dimiseramus, nec satis lucebat. Cum autem luceret, ante scripta epistula ex duabus tuis prior mihi legi coepta est. Illa omnium quidem elegantissima. Ne sim salvus, si aliter scribo ac sentio. Nihil legi humanius. Itaque veniam, quo vocas, modo adiutore te. Sed nihil tam ἀπροσδιόνυσον mihi primo videbatur quam ad eas litteras, quibus ego a te consilium petieram, te mihi ista rescribere. Ecce tibi altera, qua hortaris " παρ'

Odyssey, iii. 171 ἠνεμόεντα Μίμαντα, νῆσον ἐπὶ Ψυρίης," Appiam scilicet " ἐπ' ἀριστέρ' ἔχοντα." Itaque eo die mansi Aquini. Longulum sane iter et via mala. Inde postridie mane proficiscens has litteras dedi.

XIIIb

CICERO ATTICO SAL.

Scr. in
Arpinati III
Id. Nov. a.
710

... et quidem, ut a me dimitterem invitissimus, fecerunt Erotis litterae. Rem tibi Tiro narrabit. Tu, quid faciendum sit, videbis. Praeterea, possimne

XIIIa
CICERO TO ATTICUS, GREETING.

What a strange coincidence! On the 9th I got up before daybreak to go on from Sinuessa, and before dawn I had reached the Tirenian bridge at Menturnae, where the road for Arpinum branches off, when I met a messenger, who found me "on a far journey bent." I at once enquired: "Pray, is there anything from Atticus?" I could not read as yet, for I had dismissed the link-bearers and it was not yet light enough. But, when it got light, I began to read the first of your two letters, having already written one to you. Your note was a model of elegance. Upon my life I am not saying more than I mean. I never read a kinder. So I will come, when you call me, provided you will assist me. But at first sight I thought nothing could be more *mal à propos* than such an answer to a letter in which I had asked for your advice. Then there is your other letter, in which you advise me to go "by windy Mimas towards the Psyrian isle,"[1] that is keeping the Appian way on the left side. So I have stayed the day at Aquinum. It was rather a wearisome journey and the road was bad. This letter I am sending the next morning as I am leaving.

Aquinum,
Nov. 10, B.C.
44

XIIIb
CICERO TO ATTICUS, GREETING.

... and indeed Eros' letter made me dismiss him most unwillingly. Tiro will explain it to you. Pray see what can be done. Besides let me know whether

Arpinum,
Nov. 11, B.C.
44

[1] By Mimas Cicero means the Apennines, and by νῆσος Ψυρίης the *insula Arpinas*.

propius accedere (malo enim esse in Tusculano aut uspiam in suburbano), an etiam longius discedendum putes, crebro ad me velim scribas. Erit autem cotidie, cui des. Quod praeterea consulis, quid tibi censeam faciundum, difficile est, cum absim. Verum tamen, si pares aeque inter se, quiescendum, sin, latius manabit et quidem ad nos, deinde communiter.

XIIIc

CICERO ATTICO SAL.

Scr. in Arpinati III Id. Nov. a. 710

Avide tuum consilium exspecto. Timeo, ne absim, cum adesse me sit honestius; temere venire non audeo. De Antoni itineribus nescio quid aliter audio, atque ut ad te scribebam. Omnia igitur velim explices et ad me certa mittas.

De reliquo quid tibi ego dicam? Ardeo studio historiae (incredibiliter enim me commovet tua cohortatio); quae quidem nec institui nec effici potest sine tua ope. Coram igitur hoc quidem conferemus. In praesentia mihi velim scribas, quibus consulibus C. Fannius M. f. tribunus pl. fuerit. Videor mihi audisse P. Africano, L. Mummio censoribus. Id igitur quaero. Tu mihi de iis rebus, quae novantur, omnia certa, clara. III Idus ex Arpinati.

LETTERS TO ATTICUS XVI. 13b-13c

you think I can come nearer Rome—for I should prefer to be at Tusculum or somewhere in the neighbourhood of Rome—or whether I ought to go further off. Write frequently about it. There will be someone to give a letter to every day. You ask my advice too as to what I think you ought to do. It is difficult to say, when I am not at Rome. However, if the two[1] seem equal, keep quiet; if not, the news will spread even here; then we will take common counsel.

XIIIc

CICERO TO ATTICUS, GREETING.

I am expecting your advice eagerly. I fear I may be absent, when honour demands my presence; yet I dare not come rashly. About Antony's march I hear now rather a different tale from what I wrote. So I wish you would unravel the whole mystery and send me certain news.

Arpinum, Nov. 11, B.C. 44

For the rest what can I say? I have a burning passion for history—for your suggestion has had a wonderful effect upon me—but it is not easy to begin or to carry it out without your assistance. So we will discuss it when we meet. At the present moment I wish you would tell me in what year C. Fannius, son of Marcus, was tribune. I think I have been told it was in the censorship of Africanus and Mummius. So that is what I want to know. Please send me clear and certain details of all the changes in the constitution. Arpinum, Nov. 11.

[1] Antony and Octavian.

MARCUS TULLIUS CICERO

XIV

CICERO ATTICO SAL.

Scr. in Arpinati medio mense Novembri a. 710

Nihil erat plane, quod scriberem. Nam, cum Puteolis essem, cotidie aliquid novi de Octaviano, multa etiam falsa de Antonio. Ad ea autem, quae scripsisti (tres enim acceperam III Idus a te epistulas), valde tibi adsentior, si multum possit Octavianus, multo firmius acta tyranni comprobatum iri quam in Telluris, atque id contra Brutum fore. Sin autem vincitur, vides intolerabilem Antonium, ut, quem velis, nescias. O Sesti tabellarium hominem nequam! Postridie Puteolis Romae se dixit fore. Quod me mones, ut pedetemptim, adsentior; etsi aliter cogitabam. Nec me Philippus aut Marcellus movet. Alia enim eorum ratio est et, si non est, tamen videtur. Sed in isto iuvene, quamquam animi satis, auctoritatis parum est. Tamen vide, si forte in Tusculano recte esse possum, ne id melius sit. Ero libentius; nihil enim ignorabo. An hic, cum Antonius venerit?

Sed, ut aliud ex alio, mihi non est dubium, quin, quod Graeci καθῆκον, nos "officium." Id autem quid dubitas quin etiam in rem publicam praeclare quadret? Nonne dicimus "consulum officium, sena-

---

[1] Where the Senate met on March 17, two days after the murder of Caesar. Cf. *Att.* XIV. 10.

LETTERS TO ATTICUS XVI. 14

XIV

CICERO TO ATTICUS, GREETING.

Arpinum, middle of Nov., B.C. 44

I have nothing whatever to write about. For, when I was at Puteoli, there was something fresh about Octavian every day, and plenty of false reports about Antony. However, I had three letters from you on the fifth, and I strongly agree with what you said, that if Octavian has much success, the tyrant's proposals will receive stronger confirmation than they did in the temple of Tellus,[1] and that will be against the interests of Brutus. But if, on the other hand, he is conquered, you see Antony will be intolerable; so you don't know which you want. What a rascal Sestius' messenger is! He said he would be in Rome the day after he left Puteoli! You advise me to move slowly, and I agree, though once I thought differently. I am not influenced by Philippus or Marcellus; for their position is different, or, if it is not, it looks as though it were.[2] But that youth, though he has plenty of spirit, has little influence. However, see whether it would not be better for me to be at Tusculum, if I should do right in being there. I would rather be there; for I should get all the news. Or had I better be here when Antony comes?

But, as one thing suggests another,[3] I know that what the Greeks call καθῆκον (duty), we call *officium*. But why should you doubt whether the word fits appropriately in political affairs? Don't we say the

[2] Marcellus was Octavian's brother-in-law; Philippus his stepfather.

[3] Apparently the idea of "duty" was suggested by *recte* just above, though it hardly bears that meaning in this case.

tus officium, imperatoris officium"? Praeclare convenit; aut da melius. Male narras de Nepotis filio. Valde mehercule moveor et moleste fero. Nescieram omnino esse istum puerum. Caninium perdidi, hominem, quod ad me attinet, non ingratum. Athenodorum nihil est quod hortere. Misit enim satis bellum ὑπόμνημα. Gravedini, quaeso, omni ratione subveni. Avi tui pronepos scribit ad patris mei nepotem se ex Nonis iis, quibus nos magna gessimus, aedem Opis explicaturum idque ad populum. Videbis igitur et scribes. Sexti iudicium exspecto.

XV

CICERO ATTICO SAL.

Scr. in Arpinati ante V. Id. Dec. a. 710

Noli putare pigritia me facere, quod non mea manu scribam, sed mehercule pigritia. Nihil enim habeo aliud, quod dicam. Et tamen in tuis quoque epistulis Alexim videor adgnoscere. Sed ad rem venio.

Ego, si me non improbissime Dolabella tractasset, dubitassem fortasse, utrum remissior essem an summo iure contenderem. Nunc vero etiam gaudeo mihi causam oblatam, in qua et ipse sentiat et reliqui omnes me ab illo abalienatum, idque prae me feram, et quidem me mea causa facere et rei publicae, ut

---

[1] For Cicero's defence of him in 55 B.C.
[2] Young Quintus Cicero to Cicero's son.

LETTERS TO ATTICUS XVI. 14-15

officium of consuls, of the Senate, of generals? It is quite appropriate; if not, suggest a better word. That is bad news about Nepos' son. I am much disturbed and distressed. I had no idea he had such a son. I have lost Canidius, a man who, so far as I was concerned, has not been ungrateful.[1] There is no necessity for you to stir up Athenodorus. He has sent me quite a good memorandum. Pray do all you can for your cold. Your grandfather's great-grandson writes to my father's grandson[2] that after the 5th of December, the day of my great achievement,[3] he means to explain about the temple of Ops,[4] and that in public. Keep your eyes open then and let me know. I am anxious to hear what Sextus has to say.

XV

CICERO TO ATTICUS, GREETING.

Don't think it is laziness that prevents my writing myself; and yet, to be sure, it is nothing but laziness, for I have no other excuse to make. However, I seem to recognize Alexis' hand in your letters too. But to come to the point.

Arpinum, before Dec. 9, B.C. 44

If Dolabella had not treated me most disgracefully, I should perhaps have had some doubt whether to let him down lightly or to claim my full rights. But, as it is, I am glad to have some reason for showing him and other people that I have quarrelled with him; and I will make it clear that I detest him both on my own account and on that of the Republic, because, when at my in-

---

[3] The arrest of the Catilinarian conspirators in 63 B.C.

[4] Antony's seizure of the public funds deposited in that temple. Cf. XIV. 14.

MARCUS TULLIUS CICERO

illum oderim, quod, cum eam me auctore defendere coepisset, non modo deseruerit emptus pecunia, sed etiam, quantum in ipso fuerit, everterit. Quod autem quaeris, quo modo agi placeat, cum dies venerit, primum velim eius modi sit, ut non alienum sit me Romae esse; de quo ut de ceteris faciam, ut tu censueris. De summa autem agi prorsus vehementer et severe volo. Etsi sponsores appellare videtur habere quandam δυσωπίαν, tamen, hoc quale sit, consideres velim. Possumus enim, ut sponsores appellemus, procuratorem introducere; neque enim illi litem contestabuntur. Quo facto non sum nescius sponsores liberari. Sed et illi turpe arbitror eo nomine, quod satisdato debeat, procuratores eius non dissolvere et nostrae gravitatis ius nostrum sine summa illius ignominia persequi. De hoc quid placeat, rescribas velim; nec dubito, quin hoc totum lenius administraturus sis.

Redeo ad rem publicam. Multa mehercule a te saepe in πολιτικῷ genere prudenter, sed his litteris nihil prudentius: "Quamquam enim potest et[1] in praesentia belle iste puer retundit Antonium, tamen exitum exspectare debemus." At quae contio! nam est missa mihi. Iurat, ita sibi parentis honores consequi liceat, et simul dextram intendit ad statuam. Μηδὲ σωθείην ὑπό γε τοιούτου! Sed, ut scribis, certissimum esse video discrimen Cascae nostri tribunatum, de quo quidem ipso dixi Oppio, cum me hortaretur,

[1] potest et *Gronovius*: postea *MSS.*

---

[1] Or "is capable of holding and at present does hold."
[2] A *contio* delivered by Octavian.

stigation he had begun to defend it, he not only accepted a bribe to desert it, but did his best to overthrow it. You ask how I want things to be managed when the day comes. First, I should like them to be so arranged that it may appear natural for me to come to Rome. But about that, and indeed about the rest, I will do as you advise. On the main point, however, I want really active and serious steps to be taken. Though it is counted bad form to call upon the sureties for payment, still consider how that method would do. We can bring his agents into the case in order to call upon the sureties, for the agents will not dispute the suit, though, if they do, I know of course the sureties will escape. But I think it will be a disgrace for him, if his agents do not pay up a debt for which he gave security, and my position demands that I should prosecute my case without extreme humiliation to him. Please write and tell me what you think best; I have no doubt you will carry it through with reasonable moderation.

I return to public affairs. You have often said many a wise thing about politics, but never anything wiser than this letter: "For though the youth is strong and at present holds [1] Antony well in check, still we must wait and see." But what a speech! [2] For it has been sent to me. He swears by his hopes of attaining to the honours of his father, and at the same time stretches out his hand towards the statue. Be hanged to salvation with a saviour like that! But, as you say, I see Casca's tribuneship will afford the best criterion of his policy.[3] It was *apropos* of that that I said to Oppius, when he wanted me to

[3] Casca was one of the murderers of Caesar, and tribune elect.

ut adulescentem totamque causam manumque veteranorum complecterer, me nullo modo facere posse, ni mihi exploratum esset eum non modo non inimicum tyrannoctonis, verum etiam amicum fore. Cum ille diceret ita futurum, "Quid igitur festinamus?" inquam. "Illi enim mea opera ante Kal. Ian. nihil opus est, nos autem eius voluntatem ante Idus Decembr. perspiciemus in Casca." Valde mihi adsensus est. Quam ob rem haec quidem hactenus. Quod reliquum est, cotidie tabellarios habebis, et, ut ego arbitror, etiam quod scribas, habebis cotidie. Leptae litterarum exemplum tibi misi, ex quo mihi videtur Στρατύλαξ ille deiectus de gradu. Sed tu, cum legeris, existumabis.

Obsignata iam epistula litteras a te et a Sexto accepi. Nihil iucundius litteris Sexti, nihil amabilius. Nam tuae breves, priores erant uberrimae. Tu quidem et prudenter et amice suades, ut in his locis potissimum sim, quoad audiamus, haec, quae commota sunt, quorsus evadant. Sed me, mi Attice, non sane hoc quidem tempore movet res publica, non quo aut sit mihi quicquam carius aut esse debeat, sed desperatis etiam Hippocrates vetat adhibere medicinam. Quare ista valeant; me res familiaris movet. Rem dico; immo vero existimatio. Cum enim tanta reliqua sint, ne Terentiae quidem adhuc quod solvam expeditum est. Terentiam dico; scis nos pridem iam constituisse Montani nomine HS $\overline{XXV}$ dissolvere. Pudentissime hoc Cicero petierat ut fide sua. Liberalissime, ut tibi quoque placuerat, pro-

open my arms to the youth, the whole cause, and the troop of veterans, that I could not do anything of the kind, until I had made sure that he would not only not be an enemy, but would be a friend to the tyrannicides. He said that would be so, and I replied: "Then, what is the hurry? He does not want my assistance before the 1st of January, and we shall see what he intends before the middle of December in Casca's case." He quite agreed with me. So that's enough of that. For the rest you will have messengers every day, and I think you will have something to write every day too. I am sending a copy of Lepta's letter, and from it you will see that that toy captain[1] has had a fall. But you will judge for yourself when you have read it.

When I had already sealed this letter, I got one from you and one from Sextus. Nothing could have been pleasanter or more amiable than Sextus' letter. For yours was a short note, the earlier one having been very full. It is wise and friendly advice you give me to stay here by preference, till we hear how this disturbance is going to end. But just at this minute, my dear Atticus, it is not the Republic that I am bothered about—not that any thing is or ought to be dearer to me, but even Hippocrates admits it is useless to apply medicine in desperate cases. So let that go hang—it is my private concerns that bother me. Concerns, do I say? Nay, rather my credit; for, though I have such big balances, I have not even enough money on hand yet to pay Terentia. Do I speak of Terentia? You know we arranged long ago to pay Montanus' debt of £250.[2] My son very considerately begged me to do it out of his credit. As you also agreed, I promised quite freely,

[1] Antony.   [2] 25 sestertia.

miseram, Erotique dixeram, ut sepositum haberet.
Non modo non fecit sed iniquissimo faenore versuram
facere Aurelius coactus est. Nam de Terentiae
nomine Tiro ad me scripsit te dicere nummos a
Dolabella fore. Male eum credo intellexisse, si quisquam male intellegit, potius nihil intellexisse. Tu
enim ad me scripsisti Coccei responsum et isdem
paene verbis Eros. Veniendum est igitur vel in
ipsam flammam. Turpius est enim privatim cadere
quam publice. Itaque ceteris de rebus, quas ad me
suavissume scripsisti, perturbato animo non potui, ut
consueram, rescribere. Consenti hac cura,[1] ubi sum,
ut me expediam; quibus autem rebus, venit quidem
mihi in mentem, sed certi constituere nihil possum,
prius quam te videro. Qui minus autem ego istic
recte esse possim, quam est Marcellus? Sed non id
agitur, neque id maxime curo; quid curem, vides.
Adsum igitur.

XVI

CICERO SUO SAL. DIC. ATTICO.

Scr. in Tus-culano inter a d. V et prid. Non. Quint. a. 710

Iucundissimas tuas legi litteras. Ad Plancum
scripsi, misi. Habes exemplum. Cum Tirone quid
sit locutus, cognoscam ex ipso. Cum sorore ages
attentius, si te occupatione ista relaxaris.

[1] consenti hac cura *Tyrrell*: consenti in hac cura *MSS.*:
contendo Astura *Gurlitt*.

and told Eros to set a sum apart for it. Not only did he fail to do so, but Aurelius[1] had to raise another loan at extortionate interest. For Terentia's debt Tiro tells me you said there would be money from Dolabella. I think he misunderstood you, if anyone can misunderstand anybody, or rather he did not understand at all. For you sent me Cocceius' answer, and so did Eros in nearly the same words. So I must come even into the heart of the conflagration, for private failure is even more disgraceful than public failure. So for the other matters contained in your pleasant letter, I was too perturbed in mind to answer them as usual. Combine with me in extricating me from the tiresome position I am in; how it is to be done I have some idea, but I cannot arrange things with certainty till I see you. However, how can I be less safe in Rome than Marcellus? But that is not the point, nor is it my chief anxiety; what I am anxious about you see. So I am coming.

XVI

CICERO SENDS GREETING TO HIS FRIEND ATTICUS.

I have read your delightful letter. To Plancus I have written and sent the letter. Here is a copy. What he said to Tiro I shall learn from Tiro himself. You will attend more carefully to your sister's affairs, if you have a rest from that other business of yours.

Tusculum, between July 3 and 6, B.C. 44

---

[1] Agent of Montanus.

MARCUS TULLIUS CICERO

XVIa

M. CICERO L. PLANCO PRAET. DESIG. SAL.

Scr. in Tus-   Attici nostri te valde studiosum esse cognovi, mei
culano eodem vero ita cupidum, ut mehercule paucos aeque obser-
tempore   vantes atque amantes me habere existimem. Ad
paternas enim magnas et veteres et iustas necessi-
tudines magnam attulit accessionem tua voluntas
erga me meaque erga te par atque mutua.

Buthrotia tibi causa ignota non est. Egi enim
saepe de ea re tecum tibique totam rem demonstravi;
quae est acta hoc modo. Ut primum Buthrotium
agrum proscriptum vidimus, commotus Atticus libellum composuit. Eum mihi dedit, ut darem Caesari;
eram enim cenaturus apud eum illo die. Eum libellum Caesari dedi. Probavit causam, rescripsit Attico
aequa eum postulare, admonuit tamen, ut pecuniam
reliquam Buthrotii ad diem solverent. Atticus, qui
civitatem conservatam cuperet, pecuniam numeravit
de suo. Quod cum esset factum, adiimus ad Caesarem, verba fecimus pro Buthrotiis, liberalissimum
decretum abstulimus; quod est obsignatum ab amplissimis viris. Quae cum essent acta, mirari equidem
solebam pati Caesarem convenire eos, qui agrum
Buthrotium concupissent, neque solum pati, sed etiam
ei negotio te praeficere. Itaque et ego cum illo
locutus sum et saepius quidem, ut etiam accusarer
ab eo, quod parum constantiae suae confiderem, et
M. Messallae et ipsi Attico dixit, ut sine cura essent,

LETTERS TO ATTICUS XVI. 16a

XVIa

M. CICERO TO L. PLANCUS, PRAETOR ELECT, GREETING

I know you are much attached to our friend Atticus, and to my society you are so partial that I am sure I count myself to have few friends so attentive and affectionate. For our ancestral ties, so strong and old and natural, have been strengthened by the equal and reciprocal liking we have, you for me and I for you.

Tusculum, at the same time

The case of the Buthrotians is not unknown to you. For I have often spoken to you about it and explained the whole affair to you. This is what has happened. When first we saw that the lands of Buthrotum had been confiscated, Atticus was troubled and composed a petition. That he gave to me to hand to Caesar, for I was going to dine with him that day. That petition I handed to Caesar. He approved of the case and wrote back to Atticus that his request was reasonable, but he warned him that the Buthrotians must pay the rest of the money at the proper time. Atticus, who wanted to save the city, paid the money on his own account. When that was done we approached Caesar, said a word for the Buthrotians, and obtained a most generous decree, which was signed by persons of importance. After that I was much astonished that Caesar used to let those who had coveted the land of the Buthrotians hold meetings, and not only allowed them to do so, but even put you at the head of the commission. So I spoke to him about it, and that indeed so often that he even reproached me for having so little faith in his consistency; and he told Messalla and Atticus himself not to worry about it, and admitted candidly

MARCUS TULLIUS CICERO

aperteque ostendebat se praesentium animos (erat enim popularis, ut noras) offendere nolle ; cum autem mare transissent, curaturum se, ut in alium agrum deducerentur. Haec illo vivo. Post interitum autem Caesaris, ut primum ex senatus consulto causas consules cognoscere instituerunt, haec, quae supra scripsi, ad eos delata sunt. Probaverunt causam sine ulla dubitatione seque ad te litteras daturos esse dixerunt. Ego autem, mi Plance, etsi non dubitabam, quin et senatus consultum et lex et consulum decretum ac litterae apud te plurimum auctoritatis haberent, teque ipsius Attici causa velle intellexeram, tamen hoc pro coniunctione et benevolentia nostra mihi sumpsi, ut id a te peterem, quod tua singularis humanitas suavissimique mores a te essent impetraturi. Id autem est, ut hoc, quod te tua sponte facturum esse certo scio, honoris nostri causa libenter, prolixe, celeriter facias. Mihi nemo est amicior nec iucundior nec carior Attico. Cuius antea res solum familiaris agebatur eaque magna, nunc accessit etiam existimatio, ut, quod consecutus est magna et industria et gratia et vivo Caesare et mortuo, id te adiuvante obtineat. Quod si a te erit impetratum, sic velim existimes, me de tua liberalitate ita interpretaturum, ut tuo summo beneficio me adfectum iudicem. Ego, quae te velle quaeque ad te pertinere arbitrabor, studiose diligenterque curabo. Da operam, ut valeas.

that he did not want to offend the people, while they were in Rome—for, as you know, he aimed at popularity—but when they were across the sea, he would see to it that they were transferred to some other land. That was what happened in Caesar's lifetime. But, after Caesar's death, as soon as the consuls in accordance with a decree of the Senate began to investigate cases, the facts as I have stated them were put before them. They approved of the case without any hesitation, and said they would send you letters. Now, my dear Plancus, though I have no doubt that a decree of the Senate, a statute, a decree of the consuls, and their despatch, will have the greatest weight with you, and I understand that you will wish to please Atticus himself, yet I have taken it upon myself in view of our connection and affection, to ask you for what your own exceptional amiability and your goodness of heart would win from you themselves. That is, that you should for my sake do this thing, which I am sure you will do of your own accord, freely, fully, and quickly. I have no greater and no dearer friend than Atticus. At first it was only a question of his money, and a good sum of it too; but now it concerns his credit too, that he should obtain with your assistance what he won by his great persistency and his popularity both in Caesar's lifetime and after his death. If he obtains it from you, I hope you will consider that I shall interpret your liberality as a great favour bestowed upon myself. For my part, I will show care and diligence in anything that I think you desire or that concerns you. Take care of your health.

MARCUS TULLIUS CICERO

XVIb

CICERO PLANCO PRAET. DESIG. SAL.

Scr. paulo post ep. 16a

Iam antea petivi abs te per litteras, ut, cum causa Buthrotiorum probata a consulibus esset, quibus et lege et senatus consulto permissum erat, ut de Caesaris actis cognoscerent, statuerent, iudicarent, eam rem tu adiuvares, Atticumque nostrum, cuius te studiosum cognovi, et me, qui non minus laboro, molestia liberares. Omnibus enim rebus magna cura, multa opera et labore confectis in te positum est, ut nostrae sollicitudinis finem quam primum facere possimus. Quamquam intellegimus ea te esse prudentia, ut videas, si ea decreta consulum, quae de Caesaris actis interposita sunt, non serventur, magnam perturbationem rerum fore. Equidem, cum multa, quod necesse erat in tanta occupatione, non probentur, quae Caesar statuerit, tamen otii pacisque causa acerrime illa soleo defendere. Quod tibi idem magno opere faciendum censeo; quamquam haec epistula non suasoris est, sed rogatoris. Igitur, mi Plance, rogo te et etiam atque etiam oro sic medius fidius, ut maiore studio magisque ex animo agere non possim, ut totum hoc negotium ita agas, ita tractes, ita conficias, ut, quod sine ulla dubitatione apud consules obtinuimus propter summam bonitatem et aequitatem causae, id tu nos obtinuisse non modo facile patiare, sed etiam gaudeas. Qua quidem voluntate te esse erga Atticum saepe praesens et illi ostendisti et vero

XVIb

CICERO TO PLANCUS, PRAETOR ELECT, GREETING.

I have already written to ask you to render assistance in the matter of the Buthrotians, since the consuls, who had the authority of a statute and a senatorial decree to investigate, determine, and decide on Caesar's proceedings, have approved of their case; and to relieve Atticus, whom I know you admire, and myself, who am as much concerned as he is, from trouble. For now that we have brought the whole business to an end with the expenditure of much care, much labour, and pains, it rests with you to allow us to make an end to our anxiety as early as possible. However, I am sure that you have wisdom enough to see, that, if the decisions delivered by the consuls about Caesar's proceedings are not observed, things will be thrown into great confusion. For my part, though one cannot approve of many of Caesar's arrangements—as was natural in the case of a person so busy—still I am wont to uphold them staunchly for the sake of peace and quietness: and I am strongly of the opinion that you should do the same, though I am not writing as an adviser but as a suppliant. So, my dear Plancus, I beg and beseech you —and I do assure you I could not be more anxious or more in earnest about anything—to take in hand, to conduct, and to carry through all this business in such a way, that, what we have obtained from the consuls without any hesitation solely on the justice and equity of our case, we may obtain from you not only with your kind indulgence but with alacrity on your part. How kindly disposed you are to Atticus you have often shown him and me, too, when we

Written shortly after 16a

etiam mihi. Quod si feceris, me, quem voluntate et paterna necessitudine coniunctum semper habuisti, maximo beneficio devinctum habebis, idque ut facias, te vehementer etiam atque etiam rogo.

XVIc

CICERO CAPITONI SUO SAL.

Scr. eodem tempore quo ep. 16b

Numquam putavi fore ut supplex ad te venirem; sed hercule facile patior datum tempus, in quo amorem experirer tuum. Atticum quanti faciam, scis. Amabo te, da mihi et hoc, obliviscere mea causa illum aliquando suo familiari, adversario tuo voluisse consultum, cum illius existimatio ageretur. Hoc primum ignoscere est humanitatis tuae; suos enim quisque debet tueri; deinde, si me amas (omitte Atticum), Ciceroni tuo, quem quanti facias, prae te soles ferre, totum hoc da, ut, quod semper existimavi, nunc plane intellegam, me a te multum amari. Buthrotios cum Caesar decreto suo, quod ego obsignavi cum multis amplissimis viris, liberavisset ostendissetque nobis se, cum agrarii mare transissent, litteras missurum, quem in agrum deducerentur, accidit, ut subito ille interiret. Deinde, quem ad modum tu scis (interfuisti enim), cum consules oporteret ex senatus consulto de actis Caesaris cognoscere, res ab iis in Kal. Iun. dilata est. Accessit ad senatus consultum lex, quae lata est a. d. IIII Non.

LETTERS TO ATTICUS XVI. 16 b–c

have been together. If you will do this, you will have bound me—who have always been attached to you by my own inclination and by our family friendship—to you under a heavy obligation, and I beg you earnestly and repeatedly to do so.

XVIc

CICERO TO CAPITO, GREETING

I never thought I should have to come before you as a suppliant, but upon my soul I am not sorry that I should have an occasion for testing your affection. You know how fond I am of Atticus. Pray grant me one other favour and forget for my sake that once he wished to support a friend of his, who was an enemy of yours, when his reputation was at stake. In the first place your kindly disposition should bid you forgive that, for everyone ought to look after his own friends; in the next place, leaving Atticus out of the question, if you love me—and you are always declaring how great is the respect you have for your friend Cicero—grant me that now I may know for a certainty what I have always believed, that you have a great affection for me. By a decree, which I and many important persons signed, Caesar set free the Buthrotians, and assured us that, when the land-commissioners had crossed the sea, he would send a despatch about the territory to which they should be transferred; and then it happened that he died suddenly. Then, as you know (for you were present), when the consuls ought to have decided on Caesar's proceedings in accordance with a senatorial decree, they postponed the matter till the 1st of June. On the 2nd of June a law was passed in

Written at the same time as 16 b

435

MARCUS TULLIUS CICERO

Iun., quae lex earum rerum, quas Caesar statuisset, decrevisset, egisset, consulibus cognitionem dedit. Causa Buthrotiorum delata est ad consules. Decretum Caesaris recitatum est et multi praeterea libelli Caesaris prolati. Consules de consilii sententia decreverunt secundum Buthrotios : litteras ad[1] Plancum dederunt. Nunc, mi Capito (scio enim, quantum semper apud eos, quibuscum sis, posse soleas, eo plus apud hominem facillimum atque humanissimum, Plancum), enitere, elabora vel potius eblandire, effice, ut Plancus, quem spero optimum esse, sit etiam melior opera tua. Omnino res huius modi mihi videtur esse, ut sine cuiusquam gratia Plancus ipse pro ingenio et prudentia sua non sit dubitaturus, quin decretum consulum, quorum et lege et senatus consulto cognitio et iudicium fuit, conservet, praesertim cum hoc genere cognitionum labefactato acta Caesaris in dubium ventura videantur, quae non modo ii, quorum interest, sed etiam ii, qui illa non probant, otii causa confirmari velint. Quod cum ita sit, tamen interest nostra Plancum hoc animo libenti prolixoque facere ; quod certe faciet, si tu nervulos tuos mihi saepe cognitos suavitatemque, qua nemo tibi par est, adhibueris. Quod ut facias, te vehementer rogo.

[1] litteras ad *added by Manutius.*

LETTERS TO ATTICUS XVI. 16c

addition to the decree of the Senate, granting the consuls the right of deciding on Caesar's statutes, decrees, and proceedings. The case of the Buthrotians was put before the consuls. Caesar's decree was read to them, and many other papers of Caesar's were brought forward too. By the advice of their council the consuls decided in favour of the Buthrotians, and sent a despatch to Plancus. Now, Capito, I know the influence you always have over those with whom you are, especially with so amiable and good-natured a person as Plancus; please use all your energy, or rather all your powers of persuasion, and make Plancus, who I hope will be sufficiently kindly himself, still more kindly. In any case I think this is the state of affairs: that without favouring anybody, Plancus will have sense and wisdom enough to have no hesitation in obeying the decree of the consuls, who had the right of enquiry and decision conferred upon them by law and by a senatorial decree, especially as, if this kind of decision is rendered null, Caesar's proceedings may well be called in question; and not only those who benefit by them, but even those who disapprove of them, have to give them their support for the sake of peace. Though that is the case, still it is to our interest that Plancus should do this willingly and freely; and no doubt he will if you exert your influence, which I know so well, and your persuasive power, which is unequalled: and that I beg you earnestly to do.

MARCUS TULLIUS CICERO

XVId

CICERO C. CUPIENNIO S.

Scr. eodem tempore quo ep. 16c

Patrem tuum plurimi feci, meque ille mirifice et coluit et amavit; nec mehercule umquam mihi dubium fuit, quin a te diligerer; ego quidem id facere non destiti. Quam ob rem peto a te in maiorem modum, ut civitatem Buthrotiam subleves decretumque consulum, quod ii secundum Buthrotios fecerunt, cum et lege et senatus consulto statuendi potestatem haberent, des operam ut Plancus noster quam primum confirmet et comprobet. Hoc te vehementer, mi Cupienni, etiam atque etiam rogo.

XVIe

CICERO PLANCO PRAET. DES. S.

Scr. post ep. 16b

Ignosce mihi, quod, cum antea accuratissime de Buthrotiis ad te scripserim, eadem de re saepius scribam. Non mehercule, mi Plance, facio, quo parum confidam aut liberalitati tuae aut nostrae amicitiae, sed, cum tanta res agatur Attici nostri, nunc vero etiam existimatio, ut id, quod probavit Caesar nobis testibus et obsignatoribus, qui et decretis et responsis Caesaris interfueramus, videatur obtinere potuisse, praesertim cum tota potestas eius rei tua sit, ut ea, quae consules decreverunt secundum

XVId

CICERO TO C. CUPIENNIUS, GREETING.

Written at the same time as 16c

I was a great admirer of your father, and he was exceedingly attentive and affectionate to me; and I am sure I have never had any doubt that you have a regard for me. Certainly I have never ceased to have one for you. So I beg you with more than usual earnestness to assist the city of Buthrotum, and to make it your business that our friend Plancus should confirm and verify the decree which the consuls made in favour of the Buthrotians, when they had been granted the power of settling the question both by a statute and by a senatorial decree. This I do most earnestly beg and entreat you, my dear Cupiennius.

XVIe

CICERO TO PLANCUS, PRAETOR ELECT, GREETING.

Written after 16b

Pardon me for writing again on the same subject, when I have already written very fully to you about the Buthrotians. I do assure you, my dear Plancus, that I do not do so because I have little faith in your generosity or your friendship for me. But my friend Atticus has so great a monetary stake in the matter; and now, what is more, his very reputation is involved in showing that he can obtain what Caesar approved of, and we, who were present when Caesar made his decrees and gave his answer, witnessed and sealed. And I appeal to you especially, because it is a case where the whole power, I will not say of confirming, but of confirming freely and willingly

MARCUS TULLIUS CICERO

Caesaris decreta et responsa, non dicam comprobes, sed studiose libenterque comprobes. Id mihi sic erit gratum, ut nulla res gratior esse possit. Etsi iam sperabam, cum has litteras accepisses, fore ut ea, quae superioribus litteris a te petissemus, impetrata essent, tamen non faciam finem rogandi, quoad nobis nuntiatum erit te id fecisse, quod magna cum spe exspectamus. Deinde enim confido fore ut alio genere litterarum utamur tibique pro tuo summo beneficio gratias agamus. Quod si acciderit, velim sic existimes, non tibi tam Atticum, cuius permagna res agitur, quam me, qui non minus laboro quam ille, obligatum fore.

XVIf

CICERO CAPITONI SAL.

Scr. paulo post ep. 16c

Non dubito, quin mirere atque etiam stomachere, quod tecum de eadem re agam saepius. Hominis familiarissimi et mihi omnibus rebus coniunctissimi permagna res agitur, Attici. Cognovi ego tua studia in amicos, etiam in te amicorum. Multum potes nos apud Plancum iuvare. Novi humanitatem tuam; scio, quam sis amicis iucundus. Nemo nos in hac causa plus iuvare potest quam tu. Et res ita est firma, ut debet esse, quam consules de consilii sententia decreverunt, cum et lege et senatus consulto cognoscerent Tamen omnia posita putamus in Planci

what the consuls decreed in accordance with Caesar's decrees and promises, lies in your hands. It will be doing me a favour than which none could be greater. Although I hope that by the time you receive this letter you will have granted me the petition I made in my former letter, still I shall not cease from asking until I have news that you have done what I am looking forward to with great hope. Then I trust I shall write a different kind of letter, and pay my thanks for your exceeding kindness. If that comes to pass I would have you think that you have not so much put Atticus, in spite of the huge sum of money he has at stake, under an obligation, as myself, who take an equal interest in the matter.

XVIf

CICERO TO CAPITO, GREETING.

Written shortly after 16c

I have no doubt you are astonished and even annoyed with me for approaching you twice on the same subject. Atticus, my greatest friend and my closest intimate in every way, has grave interests at stake. I know the willingness with which you help your friends and your friends help you. You can render us much assistance with Plancus. I know the kindness of your heart; I know how welcome you are to your friends. There is no one who can help us more than you in this case. And the case is as sound as a case ought to be which the consuls have decided on the advice of their council, when they had the right of decision conferred on them by statute and by senatorial decree. Still to us the whole case seems to lie in the generosity of your

tui liberalitate; quem quidem arbitramur cum officii sui et rei publicae causa decretum consulum comprobaturum tum libenter nostra causa esse facturum. Adiuvabis igitur, mi Capito. Quod ut facias, te vehementer etiam atque etiam rogo.

friend Plancus; and, indeed, we think he will ratify the consuls' decree both for duty's sake and for the sake of the constitution, and that he will do so willingly for our sake. So please help us, my dear Capito. I entreat and beseech you earnestly to do so.

CHRONOLOGICAL ORDER OF THE LETTERS
based on the order fixed in R. Y. Tyrrell and L. C. Purser, *The Correspondence of M. Tullius Cicero*, vol. vii., Dublin, 1901 (by kind permission of the Board of Trinity College, Dublin).

ABBREVIATIONS

A = *Epistulae ad Atticum.*
F = *Epistulae ad Familiares.*
Q.Fr. = *Epistulae ad Quintum Fratrem.*
Br. = *Epistulae ad M. Brutum.*

B.C.
68 A i. 5, 6, 7 ?
67 A i. 9, 8, 10, 11
66 A i. 3, 4
65 A i. 1, 2
64 [*Q. Cic. de petit. consul.*]
63 F xiii. 76 ?
62 F v. 7, 1, 2, 6
61 A i. 12, F v. 5, A i. 13, 14, 15, 16, 17
60 A i. 18, 19, 20, ii. 1, 2, 3, Q.Fr. i. 1
59 A ii. 4, 5, 6, 7, 8, 9, 12, 10, 11, 13, 14, 15, 16, 17, 18, 19, 20, 21, 22, 23, 24, 25, Q.Fr. i. 2, F xiii. 42, 41; also 43 ? (before 58 B.C.; so also xiii. 44, 45, 46)
58 A iii. 3, 2, 4, 1, 5, 6, F xiv. 4, A iii. 7, 8, 9, Q.Fr. i. 3, A iii. 10, 11, 12, 14, 13, Q.Fr. i. 4, A iii. 15, 16, 17, 18, 19, 20, F xiv. 2, A iii. 21, 22, F xiv. 1, A iii. 23, F xiv. 3, A iii. 24, 25
57 A iii. 26, 27, F v. 4, A iv. 1, 2, 3, Q.Fr. ii. 1, F vii. 26; also xiii. 51 ?
56 F i. 1, 2, 3, 4, 5a, Q.Fr. ii. 2, A iv. 4, Q.Fr. ii. 3, F i. 5b, 6, Q.Fr. ii. 4, 5, A iv. 4a, 5, F v. 12, A iv. 6, 7, 8, F v. 3, i. 7, xiii. 6a, 6b, Q.Fr. ii. 8 (=6), A iv. 8a

445

ORDER OF THE LETTERS

B.C.

55 F i. 8, Q.Fr. ii. 9 (= 7), A iv. 10, 9, Q.Fr. ii. 10 (= 8),
   A iv. 11, 12, F vii. 2, 3, 1, xiii. 74, 40, A iv. 13

54 F v. 8, Q.Fr. ii. 11 (= 9), 12 (= 10), F. vii. 5, Q.Fr. ii.
   13 (= 11), F vii. 6, 7, A iv. 14, Q.Fr. ii. 14 (= 12),
   F vii. 8, Q.Fr. ii. 15a (= 13), 15b (= 14), A iv. 15, 16,
   Q.Fr. ii. 16 (= 15), iii. 1, A iv. 17 (part) plus 18
   (part), F vii. 9, 17, Q.Fr. iii. 2, 3, 4, A iv. 18 (part),
   Q.Fr. iii. 5 plus 6, 7, F vii. 16, Q.Fr. iii. 8, A iv. 19
   (part), 17 (part), Q.Fr. iii. 9, F i. 9, vii. 10, i, 10,
   xiii. 49, 60, 73

53 F ii. 1, vii. 11, ii. 2, 3, vii. 12, 13, 14, 18, 15, ii. 4, 5, 6,
   xiii. 75; also xvi. 13 ?, 14 ?, 15 ?, 10 ?, 16 ?

52 F v. 17, 18, iii. 1, vii. 2

51 F iii. 2, A v. 1, 2, 3, 4, 5, 6, 7, F iii. 3, viii. 1, A v. 8,
   F iii. 4, A v. 9, F viii. 2, 3, A v. 10, F xiii. 1, A v. 11,
   F ii. 8, A v. 12, 13, 14, F iii. 5, viii. 4, A v. 15, 16, 17,
   F viii. 5, 9, xv. 3, iii. 6, xv. 7, 8, 9, 12, A v. 18, F xv. 2,
   A v. 19, F xv. 1, iii. 8, viii. 8, ii. 9, 10, viii. 10, ii. 7,
   A v. 20, F vii. 32, xiii. 53, 56, 55, 61, 62, 64, 65, 9;
   also 47 ?

50 F xv. 4, 10, 13, 14, viii. 6, 7, iii. 7, ii. 14, ix. 25, xiii. 59,
   58, iii. 9, A v. 21, F xiii. 63, A vi. 1, F xiii. 54, 57,
   ii. 11, A vi. 2, F iii. 13, 18, xiii. 2, 3, iii. 10, ii. 19, 12,
   A vi. 3, F iii. 11, xv. 5, viii. 11, A vi. 4, 5, 7, F viii. 13,
   ii. 17, 15, xv. 11, iii. 12, A vi. 6, F iii. 13, xv. 6,
   viii. 12, 14, A vi. 8, 9, F xiv. 5, A vii. 1, F xvi. 1, 2,
   3, 4, 5, 6, 7, 9, A vii. 3, 4, 5, 6, 7, 8, 9

49 F xvi. 11, v. 20, A vii. 10, 11, 12, F xiv. 18, A vii. 13,
   13a, F xiv. 14, A vii. 14, 15, F xvi. 12, A vii. 16,
   F xvi. 8, A vii. 17, 18, 19, 20, 21, 22, 23, 24, viii. 11a,
   vii. 25, viii. 12b, viii. 26, viii. 1, 11b, 12c, 12d, 2, 12a,
   3, 11c, 6, 4, 5, 7, 8, 9, 10, 11, 11d, 12, F viii. 15,
   A viii. 15a, 13, 14, 15, 16, ix. 1, 2, 12a, 3, 5, 7a, 6, 6a,
   7c, 7b, 4, 7, 8, 9, 10, 11a, 11, 12, 13a, 13, 14, 15, 16,
   17, 18, 19, x. 1, 2, 3, 3a, 4, 9a (= F viii. 16),
   A x. 5, 8a, 8b, 6, F iv. 1, A x. 7, F iv. 2, 19, A x. 8, 9,
   F ii. 16, A x. 10, 11, 12, 12a, 13, 14, 15, 16, 17, 18,
   F xiv. 7

48 A xv. 1, 2, F viii. 17, ix. 9, xiv. 8, A xi. 3, F xiv. 21,
   A xi. 4, F xiv. 6, 12, A xi. 5, F xiv. 19, A xi. 6, F xiv.
   9, A xi. 7, F xiv. 17, A xi. 8

47 A xi. 9, F xiv. 6, A xi. 10, 11, 12, 13, 14, 15, 16, 17,

446

ORDER OF THE LETTERS

B.C.

F xiv. 11, A xi. 18, F xiv. 15, A xi. 25, 23, F xiv. 10, 13, A xi. 19, 24, F xiv. 24, 23, A xi. 20, 21, 22, F xiv. 22, xv. 15, xiv. 20, 21; also xiii. 48 ?

46 F xiii. 10, 11, 12, 13, 14, xi. 1, xiii. 29, v. 21, A xii. 2, F ix. 3, 2, 7, 5, vii. 3, vi. 22, ix. 4, A xii. 5c, 3, 4, F ix. 6. A xii. 5, F ix. 16, 18, vii. 33, ix. 20, vii. 27, 28, ix. 19, 26, 17, 15, xiii. 68, iv. 13, 15, 8, 7, 9, vi. 6, 13, 12, 10a, 10b, xii. 17, iv. 3, 4, 11, ix. 21, vi. 14, A xii. 6a, 6b, 7, 8, 11, F vii. 4, ix. 23, A xii. 1, F xiii. 66, 67, 69, 70, 71, 72, 17, 18, 19, 20, 21, 22, 23, 24, 25, 26, 27, 28a, 28b, 78, 79, vi. 8, 9, v. 16, xv. 18; also xii. 20 ?, xiii. 52 ?

45 F xv. 16, vi. 7, 5, 18, iv. 14, 10, ix. 10, vi. 1, 3, 4, xv. 17, 19, ix. 13, xiii. 16, A xii. 13, 14, 15, 16, 18, 17, 18a, 19, 20, xiii. 6, F iv. 5, A xiii. 12, 21, 22, 23, 24, 25, 26, 27, 28, 29, 33, 30, 32, 31, 34, 35 ?, F xiii. 15, v. 13, vi. 21, iv. 6, vi. 2, ix. 11, 36, 37, 37a, 38, 38a, 39, 40, F v. 14, A xii. 42, F v. 15, A xii. 41, 43, 44, 45 ?, xiii. 26, xii. 46, 47, 48, 50, 49, 51, 52, 53, xiii. 1, 2, 27, 28, 29, 2a, 30, 31, 32, xii. 5a, F iv. 12, A xiii. 4, 5, 33, 6a, 8, 7, 7a, xii. 5b, F vi. 11, A xiii. 9, 10, 11, 12, 13, 14, 15, 16, 17, 18, 19, 21a, F ix. 22, A xiii. 20, 22, 33a, 23, F xiii. 77, v. 9, A xiii. 24, 25, F ix. 8, A xiii. 35, 36, 43, F vi. 20, A xiii. 44, 34, F vi. 19, A xii. 9, F xvi. 22, A xii. 10, xiii. 21, F xvi. 17, A xiii. 47a, F xvi. 19, A xiii. 48, 37, 38, 39, 40, 41, 45, 46, 47, F vii. 24, A xiii. 49, 50, F vii. 35, A xiii. 51, F xii. 18, 19, xiii. 4, 5, 7, 8, v. 11, vii. 29, v. 10b, A xiii. 52, F ix. 12, A xiii. 42, F xiii. 30, 31, 32, 33, 34, 35, 36, 37, 38, 39, xvi. 18, 20

44 F vii. 30, viii. 50, v. 10a, vii. 31, xii. 21, vi. 15, xi. 1, vi. 16, xv. 20, A xiv. 1, 2, 3, 4, 5, 6, 7, 8, F vi. 17, A xiv. 9, 10, 11, 12, 13a, 13b, 13, 14, 15, 16, 17a (= F ix. 14), F xii. 1, A xiv. 17, 19, 18, 20, 21, 22, xv. 1, 1a, 2, 3, 4, 4a, F xii. 16, A xv. 6, 5, 7, F xi. 2, A xv. 8, 9, 10, 11, 12, 16, 16a, 15, 17, 18, 19, 20, 21, F xvi. 23, A xv. 22, 23, 24, 14, 25, F vii. 21, 22, xi. 29, A xv. 26, 27, 28, xvi. 16, 16a, xv. 29, xvi. 1, 5, 4, 2, 3, F vii. 20, A xvi. 6, F vii. 19, A xvi. 16b, 16c, 16d, 16e, 16f, F xi. 3, A xvi. 7, F xi. 27, 28, xvi. 21, x. 1, 2, xii. 22 (1–2) = xiia, 2, xvi. 25, xi. 4, 6 (1) = 6a, xii. 3, 23, A xv. 13, xvi. 8, 9, 11, 12, 10, 13a, 13b, 13c,

447

ORDER OF THE LETTERS

B.C.

    14, F xvi. 24, A xvi. 15, F xi. 5, x. 3, xi. 7, 6 (2–3) = 6b, xii. 22 (3–4) = xiib, xvi. 26, 27, x. 4

43 F x. 5, xi. 8, xii. 24, 4, x. 28, ix. 24, xii. 5, 11, 7, x. 31, xii. 25a, x. 6, 27, xii. 28, 26, 27, 29, x. 7, 8, 10, xii. 6, Br. ii. 1, 3, 2, F x. 12, Br. ii. 4, F x. 30, Br. ii. 5, i. 2, sects. 4–6, i. 3, sects. 1–3, F x. 9, Br. i. 3, sect. 4, F xi. 9, 11, 13b, xii. 25b, Br. i. 5, F x. 14, xi. 10, 11, xii. 12, x. 13, xi. 13a, 15, 21, sects. 1–6, Br. i. 4, sects. 1–3, F x. 21, Br. i. 4, sects. 4–6, F xi. 12, x. 34a, 18, xi. 18, Br. i. 6, 1, 2, sects. 1–3, F x. 17, xi. 19, x. 34, sects. 3–4, xi. 20, 23, x. 19, 25, 16, xii. 15, sects. 1–6, 14, x. 20, 35, Br. i. 8, F xi. 16, 17, x. 33, Br. i. 11, 17, F xii. 15, xi. 26, 21, 24, x. 23, 32, Br. i. 10. F xii. 8, 30, xi. 13, sects. 4–5, xii. 13, Br. i. 9, F xi. 25, xii. 9, Br. i. 7, F xi. 15, x. 22, 26, Br. i. 13, F xii. 10, x. 29, xi. 32, Br. i. 12, 14, 15, 16, 18, F x. 24

INDEX OF NAMES.

[*The references are to the pages of Latin text.*]

ACADEMIA, 130, 160
Academica, 140; -ca quaestio 138; -cus 132, 392
Accius, 374, 384
Achaia, 116, 322
Acidinus, 68
Acilius Balbus (M'.), 12
Acilius Glabrio (M'.), 42
Aebutius, 374
Aeculanum, 374
Aegypta, 74, 114
Aelius (M.), 360, 364
Aelius Lamia, *see* Lamia (L. Aelius)
Aelius Tubero (L.), 142
Aemilius Lepidus, *father of Regillus*, 52
Aemilius Lepidus (M'.), 42
Aemilius Lepidus (M.), 194, 200, 204, 216, 388
Aemilius Paulus (L.), 226, 228
Africa, 52, 176
Africanus, *see* Cornelius Scipio Africanus
Agamemnon, 202, 268
Ahala, *see* Servilius Ahala
Ἀκαδημική (σύνταξις), 130, 134
Alaudae (legio), 400
Albanius (C.), 172
Albanum (negotium), 272, 280
Albinus, *see* Postumius Albinus
Albius Sabinus, 132
Aledius, 8, 50, 52, 58, 60
Alexander, *letter carrier*, 112
Alexander Magnus, 82, 164
Alexandrinae legiones, 330
Alexio, 158, 292, 300, 302
Alexis, 22, 420
Ἀλφειός, 10
Alsius, 210
Ammonius, 336
Amyntas, 20
Anagnia, 400
Anagninum (praedium), 2, 358; -ni, 408
Andromenes, 156

Annianus, 330
Annius (*i.e.* Asinius Pollio), 222
Antaeus, 198
Antiates, 326
Antiochia, 130, 140; ratio, 136; -ius, 158
Antiochus, *philosopher*, 138, 140
Antiochus, *slave*, 178
Antisthenes, 80
Antistius Vetus (C.), 232
Antium, 38, 204, 320, 322, 324
Antonius (C.), *brother of the triumvir*, 346
Antonius (L.), *brother of the triumvir*, 280, 284, 298, 312, 326, 336, 342
Antonius (M.), *orator*, 140
Antonius (M.), *the triumvir*, 36, 40, 220, 222, 224, 228, 236, 240, 246, 250, 256, 258, 264, 276, 278, 280, 282, 284, 292, 294, 302, 304, 308, 310, 312, 316, 326, 328, 342, 346, 348, 350, 354, 374, 376, 394, 398, 400, 402, 404, 410, 416, 418, 422; *letter from*, 246; *letter to*, 250. *See also* Cytherius
Antro, 342
Apella, 38
Apollinares ludi, 380
Apollo, 302
Apollodorus, 50
Appia via, 404, 414
Appuleius, *estate agent*, 28, 32
Appuleius (M.), *augur*, 26, 30, 32, 36
Aquilia, 244, 264
Aquinum, 404, 414
Arabio, 342
Arcanum (praedium), 404
Archilochus, 406
Ἀρχιμήδειον πρόβλημα, 8, 166
Argiletum, 68
Ariarathes, 110
Ariobarzanes, 110
Aristophanes, 16, 406

449

INDEX OF NAMES

Ἀριστοτέλειος, 140
Aristoteles, 82, 166
Aristoxenus, 174
Arpinas insula; 24; iter, 404; (praedium), 300, 362; -ates, 336, 414, 416
Arpinum, 90, 122, 202, 288, 296, 402
Asia, 318, 324, 326
Asiatica curatio, 322
Asinius Pollio (C.), 4, 78, 82, 146, 222, 366
Astura, 84, 94, 160, 180, 188, 220, 224, 236, 260, 278, 326, 328
Ateius Capito (C.), 178, 180, **436**, 442; *letter to*, 434, 440
Athamas, 22
Athenae, 50, 52, 68, 338, 390
Athenodorus, 408, 420
Atilius (M.)
Atilius Regulus (A.), 408
Atilius Serranus (Sex.), 12
Attica *or* Atticula, 2, 8, 18, 20, 24, 26, 30, 32, 50, 52, 56, 58, 60, 66, 70, 74, 86, 94, 98, 128, 132, 134, 138, 148, 152, 164, 196, 206, 212, 220, 262, 278, 284, 362, 364, 370, 380, 392, 398, 412
Atticus, *see* Pomponius Atticus
Ἄτυπος (*i.e.* Balbus), 8
Aurelius, 426
Aurelius, *legate of Hirtius*, 232
Aurelius Cotta (C.), 42, 138, **140**, 196
Aurelius Cotta (L.), 42, 50, 56
Aurelius Cotta (M.), 48, 196
Aventinum, 68
Avius, 10, 114
Axianus (M.), 364
Axius (Q.), 2

Babullius, 206
Bacchus, 362
Baebius, 198
Baiae, 84, 214, 332
Baiana negotia, 228
Balbilius, 330
Balbinus, 146
Balbus, *see* Cornelius Balbus
Baliares, 4
Barba, *see* Cassius Barba
Barea, 382
Barnaeus, 274
Bassus, *see* Caecilius Bassus *and* Lucilius Bassus
Bibulus, *see* Calpurnius Bibulus

Blesamius, 380
Brinniana auctio, 130; -nus fundus, 210
Brinnius, 132
Brundisium, 328, 348, 352, 374, 376, 382
Brutus, *see* Iunius Brutus
Bucilianus, 342, 382
Bursa, *see* Munatius Plancus Bursa
Buthrotia civitas, 438; res (*or* causa), 236, 240, 334, 428; -um negotium, 264; -us ager, 428
Buthrotii, 238, 256, 304, 306, 336, 344, 366, 372, 382, 428, 432, 434, 436, 438
Buthrotius (*sc.* Plancus), 366
Buthrotum, 278, 280, 298, 326, 348, 368
Byzantii, 228

Caecilius Bassus (Q.), 232, 330
Caecilius Metellus (L.), *consul* 142 B.C., 12
Caecilius Metellus (L.), *tribune* 49 B.C., 146
Caecina, 400
Caeliani, 122
Caelius, 10, 14, **112**
Caelius Rufus (M.), 176
Caepio, *see* Servilius Caepio
Caerellia, 104, 148, 150, 276, 294, 360
Caerellianum nomen, 104
Caesar, *see* Iulius Caesar
Caesaris filius (*i.e.* Caesarion), 280
Caesariana celeritas, 404
Caesonius (M.), 22
Caieta, 226
Calatia, 400
Calenus, 406
Cales, 410
Calpurnius Bibulus (M.), **68**
Calpurnius Piso (C.), 42
Calpurnius Piso Caesoninus (L.), 358, 396, 398
Calva, 302
Calvena, *see* Matius
Calvus Athenodorus, *see* Athenodorus
Camillus, *see* Furius Camillus
Cana, 192
Caninianum naufragium, 94
Caninius Gallus, 330, 420
Caninius Rebilus (C.), 76, 88
Canus, *see* Gellius Canus

450

INDEX OF NAMES

Capito, *see* Ateius Capito
Capitolina contio, 296; sessio, 254; -nus dies, 234
Capitolium, 176, 234
Capua, 264, 400, 402, 410
Carfulenus (D.), 304
Carneades, 50, 146
Carrinas (T.), 178
Carteia, 94, 348
Casca, *see* Servilius Casca
Cascellius (A.), 360
Casilinum, 400, 404
Cassiani (horti), 44
Cassii, 280
Cassius Barba, 212
Cassius Longinus (C.), 150, 264, 274, 282, 284, 310, 312, 314, 316, 318, 322, 324, 326, 330, 346, 354, 366, 374, 380 384, 392
Cassius Longinus (L.), 218
Castriciana mancipia, 64; -num negotium, 60
Castricius, 60
Catina, 410
Cato, *see* Porcius Cato
Cato (*i.e. Cicero's book on Cato*), 10, 162, 200
Cato maior (*i.e. the De Senectute*), 236
Catulus, *see* Lutatius Catulus
Catulus (*i.e. Cicero's Academica, Book I*), 174
Celer, *see* Pilius Celer
Censorinus, *see* Marcius Censorinus
Chremes, 16
Chrysippus, 166, 230
Cicero, *see* Tullius Cicero
Circeii, 38, 320
Cispiana (negotia), 52
Cispius, 176
Claudius, 42
Claudius Marcellus (C.), 302, 328, 330, 418, 426
Claudius Marcellus (M.), 124, 126, 150
Clodia, 46, 80, 88, 90, 92, 98, 104, 160, 168, 228
Clodiani (horti), 104
Clodius (L.), 64, 330
Clodius (Sex.), 246, 248, 254, 276
Clodius Hermogenes, 156
Clodius Patavinus, 94
Clodius Pulcher (P.), 250, 252
Clodius Pulcher (P.), *the younger*, 248, 250
Cluatius, 34, 70

Cluviana (negotia), 230; -ni horti, 202, 260, -num, 236, 238, 390
Cluvius (M.), 200, 202
Cocceius, 26, 36, 40, 426
Coponiana villa, 66
Corcyra, 156, 390
Corduba, 76
Corfidius (L.), 196
Corinthus, 116, 118, 178
Cornelius (Cn.), 178
Cornelius Balbus (L.), 4, 20, 24, 26, 40, 62, 94, 110, 138, 146, 150, 176, 184, 198, 200, 202, 204, 210, 214, 220, 234, 238, 282, 284, 300, 308, 310, 314, 316, 318, 380, 412; *See also* Ἄτυπος
Cornelius Balbus (L.), *son of the former*, 184, 208
Cornelius Dolabella (P.), 14, 20, 78, 122, 124, 132, 144, 148, 166, 168, 198, 204, 210, 214, 232, 258, 260, 264, 266, 268, 272, 276, 278, 280, 282, 284, 286, 302, 308, 316, 324, 326, 332, 334, 344, 346, 348, 380, 406, 420, 426; *letters to*, 266, 322, 334.
Cornelius Lentulus (Cn.), 178
Cornelius Lentulus Crus (L.), 302
Cornelius Lentulus Niger (L.), 18
Cornelius Lentulus Spinther (P.), 86, 104, 120, 126, 238
Cornelius Nepos, 388, 420
Cornelius Scipio Africanus Aemilianus (P.), 406, 416
Cornificia, 166
Cornificius (Q.), 28, 32, 40, 166
Corumbus, 220
Cosanum (praedium), 362
Cosianus, 364
Cossinius (L.), 202
Cotta, *see* Aurelius Cotta
Crassus, *see* Licinius Crassus
Craterus, 26, 30
Crispus, 10, 114, 116
Critonius, 146
Cumae, 296
Cumana regna, 260
Cumanum (praedium), 74, 164, 170, 236, 262, 278, 296, 370
Cupiennius (C.), 438; *letter to*, 438
Curio, *see* Scribonius Curio
Curius (M'.), 378
Curtilius, 224, 234
Curtius Postumus (M.), 100, 122, 232, 234, 300
Cusinius, 80, 88
Cytherius, 354

451

INDEX OF NAMES

Damasippus, *see* Licinius Damasippus
Deiotarus, 216, 240, 276, 380
Demea, 170
Demetrius, 262
Demonicus, 332
Δημοσθένης, 296
Dicaearchus, 170, 172, 174, **178**
Dida, 210
Dio, 320
Diocharinae epistulae, 198
Dionysius, 112, 180
Dolabella, *see* Cornelius Dolabella
Domitius Ahenobarbus (Cn.), 184, 206, 382
Drusiani (horti), 54
Drusus, *see* Livius Drusus
Dymaei, 368

Egnatius (L.), 36, 64, 66, 198
Egnatius (Q.), 246
Egnatius Maximus, 180
Ennius (M.), 362
Epicureus, 50, 140, 186
Epicurus, 24, 282
Epirotica (nomina), 184; -cae litterae, 106
Epirus, 158, 376, 380, 390
Ἡρακλείδειον, 306, 328, 362, 376, 408, 412
Ἑρμόδωρος, *see* Hermodorus
Eros, 18, 36, 44, 110, 130, 170, 210, 274, 336, 338, 342, 350, 368, 370, 372, 376, 414, 426
Eupolis, 16
Eurotas, 318
Eutrapelus, *see* Volumnius Eutrapelus

Faberiana (causa or res), 86, 168; -num (nomen), 62, 66, 98, 168; negotium, 170
Faberius (Q.), 44, 54, 104, 110, 112, 164, 168, 174, 176, 272, 330
Fabius Maximus (Q.), 12
Fadius (C.), 406
Fadius (L.), 336, 342, 350
Fadius Gallus (M.), 206, 208
Fanniani libri, 12
Fannius (C.), 12, 14, 416
Favonius (M.), 322, 366
Ficulensis (fundus), 70
Figulus, *see* Marcius Figulus

Flaminius Flamma (T), 104, 262, 266, 294, 300
Flavius, 32
Formianum (praedium) 226, 332; -ni, 366
Frangones, 234
Fufius Calenus (Q.), 304, 406
Fulvia, 240
Fulviaster, 94
Fundi, 224
Furius Camillus (C.), 116, 180
Furius Philus (L.), 12

Galba, *see* Sulpicius Galba
Galli, 230
Gallia, 222, 232, 256
Gallica bella, 222; -us tumultus, 216
Gallus, *see* Caninius *and* Fadius Gallus
Gamala, 50
Gellius Canus (Q.), 172, 352
Gellius Poplicola (L.), 42
Germani, 232
Glabrio (M'.), 42
Graeceius, 316
Graeci, 34, 132, 418; ludi, 384
Graecia, 228, 244, 260, 274

Hegesias, 14
Heles, 396
Heraclides, 140
Herenniani coheredes, 116
Hermodorus, 146
Hermogenes, 54, 66
Hermogenes (Clodius), *see* Clodius Hermogenes
Herodes, *agent for Atticus,* 364
Herodes, *of Athens,* 262, 274, 340, 378
Hesiodus, 128
Hetereius, 210
Hieras, 380
Hilarus, *freedman of Cicero*, 74, 138
Hilarus, *freedman of Libo*, 382
Hippocrates, 424
Hirtius (A.), 4, 70, 76, 82, 88, 92, 96, 98, 144, 184, 232, 238, 282, 286, 292, 294, 302, 310, 312, 314, 316, 328, 354; letter of, 312
Hispalis, 142
Hispani, 230
Hispania, 18, 48, 76, 222
Hispaniensis res, 330
Hordeonius (T.), 202

452

INDEX OF NAMES

Hortensius (Q.), *orator*, 14, 118, 134, 136, 140, 170, 174, 178
Hortensius (Q.), *son of the last*, 10, 372
Hostilius Tubulus (L.), 12
Hydrus, 352, 386

Isthmus, 176
Italia, 324, 376, 390
Iulia, 366
Iulia lex, 326; -ae nonae, 368, 380
Iulius Caesar (C.), *dictator*, 6, 16, 20, 42, 82, 88, 96, 98, 100, 102, 104, 108, 110, 120, 124, 126, 132, 136, 138, 142, 148, 152, 160, 162, 164, 172, 180, 184, 188, 196, 198, 202, 204, 210, 212, 216, 218, 222, 224, 232, 234, 238, 240, 242, 244, 246, 248, 264, 266, 276, 302, 382, 428, 430, 432, 434, 436, 438, 440
Iulius Caesar (L.), 42, 264, 268, 308
Iulius Caesar Octavius (*i.e.* Octavianus), 224, 234, 238, 240, 284, 286, 298, 328, 400, 402, 410, 418
Iulius Caesar Strabo Vopiscus (C.), 140
Iunia, 228
Iunius, 28
Iunii Bruti, 254, 280, 328
Iunius Brutus (D.), 242, 304, 312, 320, 322, 340, 366, 410
Iunius Brutus (D.), *consul* 510 B.C., 46
Iunius Brutus (L.), 190
Iunius Brutus (M.), *murderer of Caesar*, 12, 14, 26, 28, 30, 34, 40, 42, 58, 62, 74, 78, 114, 116, 118, 120, 122, 126, 128, 130, 132, 134, 136, 146, 150, 152, 154, 156, 158, 180, 182, 184, 186, 188, 190, 192, 200, 206, 216, 218, 222, 226, 228, 230, 232, 236, 242, 258, 260, 264, 268, 274, 276, 278, 280, 282, 288, 294, 296, 302, 306, 308, 310, 312, 314, 318, 320, 322, 326, 330, 340, 342, 344, 346, 348, 350, 354, 356, 358, 360, 362, 364, 368, 370, 372, 374, 380, 382, 384, 386, 392, 396, 398, 402, 418
Iunius Silanus (D.), 42
Iuventius Laterensis, 32
Iuventius Talna, 166, 388

Καλλιπίδης, 128
Karthago, 382
Κῦρος, *a book by Antisthenes*, 80

Labeo, 130
Lacedaemon, 318
Laco, 408
Laelius Sapiens (C.), 12
Laenas, *see* Popilius Laenas
Lamia (L. Aelius), 48, 62, 198, 252
Lamiani (horti), 44
Lanuvinum (praedium), 90, 210
Lanuvium, 86, 94, 96, 160, 162, 180, 200, 220, 226, 232, 280, 284, 306, 318, 342, 348
Laterensis, *see* Iuventius Laterensis
Latina lingua, 106; -ni, 34
Latinitas, 240
Lentulus, *son of Dolabella and Tullia*, 60, 64
Lentulus, *see also* Cornelius Lentulus
Leonides, 262, 274, 340
Lepidianae feriae, 412
Lepidus, *see* Aemilius Lepidus
Lepta, 184, 200, 206, 358, 424
Leucopetra, 390, 392
Liberalia, 234, 254
Libo, *see* Scribonius Libo
Licinius Crassus (L.), 100, 140, 228
Licinius Crassus (P.), 52
Licinius Damasippus, 62, 68
Licinius Lucullus (Cn.), 296
Licinius Lucullus Ponticus (L.), 130, 134, 140, 176
Licinius Lucullus (L.), *son of Ponticus*, 116, 278
Licinius Lucullus (M.), 118
Licinius Murena (L.), 42, 118, 210
Ligariana (*oratio*), 128, 138, 142, 196
Ligarii, 196
Ligarius (T.), 196
Ligus (L.), 50, 224
Livius Drusus, 6, 44, 50, 66, 68, 76, 80, 82, 88, 92, 160
Lollius (C.), 44
Lucceius (Cn.), 386
Lucilianus φαλλός, 406
Lucilius (C.), 146
Lucilius Bassus, 10
Lucrinus (lacus), 260
Luculli, 42
Lucullus, *see* Licinius Lucullus
Lucullus (*i.e. the 2nd book of Cicero's Academica*), 174
Lupercus, 10
Lutatius Catulus (Q.) *consul* 78 B.C., 42, 130, 134, 140
Lutatius Catulus (Q.), *consul* 102 B.C., 140

453

INDEX OF NAMES

Macedonicae legiones, 400
Madarus (*i.e.* Matius), 218
Magius Cilo (P.), 126
Mamurra. 214
Manilius (M'.), 12
Manlius Torquatus, 32, 116, 122, 142, 144, 198, 202, 412
Manlius Torquatus (L.), *consul* 65 B.C., 42
Manlius Torquatus (L.), *son of the last*, 140
Marcellus, *see* Claudius Marcellus
Marcianus, *see* Tullius Marcianus
Marcius Censorinus, 234
Marcius Censorinus (L.), 12
Marcius Figulus (C.), 42
Marcius Philippus (C.), 32, 34, 212, 214, 238, 240, 418
Marius, (C.), 100
Marius (C.), *impostor*, 100, 224, 226, 228
Mars, 322
Martius campus, 20, 180
Massilienses, 256
Matius (C.), 210, 220, 222, 300, 306. *See also* Madarus
Maximus, *see* Fabius Maximus
Menedemus, 298, 308, 346
Menturnae, 404, 414
Messalla, *see* Valerius Messalla
Metella, 120
Metellus, *see* Caecilius Metellus
Meto, 104
Meto, *astronomer*, 6
Metrodorus, 298
Mettius, 364
Μίκυλλος, 212
Μίμας, 414
Misenum, 280, 292
Montanus, *see* Tullius Montanus
Mucius Scaevola (P.), 12
Mulvius pons, 180
Mummius (L.), 114, 170, 178, 416
Mummius (Sp.), *brother of L. Mummius*, 116, 118, 170
Mummius (Sp.), *grandson of the last*, 118
Munatius Plancus, 104, 176, 362, 366, 368, 374, 382, 426, 430, 432, 436, 438, 440; *letters to*, 428, 432, 438
Munatius Plancus Bursa (T.), 234
Mundus, 362, 366
Murcus, *see* Statius Murcus
Murena, *see* Licinius Murena
Musca, 82

Mustela, 10, 92, 98, 114, **116, 120**
Mustela, *of Anagnium*, 408
Myrtilus, 332, 410

Narbo, 76
Naso (P.), 32
Neapolis, 234, 264, 268, 294, 380, 392
Neapolitanum (praedium), 286
Nemus, 308
Nepos, *see* Cornelius Nepos
Nesis, 368, 374, 380
Nestor, 268
Nicasiones, 14
Nicaea, 216
Nicias Curtius, 56, 102, 106, 108, 122, 166, 214, 232, 348
Nolanus (ager), 122

Ocella (Cn.), 412
Octavianus, *see* Iulius Caesar Octavius
Octavii pueri, 208
Octavius, *see* Iulius Caesar Octavius
Offilius (A.), 184
Ollius, 206
Olympia, 170, 398
Oppius (C.), 26, 40, 62, 94, 110, 138, 204, 210, 216, 310, 374, 376, 422
Ops, 256, 272, 420
Orator, *a book by Cicero*, 16
Oropus, 50
Ostiense (praedium), 50, 62
Otho, *see* Roscius Otho
Othones, 366
Ovia, 44, 52, 64, 150, 372
Ovius, 370

Paciaecus, 4
Pacorus, 232
Paestanus sinus, 388
Paetus, *see* Papirius Pactus
Παλλάς, 188
Panaetius, 122, 408
Pansa, *see* Vibius Pansa
Papirius Paetus (L.), 260
Parilia, 252
Parthenon, 190
Parthi, 172; -us, 232
Parthicum bellum, 162
Patavinus, 94
Patrae, 390

INDEX OF NAMES

Patulcianum nomen, 272
Paulus, *see* Aemilius Paulus (L.)
Peducaeus (Sex.), 102, 108, 112, 314, 330, 406, 420, 424
Πειρήνη, 10
Pelopidae, 240, 324
Pelops, 228
Περσική porticus, 318
Φαῖδρος (Phaedrus), 188, 396
Phamea, 206, 208
Pharnaces, 170, 198
Pheriones, 254
Philippus, *see* Marcius Philippus
Philo, 382
Philotimus, *copyist*, 176
Philotimus, *freedman of Terentia*, 10, 94, 98
Philoxenus, 122
Philus, *see* Furius Philus
Pilia, 2, 8, 20, 30, 32, 52, 56, 58, 60, 66, 74, 86, 98, 152, 204, 206, 220, 260, 262, 278, 234, 288, 296, 370, 380, 392, 398
Pilius (M.), 172
Pilius Celer (Q.), 20
Pindarus, *poet*, 188
Pindarus, *slave*, 370
Piso, *banker*, 12, 110, 114, 116, 128, 130, 136, 178
Piso, *see also* Calpurnius *and* Pupius Piso
Plaetorius (M), 342
Plancus, *see* Munatius Plancus
Plato, 146
Plotius, 202
Polla, 154
Pollex, 200, 202, 204, 206
Pollio, *see* Asinius Pollio
Polybius, 170
Pompeia lex, 208
Pompeianum (praedium), 122, 260, 262, 264, 272, 274, 278, 280, 232, 292, 332, 374, 380, 388, 396, 410
Pompeius Magnus (Cn.), 22
Pompeius (Cn.), *son of the last*, 4, 76
Pompeius (Q.), 12
Pompeius (Sex.), 76, 94, 216, 222, 230, 242, 288, 348, 352, 354, 368, 380, 382
Pomponius Atticus (T.), 6, 40, 48, 172, 186, 238, 260, 264, 280, 334, 336, 338, 348, 372, 390, 394, 414, 424, 428, 430, 432, 434, 438, 440
Pontianus, 92
Pontius Aquila (L.), 286

Popilius (P.), 174
Popilius Laenas, 26, 28, 32
Porcia, *daughter of Cato*, 322
Porcia, *sister of Cato*, 184, 206
Porcius Cato (M.) (*i.e.* "*Cato of Utica*"), 8, 42, 44, 82, 88, 92, 96, 134, 140, 370, 392
Porcius Cato (M.), *son of the last*, 116
Posidonius, 408
Postumia, 22, 48
Postumius Albinus (A.), 170, 176
Postumus, *see* Curtius Postumus
Praeneste, 4
Preciana (negotia), 52
Prognostica, *a work by Cicero*, 340
Ψυρίη, 414
Publicianus locus, 80
Publilia, 66
Publilius, 18, 38, 52, 60, 66, 182, 204, 276, 372, 390
Publilius Syrus, 218
Pupius Piso Frugi Calpurnianus (M.), 140
Puteolana regna, 260 ; -num (praedium), 226, 278, 292, 294, 296, 364, 368 ; -nus mos, 254
Puteoli, 198, 204, 210, 214, 348, 360, 362, 366, 418

Quinctius Flamininus (T.), 12
Quinctius Scapula (T.), 76, 80, 86
Quinti (*sc.* Cicerones), 280
Quirinus, 96, 166

Regillus, 52
Regini, 392
Regium, 390
Regulus, *see* Atilius Regulus
Roma, 14, 20, 22, 34, 46, 50, 68, 76, 84, 86, 90, 98, 102, 110, 120, 130, 136, 148, 154, 156, 158, 162, 174, 184, 188, 196, 203, 212, 220, 236, 240, 262, 280, 288, 302, 308, 310, 322, 324, 326, 338, 350, 358, 370, 372, 392, 394, 400, 402, 404, 410, 412, 418
Romani cives, 240 ; ludi, 198, 200 ; -nus populus, 254, 374
Roscius Otho (L.), 76, 80, 82, 86, 90, 92, 168, 172, 176
Rubriana, 406
Rufio (Vestorianus), 254
Rupilius (P.), 174
Rutilia, 42, 46

455

INDEX OF NAMES

Sabinus, *see* Albius Sabinus
Sallustius (Cn.), 210
Salus, 96
Salvius, 198, 376
Samnium, 280, 410
Sara, 336
Saserna, 300
Saturnalia, 212, 214
Satyrus, 48
Saufeius (L.), 274, 306
Saxa, 190
Scaeva, 154
Scaevae, 234
Scaevola, *see* Mucius Scaevola
Scaptius (M.), 330
Scapula, *see* Quinctius Scapula
Scapulani (horti), 76, 84, 104, 130, 180
Scipio, *see* Cornelius Scipio
Scribonius Curius (C.), 42
Scribonius Libo (L.), *tribune, 56 B.C.*, 36, 40, 380, 382
Scribonius Libo (L.), *writer of annals*, 12, 170, 174, 196
Scrofa, *see* Tremellius Scrofa
Seius (M.), 22
Sempronius Tuditanus (C.), 118
Sempronius Tuditanus (C.), *son of the last*, 114, 170, 174, 178
Septimia, 404
Septimius (C.), 26
Serranus, *see* Atilius Serranus
Servilia, *mother of Brutus*, 128, 136, 314, 322, 324, 326, 330, 342, 356
Servilia, *wife of Claudius*, 42
Servilius Ahala (C.), 190
Servilius Caepio (Cn.), *consul 141 B.C.*, 12
Servilius Caepio (Cn.), 42
Servilius Casca (P.), 196, 422, 424
Servilius Vatia (P.), 42
Servius, *see* Sulpicius Rufus
Sestius (P.), 110, 120, 208, 218, 332, 342, 362, 374, 382, 418
Sextilianus fundus, 234
Sextilius Rufus (C.), 224
Sextus, *see* Peducaeus *and* Pompeius
Sicca, 50, 52, 54, 56, 58, 64, 70, 276, 342, 388, 404, 406
Sicilia, 60, 318
Siculi, 238, 240
Silanus, *see* Iulius Silanus
Siliana villa, 56 ; -ni (horti), 66 ; -num negotium, 56

Silius, (A.), 38, 52, 54, 58, 62, 64, 66, 68, 70, 82, 88, 92, 104, 116, 120, 210
Silius Nerva (P.), 354, 356
Sinuessanum (devorsoriolum), 228, 296, 298, 404, 414
Siregius, 342
Sittius (P.), 342
Socrates, 230
Socratici viri, 230
Spintharus, 160
Spinther, *see* Cornelius Lentulus Spinther
Staberius (Q.), 122
Statilius (L.), 26, 28
Statius, 10, 336, 340, 346, 352
Statius Murcus (L.), 4
Stoica, 140
Strabo, *augur*, 22
Strenia, 360
Sulpicius Galba (Ser.), 12
Sulpicius Rufus (P.), 140
Sulpicius Rufus (Ser.), 22, 36, 124, 150, 276, 278, 314
Syracusae, 390, 410
Syrus, *slave*, 48, 342

Talna, *see* Iuventius Talna
Tarentini, 390
Tauromenium, 410
Teanum Sidicinum, 410
Tebassi, 234
Tellus, 418
Terentia, 36, 40, 42, 44, 46, 50, 76, 202, 390, 424, 426
Terentius Varro (M.), 14, 128, 130, 134, 138, 140, 144, 148, 150, 154, 156, 158, 160, 178, 182, 196, 310, 328, 332, 362, 406, 412
Terentius Varro Gibba (M.), 206
Tereus, *a play*, 374, 384
Tertulla (Tertia), 280, 322
Theophanes, 346
Theopompus, 82, 120
Tiberis, 38, 180, 338
Tibur, 376
Tigellius, 206, 210
Tirenus pons, 414
Tiro, *see* Tullius Tiro
Tisamenos, 22
Tite, O (*i.e. the De Senectute*), 376, 408
Torquatus (*i.e. Cicero's De Finibus, Book I*), 174

INDEX OF NAMES

Torquatus, *see also* Manlius Torquatus
Transtiberini (horti), 50
Trebatius Testa (C.), 122, 154
Treboniani (horti), 92
Trebonius (C.), 80, 88, 232
Tremellius Scrofa (Cn.), 148
Triarius, *see* Valerius Triarius
Τρῶες, 132, 156
Tubero, *see* Aelius Tubero
Tubulus, *see* Hostilius Tubulus
Tuditanus, *see* Sempronius Tuditanus
Tullia (Tulliola), 2, 8, 14
Tullianum caput, 360; semis 364; -nae aedes 360
Tullii (*i.e.* Marcianus *and* Montanus) 108
Tullius, *scribe*, 152
Tullius Cicero (M.), *the orator*, 218, 248, 268, 312, 314, 434
Tullius Cicero (M.), *son of the orator*, 18, 20, 40, 52, 56, 58, 68, 100, 104, 108, 156, 184, 226, 238, 244, 260, 262, 264, 280, 332, 338, 340, 342, 350, 370, 378, 408, 424
Tullius Cicero (Q.), *brother of the orator*, 2, 10, 60, 144, 192, 202, 204, 212, 236, 244, 298, 350, 352, 358, 380
Tullius Cicero (Q.), *son of the last*, 122, 168, 184, 244, 252, 264, 294, 302, 346, 354, 366, 370, 378, 384
Tullius Marcianus, 32, 106
Tullius Montanus (L.), 104, 106, 262, 266, 274, 424; *see also* Tullii
Tullius Tiro (M.), 8, 14, 22, 40, 70, 98, 100, 102, 104, 122, 160, 308, 316, 326, 338, 342, 344, 350, 388, 414, 426
Tullus, *see* Volcatius Tullus
Tusculana disputatio, 300, 306; -num (praedium), 2, 6, 74, 86, 88, 90, 92, 94, 96, 98, 114, 120, 126, 128, 130, 132, 136, 144, 154, 158, 168, 186, 204, 206, 214, 220, 300, 306, 310, 312, 314, 316, 326, 334, 340, 342, 360, 362, 392, 416, 418
Tutia, 374
Tyndaritani, 300
Tyrannio, 6, 16

Utica, 4

Valerius, *friend of Cicero*, 392
Valerius, *interpreter*, 410
Valerius (P.), 102, 106, 134
Valerius Messalla (M.), 124, 234, 342
Valerius Messalla Corvinus (M.), 68, 428
Valerius Triarius (C.), 60
Varro, *see* Terentius Varro
Vaticani montes, 180; -nus campus 180
Velia, 388, 396
Vennonius, 6
Ventidius (P.), 368
Venuleia, 52
Venusia, 386
Vergilius, 102, 160, 176
Verginius, 10
Vescianum (praedium), 298
Vestoriana haeresis, 254
Vestorianus, *see* Rufio
Vestorius (C.), 120, 130, 170, 184, 200, 202, 210, 230, 242, 284, 286, 306
Vettienus, 8, 330, 332, 348
Vettius (Sex.), 130
Vetus, *see* Antistius Vetus
Vibius Pansa, 30, 32, 40, 58, 146, 238, 276, 282, 294, 328, 354, 370, 402
Vibo, 388
Vibonensis sinus, 388
Victor, 254
Victoria, 196
Visellia, 330
Volaterranus Caecina, 400
Volcatius Tullus (L.), 42
Volcatius Tullus (L.), *praetor*, 46 B.C., 232
Volumnius Eutrapelus, 316

Xeno, 184, 262, 352, 370, 378

457

PRINTED IN GREAT BRITAIN BY
RICHARD CLAY AND COMPANY, LTD.,
BUNGAY, SUFFOLK

THE LOEB CLASSICAL LIBRARY

VOLUMES ALREADY PUBLISHED

Latin Authors

AMMIANUS MARCELLINUS. Translated by J. C. Rolfe. 3 Vols.

APULEIUS: THE GOLDEN ASS (METAMORPHOSES). W. Adlington (1566). Revised by S. Gaselee.

ST. AUGUSTINE: CITY OF GOD. 7 Vols. Vol. I. G. H. McCracken. Vol. VI. W. C. Greene.

ST. AUGUSTINE, CONFESSIONS OF. W. Watts (1631). 2 Vols.

ST. AUGUSTINE, SELECT LETTERS. J. H. Baxter.

AUSONIUS. H. G. Evelyn White. 2 Vols.

BEDE. J. E. King. 2 Vols.

BOETHIUS: TRACTS and DE CONSOLATIONE PHILOSOPHIAE. Rev. H. F. Stewart and E. K. Rand.

CAESAR: ALEXANDRIAN, AFRICAN and SPANISH WARS. A. G. Way.

CAESAR: CIVIL WARS. A. G. Peskett.

CAESAR: GALLIC WAR. H. J. Edwards.

CATO: DE RE RUSTICA; VARRO: DE RE RUSTICA. H. B. Ash and W. D. Hooper.

CATULLUS. F. W. Cornish; TIBULLUS. J. B. Postgate; PERVIGILIUM VENERIS. J. W. Mackail.

CELSUS: DE MEDICINA. W. G. Spencer. 3 Vols.

CICERO: BRUTUS, and ORATOR. G. L. Hendrickson and H. M. Hubbell.

[CICERO]: AD HERENNIUM. H. Caplan.

CICERO: DE ORATORE, etc. 2 Vols. Vol. I. DE ORATORE, Books I. and II. E. W. Sutton and H. Rackham. Vol. II. DE ORATORE, Book III. De Fato; Paradoxa Stoicorum; De Partitione Oratoria. H. Rackham.

CICERO: DE FINIBUS. H. Rackham.

CICERO: DE INVENTIONE, etc. H. M. Hubbell.

CICERO: DE NATURA DEORUM and ACADEMICA. H. Rackham.

CICERO: DE OFFICIIS. Walter Miller.

CICERO: DE REPUBLICA and DE LEGIBUS; SOMNIUM SCIPIONIS. Clinton W. Keyes.

CICERO: DE SENECTUTE, DE AMICITIA, DE DIVINATIONE. W. A. Falconer.

CICERO: IN CATILINAM, PRO FLACCO, PRO MURENA, PRO SULLA. Louis E. Lord.

CICERO: LETTERS TO ATTICUS. E. O. Winstedt. 3 Vols.

CICERO: LETTERS TO HIS FRIENDS. W. Glynn Williams. 3 Vols.

CICERO: PHILIPPICS. W. C. A. Ker.

CICERO: PRO ARCHIA POST REDITUM, DE DOMO, DE HARUSPICUM RESPONSIS, PRO PLANCIO. N. H. Watts.

CICERO: PRO CAECINA, PRO LEGE MANILIA, PRO CLUENTIO, PRO RABIRIO. H. Grose Hodge.

CICERO: PRO CAELIO, DE PROVINCIIS CONSULARIBUS, PRO BALBO. R. Gardner.

CICERO: PRO MILONE, IN PISONEM, PRO SCAURO, PRO FONTEIO, PRO RABIRIO POSTUMO, PRO MARCELLO, PRO LIGARIO, PRO REGE DEIOTARO. N. H. Watts.

CICERO: PRO QUINCTIO, PRO ROSCIO AMERINO, PRO ROSCIO COMOEDO, CONTRA RULLUM. J. H. Freese.

CICERO: PRO SESTIO, IN VATINIUM. R. Gardner.

CICERO: TUSCULAN DISPUTATIONS. J. E. King.

CICERO: VERRINE ORATIONS. L. H. G. Greenwood. 2 Vols.

CLAUDIAN. M. Platnauer. 2 Vols.

COLUMELLA: DE RE RUSTICA. DE ARBORIBUS. H. B. Ash, E. S. Forster and E. Heffner. 3 Vols.

CURTIUS, Q.: HISTORY OF ALEXANDER. J. C. Rolfe. 2 Vols.

FLORUS. E. S. Forster; and CORNELIUS NEPOS. J. C. Rolfe.

FRONTINUS: STRATAGEMS and AQUEDUCTS. C. E. Bennett and M. B. McElwain.

FRONTO: CORRESPONDENCE. C. R. Haines. 2 Vols.

GELLIUS, J. C. Rolfe. 3 Vols.

HORACE: ODES and EPODES. C. E. Bennett.

HORACE: SATIRES, EPISTLES, ARS POETICA. H. R. Fairclough.

JEROME: SELECTED LETTERS. F. A. Wright.

JUVENAL and PERSIUS. G. G. Ramsay.

LIVY. B. O. Foster, F. G. Moore, Evan T. Sage, and A. C. Schlesinger and R. M. Geer (General Index). 14 Vols.

LUCAN. J. D. Duff.

LUCRETIUS. W. H. D. Rouse.

MARTIAL. W. C. A. Ker. 2 Vols.

MINOR LATIN POETS: from PUBLILIUS SYRUS to RUTILIUS NAMATIANUS, including GRATTIUS, CALPURNIUS SICULUS, NEMESIANUS, AVIANUS, and others with "Aetna" and the "Phoenix." J. Wight Duff and Arnold M. Duff.

OVID: THE ART OF LOVE and OTHER POEMS. J. H. Mozley.

OVID: FASTI. Sir James G. Frazer.
OVID: HEROIDES and AMORES. Grant Showerman.
OVID: METAMORPHOSES. F. J. Miller. 2 Vols.
OVID: TRISTIA and EX PONTO. A. L. Wheeler.
PERSIUS. Cf. JUVENAL.
PETRONIUS. M. Heseltine; SENECA: APOCOLOCYNTOSIS. W. H. D. Rouse.
PLAUTUS. Paul Nixon. 5 Vols.
PLINY: LETTERS. Melmoth's Translation revised by W. M. L. Hutchinson. 2 Vols.
PLINY: NATURAL HISTORY. H. Rackham and W. H. S. Jones. 10 Vols. Vols. I.–V. and IX. H. Rackham. Vols. VI. and VII. W. H. S. Jones.
PROPERTIUS. H. E. Butler.
PRUDENTIUS. H. J. Thomson. 2 Vols.
QUINTILIAN. H. E. Butler. 4 Vols.
REMAINS OF OLD LATIN. E. H. Warmington. 4 Vols. Vol. I. (ENNIUS AND CAECILIUS.) Vol. II. (LIVIUS, NAEVIUS, PACUVIUS, ACCIUS.) Vol. III. (LUCILIUS and LAWS OF XII TABLES.) (ARCHAIC INSCRIPTIONS.)
SALLUST. J. C. Rolfe.
SCRIPTORES HISTORIAE AUGUSTAE. D. Magie. 3 Vols.
SENECA: APOCOLOCYNTOSIS. Cf. PETRONIUS.
SENECA: EPISTULAE MORALES. R. M. Gummere. 3 Vols.
SENECA: MORAL ESSAYS. J. W. Basore. 3 Vols.
SENECA: TRAGEDIES. F. J. Miller. 2 Vols.
SIDONIUS: POEMS and LETTERS. W. B. Anderson. 2 Vols.
SILIUS ITALICUS. J. D. Duff. 2 Vols.
STATIUS. J. H. Mozley. 2 Vols.
SUETONIUS. J. C. Rolfe. 2 Vols.
TACITUS: DIALOGUES. Sir Wm. Peterson. AGRICOLA and GERMANIA. Maurice Hutton.
TACITUS: HISTORIES AND ANNALS. C. H. Moore and J. Jackson. 4 Vols.
TERENCE. John Sargeaunt. 2 Vols.
TERTULLIAN: APOLOGIA and DE SPECTACULIS. T. R. Glover. MINUCIUS FELIX. G. H. Rendall.
VALERIUS FLACCUS. J. H. Mozley.
VARRO: DE LINGUA LATINA. R. G. Kent. 2 Vols.
VELLEIUS PATERCULUS and RES GESTAE DIVI AUGUSTI. F. W. Shipley.
VIRGIL. H. R. Fairclough. 2 Vols.
VITRUVIUS: DE ARCHITECTURA. F. Granger. 2 Vols.

Greek Authors

ACHILLES TATIUS. S. Gaselee.

AELIAN: ON THE NATURE OF ANIMALS. A. F. Scholfield. 3 Vols.

AENEAS TACTICUS, ASCLEPIODOTUS and ONASANDER. The Illinios Greek Club.

AESCHINES. C. D. Adams.

AESCHYLUS. H. Weir Smyth. 2 Vols.

ALCIPHRON, AELIAN, PHILOSTRATUS: LETTERS. A. R. Benner and F. H. Fobes.

ANDOCIDES, ANTIPHON, Cf. MINOR ATTIC ORATORS.

APOLLODORUS. Sir James G. Frazer. 2 Vols.

APOLLONIUS RHODIUS. R. C. Seaton.

THE APOSTOLIC FATHERS. Kirsopp Lake. 2 Vols.

APPIAN: ROMAN HISTORY. Horace White. 4 Vols.

ARATUS. Cf. CALLIMACHUS.

ARISTOPHANES. Benjamin Bickley Rogers. 3 Vols. Verse trans.

ARISTOTLE: ART OF RHETORIC. J. H. Freese.

ARISTOTLE: ATHENIAN CONSTITUTION, EUDEMIAN ETHICS, VICES AND VIRTUES. H. Rackham.

ARISTOTLE: GENERATION OF ANIMALS. A. L. Peck.

ARISTOTLE: METAPHYSICS. H. Tredennick. 2 Vols.

ARISTOTLE: METEROLOGICA. H. D. P. Lee.

ARISTOTLE: MINOR WORKS. W. S. Hett. On Colours, On Things Heard, On Physiognomies, On Plants, On Marvellous Things Heard, Mechanical Problems, On Indivisible Lines, On Situations and Names of Winds, On Melissus, Xenophanes, and Gorgias.

ARISTOTLE: NICOMACHEAN ETHICS. H. Rackham.

ARISTOTLE: OECONOMICA and MAGNA MORALIA. G. C. Armstrong; (with Metaphysics, Vol. II.).

ARISTOTLE: ON THE HEAVENS. W. K. C. Guthrie.

ARISTOTLE: ON THE SOUL. PARVA NATURALIA. ON BREATH. W. S. Hett.

ARISTOTLE: ORGANON—Categories, On Interpretation, Prior Analytics. H. P. Cooke and H. Tredennick.

ARISTOTLE: ORGANON—Posterior Analytics, Topics. H. Tredennick and E. S. Foster.

ARISTOTLE: ORGANON—On Sophistical Refutations.
On Coming to be and Passing Away, On the Cosmos. E. S. Forster and D. J. Furley.

ARISTOTLE: PARTS OF ANIMALS. A. L. Peck; MOTION AND PROGRESSION OF ANIMALS. E. S. Forster.

ARISTOTLE: PHYSICS. Rev. P. Wicksteed and F. M. Cornford. 2 Vols.

ARISTOTLE: POETICS and LONGINUS. W. Hamilton Fyfe; DEMETRIUS ON STYLE. W. Rhys Roberts.

ARISTOTLE: POLITICS. H. Rackham.

ARISTOTLE: PROBLEMS. W. S. Hett. 2 Vols.

ARISTOTLE: RHETORICA AD ALEXANDRUM (with PROBLEMS. Vol. II.). H. Rackham.

ARRIAN: HISTORY OF ALEXANDER and INDICA. Rev. E. Iliffe Robson. 2 Vols.

ATHENAEUS: DEIPNOSOPHISTAE. C. B. Gulick. 7 Vols.

ST. BASIL: LETTERS. R. J Deferrari. 4 Vols.

CALLIMACHUS: FRAGMENTS. C. A. Trypanis.

CALLIMACHUS, Hymns and Epigrams, and LYCOPHRON. A. W. Mair; ARATUS. G. R. Mair.

CLEMENT of ALEXANDRIA. Rev. G. W. Butterworth.

COLLUTHUS. Cf. OPPIAN.

DAPHNIS AND CHLOE. Thornley's Translation revised by J. M. Edmonds; and PARTHENIUS. S. Gaselee.

DEMOSTHENES I.: OLYNTHIACS, PHILIPPICS and MINOR ORATIONS. I.-XVII. AND XX. J. H. Vince.

DEMOSTHENES II.: DE CORONA and DE FALSA LEGATIONE. C. A. Vince and J. H. Vince.

DEMOSTHENES III.: MEIDIAS, ANDROTION, ARISTOCRATES, TIMOCRATES and ARISTOGEITON, I. AND II. J. H. Vince.

DEMOSTHENES IV.-VI.: PRIVATE ORATIONS and IN NEAERAM. A. T. Murray.

DEMOSTHENES VII.: FUNERAL SPEECH, EROTIC ESSAY, EXORDIA and LETTERS. N. W. and N. J. DeWitt.

DIO CASSIUS: ROMAN HISTORY. E. Cary. 9 Vols.

DIO CHRYSOSTOM. J. W. Cohoon and H. Lamar Crosby. 5 Vols.

DIODORUS SICULUS. 12 Vols. Vols. I.-VI. C. H. Oldfather. Vol. VII. C. L. Sherman, Vols. IX. and X. R. M. Geer. Vol. XI. F. Walton.

DIOGENES LAERITIUS. R. D. Hicks. 2 Vols.

DIONYSIUS OF HALICARNASSUS: ROMAN ANTIQUITIES. Spelman's translation revised by E. Cary. 7 Vols.

EPICTETUS. W. A. Oldfather. 2 Vols.

EURIPIDES. A. S. Way. 4 Vols. Verse trans.

EUSEBIUS: ECCLESIASTICAL HISTORY. Kirsopp Lake and J. E. L. Oulton. 2 Vols.

GALEN: ON THE NATURAL FACULTIES. A. J. Brock.

THE GREEK ANTHOLOGY. W. R. Paton. 5 Vols.

GREEK ELEGY AND IAMBUS with the ANACREONTEA. J. M. Edmonds. 2 Vols.

THE GREEK BUCOLIC POETS (THEOCRITUS, BION, MOSCHUS). J. M. Edmonds.

GREEK MATHEMATICAL WORKS. Ivor Thomas. 2 Vols.

HERODES. Cf. THEOPHRASTUS: CHARACTERS.

HERODOTUS. A. D. Godley. 4 Vols.

HESIOD AND THE HOMERIC HYMNS. H. G. Evelyn White.

HIPPOCRATES and the FRAGMENTS OF HERACLEITUS. W. H. S. Jones and E. T. Withington. 4 Vols.

HOMER: ILIAD. A. T. Murray. 2 Vols.

HOMER: ODYSSEY. A. T. Murray. 2 Vols.

ISAEUS. E. W. Forster.

ISOCRATES. George Norlin and LaRue Van Hook. 3 Vols.

ST. JOHN DAMASCENE: BARLAAM AND IOASAPH. Rev. G. R. Woodward and Harold Mattingly.

JOSEPHUS. H. St. J. Thackeray and Ralph Marcus. 9 Vols. Vols. I.–VII.

JULIAN. Wilmer Cave Wright. 3 Vols.

LUCIAN. 8 Vols. Vols. I.–V. A. M. Harmon. Vol. VI. K. Kilburn.

LYCOPHRON. Cf. CALLIMACHUS.

LYRA GRAECA. J. M. Edmonds. 3 Vols.

LYSIAS. W. R. M. Lamb.

MANETHO. W. G. Waddell: PTOLEMY: TETRABIBLOS. F. E. Robbins.

MARCUS AURELIUS. C. R. Haines.

MENANDER. F. G. Allinson.

MINOR ATTIC ORATORS (ANTIPHON, ANDOCIDES, LYCURGUS, DEMADES, DINARCHUS, HYPEREIDES). K. J. Maidment and J. O. Burrt. 2 Vols.

NONNOS: DIONYSIACA. W. H. D. Rouse. 3 Vols

OPPIAN, COLLUTHUS, TRYPHIODORUS. A. W. Mair.

PAPYRI. NON-LITERARY SELECTIONS. A. S. Hunt and C. C. Edgar. 2 Vols. LITERARY SELECTIONS (Poetry). D. L. Page.

PARTHENIUS. Cf. DAPHNIS AND CHLOE.

PAUSANIAS: DESCRIPTION OF GREECE. W. H. S. Jones. 4 Vols. and Companion Vol. arranged by R. E. Wycherley.

PHILO. 10 Vols. Vols. I.–V.; F. H. Colson and Rev. G. H. Whitaker. Vols. VI.–IX.; F. H. Colson.

PHILO: two supplementary Vols. (*Translation only.*) Ralph Marcus.

PHILOSTRATUS: THE LIFE OF APOLLONIUS OF TYANA. F. C. Conybeare. 2 Vols.

PHILOSTRATUS: IMAGINES; CALLISTRATUS: DESCRIPTIONS. A. Fairbanks.

PHILOSTRATUS and EUNAPIUS: LIVES OF THE SOPHISTS. Wilmer Cave Wright.

PINDAR. Sir J. E. Sandys.

PLATO: CHARMIDES, ALCIBIADES, HIPPARCHUS, THE LOVERS, THEAGES, MINOS and EPINOMIS. W. R. M. Lamb.

PLATO: CRATYLUS, PARMENIDES, GREATER HIPPIAS, LESSER HIPPIAS. H. N. Fowler.

PLATO: EUTHYPHRO, APOLOGY, CRITO, PHAEDO, PHAEDRUS. H. N. Fowler.

PLATO: LACHES, PROTAGORAS, MENO, EUTHYDEMUS. W. R. M. Lamb.

PLATO: LAWS. Rev. R. G. Bury. 2 Vols.

PLATO: LYSIS, SYMPOSIUM, GORGIAS. W. R. M. Lamb.

PLATO: REPUBLIC. Paul Shorey. 2 Vols.

PLATO: STATESMAN, PHILEBUS. H. N. Fowler; ION. W. R. M. Lamb.

PLATO: THEAETETUS and SOPHIST. H. N. Fowler.

PLATO: TIMAEUS, CRITIAS, CLITOPHO, MENEXENUS, EPISTULAE. Rev. R. G. Bury.

PLUTARCH: MORALIA. 15 Vols. Vols. I.–V. F. C. Babbitt. Vol. VI. W. C. Helmbold. Vol. VII. P. H. De Lacy and B. Einarson. Vol. IX. E. L. Minar, Jr., F. H. Sandbach, W. C. Helmbold. Vol. X. H. N. Fowler. Vol. XII. H. Cherniss and W. C. Helmbold.

PLUTARCH: THE PARALLEL LIVES. B. Perrin. 11 Vols.

POLYBIUS. W. R. Paton. 6 Vols.

PROCOPIUS: HISTORY OF THE WARS. H. B. Dewing. 7 Vols.

PTOLEMY: TETRABIBLOS. Cf. MANETHO.

QUINTUS SMYRNAEUS. A. S. Way. Verse trans.

SEXTUS EMPIRICUS. Rev. R. G. Bury. 4 Vols.

SOPHOCLES. F. Storr. 2 Vols. Verse trans.

STRABO: GEOGRAPHY. Horace L. Jones. 8 Vols.

THEOPHRASTUS: CHARACTERS. J. M. Edmonds. HERODES, etc. A. D. Knox.

THEOPHRASTUS: ENQUIRY INTO PLANTS. Sir Arthur Hort, Bart. 2 Vols.

THUCYDIDES. C. F. Smith. 4 Vols.

TRYPHIODORUS. Cf. OPPIAN.

XENOPHON: CYROPAEDIA. Walter Miller. 2 Vols.

XENOPHON: HELLENICA, ANABASIS, APOLOGY, and SYMPOSIUM. C. L. Brownson and O. J. Todd. 3 Vols.

XENOPHON: MEMORABILIA and OECONOMICUS. E. C. Marchant.

XENOPHON: SCRIPTA MINORA. E. C. Marchant.

IN PREPARATION

Greek Authors

ARISTOTLE: HISTORY OF ANIMALS. A. L. Peck.
PLOTINUS: A. H. Armstrong.

Latin Authors

BABRIUS AND PHAEDRUS. Ben E. Perry.

DESCRIPTIVE PROSPECTUS ON APPLICATION

London   **WILLIAM HEINEMANN LTD**
Cambridge, Mass.   **HARVARD UNIVERSITY PRESS**

DATE DUE